INTERMEDIATE ALGEBRA, 5TH EDITION, BY WOOTON AND DROOYAN

STUDY GUIDE

BERNARD FELDMAN
Los Angeles Pierce College

Wadsworth Publishing Company, Inc.
Belmont, California
A division of Wadsworth

Mathematics Editor: Richard Jones

Production: Dick Palmer

Typist: Judy Foil

Printed in the United States of America
1 2 3 4 5 6 7 8 9 10—84 83 82 81 80

CONTENTS

PREFACE

This book is a supplement to Wooton and Drooyan's *Intermediate Algebra, 5th Edition*. It conforms, section by section, to the style and development of the basic text and provides programmed material that can be used either to supplement the text and classroom instruction in a traditional lecture course or as an integral part of a self-study program.

In a traditional lecture course, students will find this Study Guide useful as:

1. a tutor for sections of the course that are found to be difficult;

2. an aid in catching up with work covered during an absence;

3. a vehicle for reviewing prior to examinations.

In a self-study program, students may wish to use this guide in accordance with the suggested self-study program immediately following this preface.

Whether this Study Guide is used in a traditional lecture course or a self-study program, it should be used as follows:

1. Thoroughly study the appropriate section of the text.

2. Turn to the corresponding section of this Study Guide and place a strip of paper over the response column (the column on the left-hand side of the page) so that the entire column is covered.

3. Read the first frame and write your responses in the blanks.

4. When all of the response blanks in a line are filled, slide the strip of paper covering the response column down to expose the correct responses. Compare your responses with these correct responses and, if necessary, correct your thinking.

5. Then go to the next frame and repeat the procedure.

6. When you have completed a chapter, you may test yourself by working the text problems at the end of the chapter. Complete solutions to these problems are supplied in the Appendix.

Frames that are shaded contain statements of definitions, axioms, theorems, etc, and their concepts are usually illustrated by those frames that immediately follow.

1

THE SET OF REAL NUMBERS

1.1

SETS AND SYMBOLISM

Answer	Frame
set { }	1. A collection of things is called a _____ and is represented symbolically by _____, used in conjunction with words or other symbols.
3	2. {natural numbers} is equivalent to {1,2,___, . . .}.
2,4,6,8	3. {even natural numbers less than 10} is equivalent to {_________}.
empty null; ∅	4. The set with no members is called the _______ set or ______ set and is represented symbolically by ___.
null (empty)	5. {odd natural numbers between 3 and 5} is the ______ set.
false	6. True or false: $0 \in \emptyset$.
equal (=) unequal (≠)	7. If two sets have the same members, they are _________. If they do not have the same members, they are _________.

In Frames 8-10, use one of the symbols = or ≠ to make a true statement.

Answer	Frame
=	8. {1,2,3} ___ {3,1,2}.
≠	9. {1,2,3} ___ {2,3,4}.
=	10. {natural numbers less than 4} ___ {1,2,3}.
infinite finite	11. If a set has no last member, it is called a(n) _________ set; if it does have a last member it is called a(n) ________ set.
is	12. {even natural numbers} _____ (is/is not) an infinite set.
is not	13. {natural numbers less than 10} ________ (is/is not) an infinite set.

rule

14. When a set is denoted by roster notation, the members of the set are listed; when the members are described verbally, the notation used is ______ notation.

roster
rule

15. {1,2,3} is an example of ________ notation and {natural numbers less than 4} is an example of ______ notation.

set-builder
all; such that
all

16. Symbolism such as $\{x|x^2 = 1, x \text{ a natural number}\}$ is called ____________ notation. The first x is read "_____ x" and the vertical bar is read "______ ______." Hence, the symbolism is read "the set of _____ x such that $x^2 = 1$ and x is a natural number."

subset (⊂)
is not (⊄)

17. If every member of set A is a member of set B, A is a ________ of B. If every member of set A is not a member of set B, A ________ (is/is not) a subset of set B.

In Frames 18-21, use one of the symbols ⊂ or ⊄ to make a true statement.

⊂; ⊄

18. {5} ___ {1,3,5,7}.

19. {0} ___ ∅.

⊂; ⊂

20. ∅ ___ {0}.

21. {1,3} ___ {1,3,5}.

In Frames 22-24, list all of the subsets of {a,b,c}.

{*a*,*b*,*c*}

22. The 3-member subset is _________.

{*b*,*c*}

23. The 2-member subsets are {*a*,*b*}, {*a*,*c*}, and _______.

{*b*}, {*c*}

24. The 1-member subsets are {*a*}, _____, and _____.

variables; replacement

25. Symbols used to represent unspecified elements of a given set containing more than one element are called __________. The given set is called the ____________ set.

variable
{1,2,3}

26. x is an element of {1,2,3} means that x represents an unspecified element of {1,2,3} and hence is a __________. The replacement set of x is _________.

constant

27. A symbol used to represent the element of a set with only one member is called a __________.

constant

28. Because, in the statement "c is a member of {1}," c represents the member of a set with only one member, c is a __________.

member
not

29. If x is in set A, $(x \in A)$, it is said to be a _________ of set A. If it is not in set A, $(x \notin A)$, it is _______ a member of set A.

In Frames 30-33, let $A = \{-3,-\sqrt{2},0,1,3,5\}$ *and* $B = \{a,-1,3,x\}$. *Supply the correct symbol to make a true statement.*

Answer	Frame
A; B	30. $-3 \in$ ___. 31. $0 \notin$ ___.
$\notin$; $\in$	32. 5 ___ B. 33. 3 ___ B.
natural	34. $\{1,2,3, . . .\}$ is the set N of ________ numbers.
whole	35. $\{0,1,2,3, . . .\}$ is the set W of _______ numbers.

In Frames 36-38, use N or W to make a true statement.

Answer	Frame
N; W; W	36. $0 \notin$ ___. 37. $0 \in$ ___. 38. $N \subset$ ___.
integers	39. $\{. . . ,-3,-2,-1,0,1,2,3, . . .\}$ is the set J of __________.
J	40. $N \subset J$ because if $a \in N$, then $a \in$ ___.
rational	41. The set whose members are all those numbers that can be represented in the form a/b, where a,b represent integers and $b \neq 0$, is the set Q of __________ numbers.
$\frac{3}{10}$	42. $.3 \in Q$ because $.3 =$ ___.
$\frac{5}{1}$	43. $5 \in Q$ because $5 =$ ___.
$\frac{0}{1}$	44. $0 \in Q$ because $0 =$ ___.
Q	45. $N \subset Q$ because if $a \in N$, then $a \in$ ___.
irrational; integers; 0	46. The set made up of those numbers whose decimal representations are nonterminating and nonrepeating numerals is the set H of __________ numbers. These numbers cannot be represented in the form a/b where a and b are __________ and $b \neq$ ___.
no	47. Q and H have ____ (how many?) members in common.
J	48. $\frac{2}{3} \notin H$ because $\frac{2}{3}$ is in the form $\frac{a}{b}$ where $a,b \in$ ___, $b \neq 0$.
H	49. Because $\sqrt{2}$ cannot be expressed in the form $\frac{a}{b}$, where $a,b \in J$, $b \neq 0$, $\sqrt{2} \in$ ___.
real	50. The collection of all rational and irrational numbers is the set R of ______ numbers.
R; R	51. $Q \subset$ ___. 50. $H \subset$ ___.
real	52. $\frac{2}{3} \in Q$ and also is a ______ number.
real	53. $\sqrt{2} \in H$ and also is a ______ number.

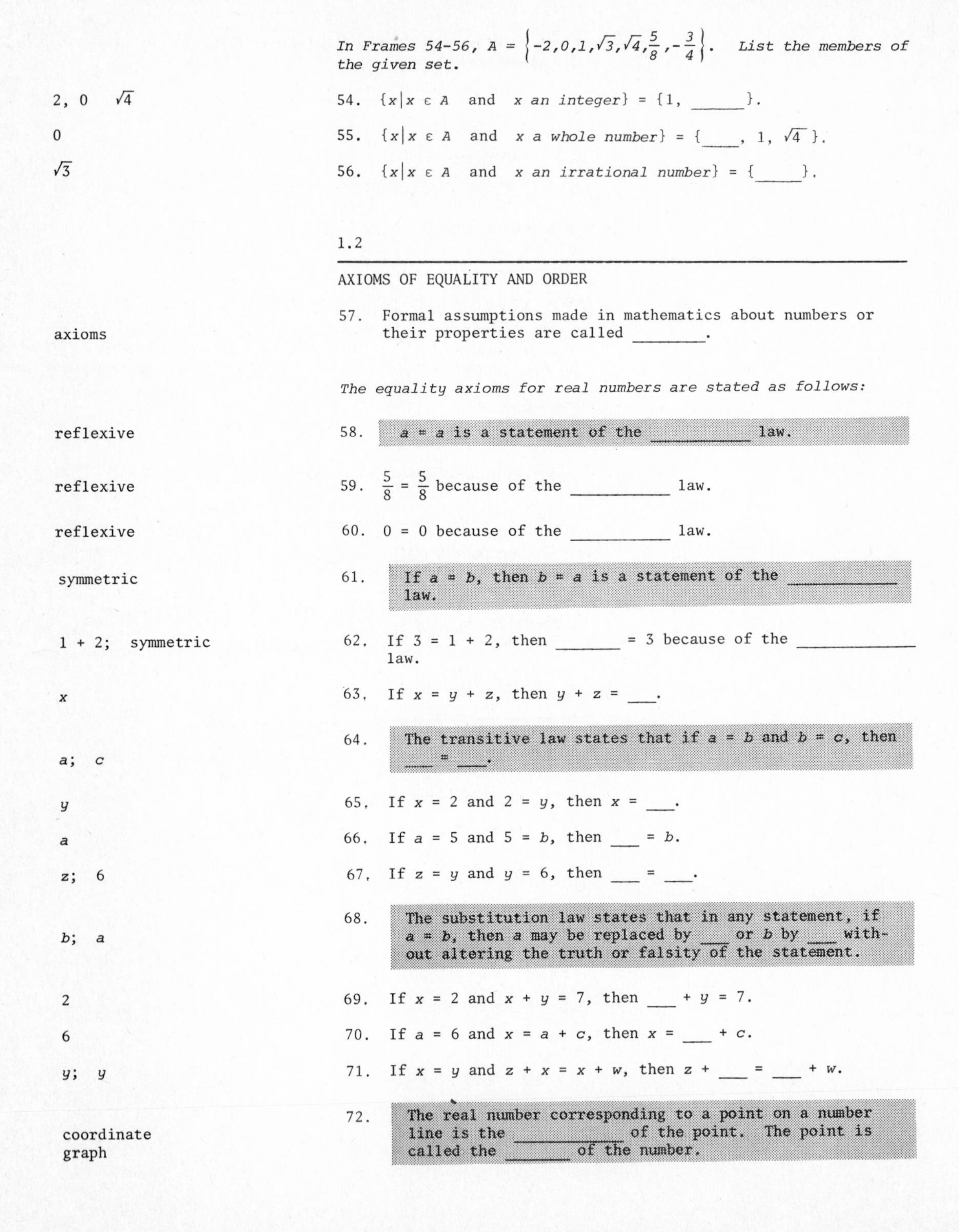

In Frames 54-56, $A = \left\{-2, 0, 1, \sqrt{3}, \sqrt{4}, \frac{5}{8}, -\frac{3}{4}\right\}$. *List the members of the given set.*

2, 0 $\sqrt{4}$

54. $\{x \mid x \in A$ and x *an integer*$\} = \{1,$ ______$\}$.

0

55. $\{x \mid x \in A$ and x *a whole number*$\} = \{$____, $1, \sqrt{4}\}$.

$\sqrt{3}$

56. $\{x \mid x \in A$ and x *an irrational number*$\} = \{$_____$\}$.

1.2

AXIOMS OF EQUALITY AND ORDER

axioms

57. Formal assumptions made in mathematics about numbers or their properties are called ________.

The equality axioms for real numbers are stated as follows:

reflexive

58. **$a = a$ is a statement of the ___________ law.**

reflexive

59. $\frac{5}{8} = \frac{5}{8}$ because of the ___________ law.

reflexive

60. $0 = 0$ because of the ___________ law.

symmetric

61. **If $a = b$, then $b = a$ is a statement of the ___________ law.**

1 + 2; symmetric

62. If $3 = 1 + 2$, then _______ $= 3$ because of the ___________ law.

x

63. If $x = y + z$, then $y + z =$ ___.

a; c

64. **The transitive law states that if $a = b$ and $b = c$, then ___ = ___.**

y

65. If $x = 2$ and $2 = y$, then $x =$ ___.

a

66. If $a = 5$ and $5 = b$, then ___ $= b$.

z; 6

67. If $z = y$ and $y = 6$, then ___ = ___.

b; a

68. **The substitution law states that in any statement, if $a = b$, then a may be replaced by ___ or b by ___ without altering the truth or falsity of the statement.**

2

69. If $x = 2$ and $x + y = 7$, then ___ $+ y = 7$.

6

70. If $a = 6$ and $x = a + c$, then $x =$ ___ $+ c$.

y; y

71. If $x = y$ and $z + x = x + w$, then $z +$ ___ = ___ $+ w$.

coordinate
graph

72. **The real number corresponding to a point on a number line is the ____________ of the point. The point is called the _______ of the number.**

Use the following number line for Frames 73-76.

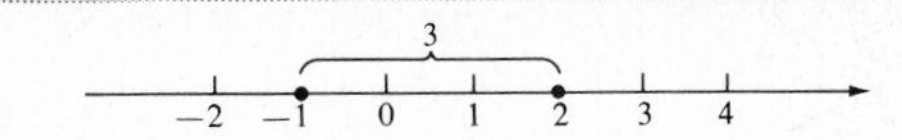

0	73. The coordinate of point *A* is ___.
E	74. The graph of 4 is point ___.
coordinate	75. 1 is the __________ of point *B*.
graph	76. Point *C* is the ______ of 2.
$b + c$	77. **The point corresponding to *b* is located on the number line and then, from this point, *c* units are counted off to the right of *b*. The point arrived at corresponds to ______.**
	78.
2	The above illustrates that -1 + 3 = ___.
	79.
6	The above illustrates that -3 + ___ = 3.
b; *a* left	80. **If *b* is less than *a*, there exists a positive number $c \in R$ such that ___ + *c* = ___. On a line graph, the graph of *b* would be to the _____ of the graph of *a*.**
-6; -4	81. Of the numbers -4 and -6, ____ is less than ____.
-2; 1	82. Of the numbers -2 and 1, ____ is less than ___.
-1; 0	83. Of the numbers 0 and -1, ____ is less than ___.
less	84. $a < b$ is read "*a* is _____ than *b*."
greater	85. $a > b$ is read "*a* is ________ than *b*."
equal	86. $a \leq b$ is read "*a* is less than or ______ to *b*."
equal	87. $a \geq b$ is read "*a* is greater than or ______ to *b*."

In Frames 88-91, supply one of the symbols <, >, or = to make a true statement.

>; >	88. -4 ___ - 6. 89. 0 ___ -1.
<	90. -2 ___ -1.
=	91. $5 \geq 5$ is true because 5 ___ 5.

Two of the order axioms for real numbers are stated as follows.

$a > b$
trichotomy

92. Exactly one of the following relationships holds: $a < b$, $a = b$, or ______. This law is called the ____________ law.

a; c
transitive

93. If $a < b$ and $b < c$ then ___ < ___. This law is called the ____________ law.

$>$

94. If $a \neq 2$ and $a \not< 2$, then a ___ 2.

y

95. If $x < 1$ and $1 < y$, then $x <$ ___.

1

96. If $y < z$ and $z < 1$, then $y <$ ___.

In Frames 97-99, graph each set of integers.

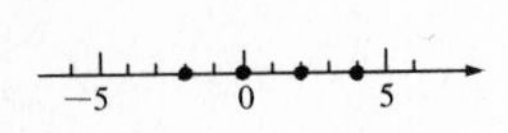

97. $\{-2,0,2,4\}$

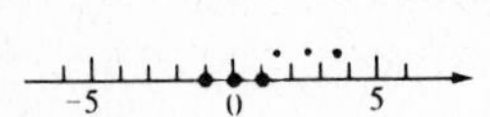

98. $\{x \mid x \geq -1\}$

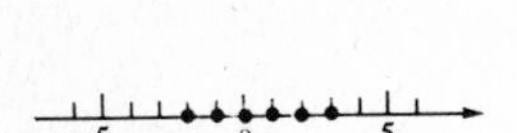

99. $\{x \mid -2 \leq x < 4\}$

1.3

INTERVALS OF REAL NUMBERS

100. In interval notation,

$\{x \mid a < x < b\}$ is written as (a,b).

$[a,b)$ — $\{x \mid a \leq x < b\}$ is written as ______.

$(a,b]$ — $\{x \mid a < x \leq b\}$ is written as ______.

$[a,b]$ — $\{x \mid a \leq x \leq b\}$ is written as ______.

In Frames 101-104, replace each ? with < or ≤ to make a true statement.

$<$ — 101. $(2,4) = \{x \mid 2 < x \; ? \; 4\}$.

$\leq$ — 102. $(2,4] = \{x \mid 2 < x \; ? \; 4\}$.

$\leq$, $<$ — 103. $[2,4) = \{x \mid 2 \; ? \; x \; ? \; 4\}$.

$\leq$, $\leq$ — 104. $[2,4] = \{x \mid 2 \; ? \; x \; ? \; 4\}$.

closed
closed

105. If an interval includes both of its endpoints it is called a(n) ________ (open/closed) interval. [2,4] is a ________ interval.

open
open

106. If an interval includes neither of its endpoints it is called a(n) _________ (open/closed) interval. (2,4) is a(n) _________ interval.

half; half

107. If an interval includes one endpoint only, it is called a ______-open interval. (2,4] and [2,4) are _______-open intervals.

less

108. The symbols $+\infty$ and $-\infty$ are used when an interval either includes all numbers greater than a given real number or all numbers _______ than a given real number.

$+\infty$

109. $\{x|x \geq 1\} = [1, \_\_)$.

$(1, +\infty)$

110. $\{x|x > 1\} =$ _______.

$-\infty$

111. $\{x|x < 1\} = (\_\_, 1)$.

$(-\infty, 1]$

112. $\{x|x \leq 1\} =$ _______.

In Frames 113-115, graph each interval of real numbers and represent the numbers of the interval, using set-builder notation.

113. $(-3,2]$.

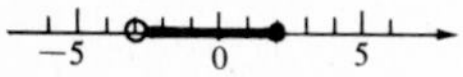

−5 0 5 ; $\leq$

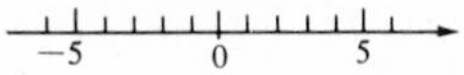

−5 0 5 ; $\{x|-3 < x$ ___ $2\}$

114. $(-2, +\infty)$

−5 0 5 ; $>$

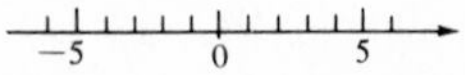

−5 0 5 $\{x|x$ ___ $-2\}$

115. $[-3,3]$

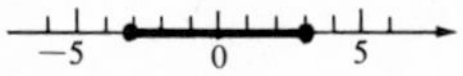

−5 0 5 ; $\leq$

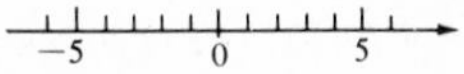

−5 0 5 $\{x|-3 \leq x$ ___ $3\}$

In Frames 116-118, graph each set of real numbers and represent the numbers of the set using interval notation.

116. $\{x|x > -3\}$

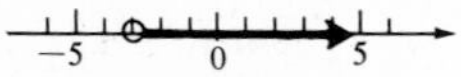

−5 0 5 ; $+\infty$

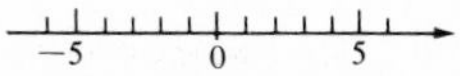

−5 0 5 ; $(-3,$ ___$)$

117. $\{t|10 < t \leq 40\}$

−10 0 50 ; $40]$

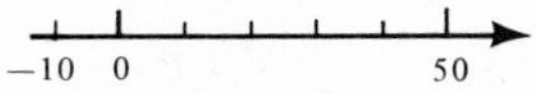

−10 0 50 ; $(10,$ ___

118. $\{z|z < 30\}$

−10 0 50 ; $(-\infty$

−10 0 50 ; ___$, 30)$

1.4

SOME PROPERTIES OF THE REAL NUMBERS

The axioms that set forth the properties of the real numbers are stated as follows.

Answer	Frame
unique	119. **The closure property of addition states that $a+b$ is a ________ real number.**
R	120. $(5{,}862 + 10{,}851) \in$ ___ and is unique.
R	121. Since $4 \in R$ and $\sqrt{2} \in R$, $4 + \sqrt{2}$ is a unique element of _____.
$b + a$	122. **The commutative property of addition states that $a+b =$ ____________.**
3; 8	123. $3 + 4 = 4 +$ ___.
3	124. $3 + (2 \cdot 3) = (2 \cdot 3) +$ ___.
$(3 + 4)$	125. $(3 + 4) + (2 \cdot 5) = (2 \cdot 5) +$ _________.
$a + (b + c)$	126. **The associative property of addition states that $(a+b) + c =$ ___________.**
4	127. $3 + (2 + 4) = (3 + 2) +$ ___.
6	128. $5 + (6 + 7) = (5 +$ ___ $) + 7$.
$2 + 3$	129. $(1 + 2) + 3 = 1 + ($_______$)$.
m	130. $(m + n) + p =$ ___ $+ (n + p)$.
real	131. **The closure property of multiplication states that $a \cdot b$ is a unique ________ number.**
real	132. $(5{,}862) \cdot (67{,}476)$ is a unique ______ number.
real	133. If $a, b \in R$ and $a \cdot b = c$, then c is a unique ______ number.
$b \cdot a$	134. **The commutative property of multiplication states that $a \cdot b =$ _________.**
3	135. $3 \cdot 4 = 4 \cdot$ ___.
3	136. $3(4 \cdot 5) = (4 \cdot 5) \cdot$ ___.
6	137. $6(4 + 7) = (4 + 7) \cdot$ ___.
$[4 + (-7)]$	138. $(-3)[4 + (-7)] =$ ___________(-3).
a	139. $a(b \cdot c) = (b \cdot c) \cdot$ ___.
$(a \cdot b)c$	140. **The associative property of multiplication states that $a(b \cdot c) =$ ___________.**

5	141. $3 \cdot (4 \cdot 5) = (3 \cdot 4) \cdot$ ___.
$7 \cdot 2$	142. $(6 \cdot 7) \cdot 2 = 6 \cdot ($_______$)$.
-2	143. $[(-2) \cdot (3)](-4) = ($____$)[(3) \cdot (-4)]$.
$c + d$	144. $a[b(c + d)] = [a \cdot b]($_______$)$.
$(a \cdot b) + (a \cdot c)$	145. **The distributive property states that $a(b + c) =$ _________.**
4	146. $2(3 + 4) = 2 \cdot 3 + 2 \cdot$ ___.
$5 \cdot 7$	147. $5(6 + 7) = 5 \cdot 6 +$ _______.
$(-3)(5)$; $(-3)(-2)$	148. $-3[5 + (-2)] =$ _________ + _________.
3	149. $3 \cdot 6 + 3 \cdot 4 =$ ___$(6 + 4)$.
$\frac{1}{x}(z)$	150. $\frac{1}{x}(y + z) = \frac{1}{x}(y) +$ ______.
0 a; a	151. **The identity element for addition assumes the existence of a unique real number ___ with the property: $a + 0 =$ _____ and $0 + a =$ _____.**
5; 0	152. $5 + 0 =$ ___. 153. $6 +$ ___ $= 6$.
$x + y$	154. $(x + y) + 0 =$ _______.
$m + n$	155. _______ $+ 0 = m + n$.
1 a; a	156. **The identity element for multiplication assumes the existence of a unique real number ___, with the property $a \cdot 1 =$ ___ and $1 \cdot a =$ ___.**
5; 1	157. $5 \cdot 1 =$ ___. 158. $6 \cdot$ ___ $= 6$.
$x + y$	159. $(x + y) \cdot 1 =$ _______.
$m + n$	160. _______ $\cdot 1 = m + n$.
$-a$; $-a$; $-a$	161. **The negative or additive-inverse axiom states that for each real number a, there exists a unique real number ____ with the property $a +$ ____ $= 0$ and ____ $+ a = 0$.**
0; -8	162. $5 + (-5) =$ ___. 163. $8 +$ ____ $= 0$.
9	164. ___ $+ (-9) = 0$.
0	165. $(m + n) + [-(m + n)] =$ ___.
$-(x + y)$	166. $(x + y) +$ __________ $= 0$.
$\frac{1}{a}$; 1	167. **The reciprocal or multiplicative-inverse property states that for each real number $a \neq 0$ there exists a unique real number ___ with the property $(a) \cdot \frac{1}{a} = \frac{1}{a} \cdot (a) =$ ___.**

Answer	Frame
1; $\frac{1}{8}$	168. $5\left(\frac{1}{5}\right) =$ ___. 169. $8 \cdot$ ___ $= 1$.
9	170 ___ $\cdot \frac{1}{9} = 1$.
$\frac{1}{-6}$; $\frac{1}{m+n}$	171. $(-6) \cdot$ ___ $= 1$. 172. $(m + n) \cdot$ ___ $= 1$.
$b + c$; $c + b$	173. **The addition property of equality states that if $a = b$, then $a + c =$ ___ and $c + a =$ ___.**
6	174. If $7 = 3 + 4$, then $7 + 6 = 3 + 4 +$ ___.
8	175. If $3 \cdot 4 = 6 \cdot 2$, then $8 + 3 \cdot 4 =$ ___ $+ 6 \cdot 2$.
p	176. If $ab = c$, then $p + ab =$ ___ $+ c$.
$r + zw$	177. If $x + y = zw$, then $r + x + y =$ ___.
bc; cb	178. **The multiplication property of equality states that if $a = b$, $ac =$ ___ and $ca =$ ___.**
6	179. If $7 = 3 + 4$, then $6 \cdot 7 =$ ___ $\cdot (3 + 4)$.
8	180. If $3 \cdot 4 = 6 \cdot 2$, then $8 \cdot (3 \cdot 4) =$ ___ $\cdot (6 \cdot 2)$.
p	181. If $a + b = c$, then ___ $\cdot (a + b) = pc$.
$r(zw)$	182. If $x + y = zw$, then $r(x + y) =$ ___.
0	183. **The zero-factor property states that for every $a \in R$, $a \cdot 0 =$ ___.**
0; 0	184. $1 \cdot 0 =$ ___. 185. $-3 \cdot$ ___ $= 0$.
0	186. $(x + y) \cdot 0 =$ ___.
a	187. **The double-negative property states that for each $a \in R$, $-(-a) =$ ___.**
2; x	188. $-(-2) =$ ___. 189. $-(-x) =$ ___.
$-y$	190. $y = -($___$)$.
a; $-a$	191. **The absolute value of a real number a is defined by the following statements: If $a \geq 0$, $\lvert a \rvert =$ ___, but if $a < 0$, then $\lvert a \rvert =$ ___.**
2; 2	192. $\lvert 2 \rvert =$ ___. 193. $\lvert -2 \rvert =$ ___.
0	194. $\lvert 0 \rvert =$ ___.
$a - b$	195. If $a - b \geq 0$, then $\lvert a - b \rvert =$ ___.
$-(a - b)$	196. If $a - b < 0$, then $\lvert a - b \rvert =$ ___.

Answer	Frame
$mp + r$	197. If $mp + r \geq 0$, then $\lvert mp + r\rvert =$ ________.
$-(mp + r)$	198. If $mp + r < 0$, then $\lvert mp + r\rvert =$ __________.

1.5

SUMS AND DIFFERENCES

Answer	Frame
0	199. If $a > 0$ and $b > 0$, then $a + b >$ ____.
0	200. $2 + 3 >$ ___.
uv	201. If $xy > 0$ and $uv > 0$, then $xy +$ ____ > 0.
$m + p$	202. If $m > 0$ and $p > 0$, then _______ > 0.
a	203. The difference $a - b$ of two positive numbers a and b is defined as the number d such that $b + d =$ ____.
5	204. $5 - 2 = 3$ because $3 + 2 =$ ___.
$5 + 2$	205. $7 - 5 = 2$ because _______ $= 7$.
$4 + 2$; 6	206. $6 - 4 = 2$ because _______ $=$ ___.
$y + z$	207. $x - y = z$ because _______ $= x$.
$a + b$	208. The theorem that states how to find the sum of two negative numbers asserts that for positive numbers a and b, $-a + (-b) = -($________$)$.
3; 5	209. $(-2) + (-3) = -(2 +$ ___$) = -$___.
7; 5; -12	210. $(-7) + (-5) = -($___ $+$ ___$) =$ _____.
-14	211. $(-6) + (-8) =$ _____.
$p + q$	212. If $p,q > 0$, then $(-p) + (-q) = -($_______$)$.
difference; greater	213. The following rule is used to find the sum of a positive and a negative number; The absolute value of the sum of a positive and a negative number is equal to the nonnegative __________ of the absolute values of the numbers; and the sum is positive or negative as the addend of the number of _________ (greater/lesser) absolute value is positive or negative.
3	214. $4 + (-1) = 4 - 1 =$ ___.
4	215. $6 + (-2) = 6 - 2 =$ ___.
7; 1	216. $8 + (-7) = 8 -$ ___ $=$ ___.
3; 7	217. $10 + (-3) = 10 -$ ___ $=$ ___.

Answer	Frame
-3	218. 2 + (-5) = -(5 - 2) = ____.
7; -3	219. 7 + (-10) = -(10 - ___) = ____.
7; 6; -1	220. 6 + (-7) = -(___ - ___) = ____.
-(20 - 12); -8	221. 12 + (-20) = ____________ = ____.
$-b$	222. **The theorem that states how to find the difference of two real numbers a and b asserts that $a - b = a + (\_\_\_\_)$.**
5	223. 9 - 4 = 9 + (-4) = ___.
7	224. 10 - 3 = 10 + (-3) = ___.
-4; 11	225. 15 - 4 = 15 + (____) = ____.
13	226. 9 - (-4) = 9 + [-(-4)] = 9 + [4] = ____.
2; 8	227. 6 - (-2) = 6 + [___] = ___.
-2; -5	228. -3 - 2 = -3 + (____) = ____.
2; -6	229. -8 - (-2) = -8 + ___ = ____.
3; -7	230. -10 - (-3) = -10 + ___ = ____.

1.6

PRODUCTS AND QUOTIENTS

Answer	Frame
product factors	231. **The number ab is called the _________ of a and b and in ab, a and b are called _________.**
2	232. 6 is the product of 3 and ___.
8	233. The product of 4 and 2 is ___.
2	234. In Frame 232 above, 3 and ___ are factors of 6.
4; 2	235. In Frame 233 above, ___ and ___ are factors of 8.
>	236. **If $a > 0$ and $b > 0$, then ab ___ 0.**
0	237. Since 3 > 0 and 2 > 0, 3 • 2 > ___.
3	238. Since 4 > 0 and 3 > 0, 4 • ___ > 0.
14; 0	239. Since 7 > 0 and 2 > 0, ____ > ___.
negative	240. **The product of two nonzero real numbers with unlike signs is a _________ (positive/negative) number and is equal in absolute value to the product of their absolute values.**

positive

241. The product of two nonzero real numbers with like signs is a _________ (positive/negative) number and is equal in absolute value to the product of their absolute values.

-6; -15

242. (3)(-2) = ____.   243. (-3)(5) = _____.

1; -2

244. (-2)(___) = -2.   245. (5)(____) = -10.

6; 6

246. (3)(2) = ___.   247. (-3)(-2) = ___.

-4; -2

248. (-2)(____) = 8.   249. (____)(-9) = 18.

positive

negative

250. If a product contains
1. an even number of negative factors, the product is ________ (positive/negative).
2. an odd number of negative factors, the product is ___________.

odd

-4

251. (-1)(2)(-2)(-1) has an (even/odd) number of negative factors, hence
(-1)(2)(-2)(-1) = ___.

even

48

252. (-1)(-2)(-3)(2)(-4) has an _______ (even/odd) number of negative factors, hence
(-1)(-2)(-3)(2)(-4) = ____.

a

253. The quotient of two real numbers a and b (written $\frac{a}{b}$; $b \neq 0$) is defined as the real number q such that $b \cdot q =$ ___.

defined

254. Division by zero is not ________.

2

255. $\frac{4}{2} = 2$ because $2 \cdot$ ___ $= 4$.

-12

256. $\frac{-12}{3} = -4$ because $(3)(-4) =$ _____.

-3

257. $\frac{-15}{-3} = 5$ because (____)(5) = -15.

-3; -3

258. $\frac{18}{-6} =$ ____ because (-6)(____) = 18.

negative

259. The quotient of two nonzero real numbers with unlike signs is a _______ (positive/negative) number and is equal in absolute value to the quotient of their absolute values.

positive

260. The quotient of two nonzero real numbers with like signs is a _________ (positive/negative) number and is equal in absolute value to the quotient of their absolute values.

-3; -2

261. $\frac{6}{-2} =$ ____.   262. $\frac{-10}{5} =$ ____.

Answers	Frames
-2; -2	263. $\frac{-12}{6} = $ ____. 264. $\frac{20}{-10} = $ ____.
3; 7	265. $\frac{-6}{-2} = $ ____. 266. $\frac{-14}{-2} = $ ____.
2; 6	267. $\frac{8}{4} = $ ____. 268. $\frac{12}{2} = $ ____.
$\frac{1}{b}$	269. The same number is represented by $\frac{a}{b}$ and $a \cdot (\_\_\_)$.
$\frac{1}{3}$; $\frac{1}{-8}$	270. $\frac{2}{3} = 2(\_\_\_)$. 271. $\frac{5}{-8} = 5(\_\_\_)$.
-3; 5	272. $\frac{-3}{5} = (\_\_\_)\frac{1}{5}$. 273. $\frac{5}{-4} = (\_\_\_)\frac{1}{-4}$.
$\frac{4}{7}$; $\frac{8}{-3}$	274. $4\left(\frac{1}{7}\right) = $ ____. 275. $8\left(\frac{1}{-3}\right) = $ ____.
$\frac{4}{5}$; $\frac{-7}{8}$	276. $\frac{1}{5}(4) = $ ____. 277. $\frac{1}{8}(-7) = $ ____.

1.7

ORDER OF OPERATIONS

Operations must be performed in the following order:

Answers	Frames
innermost	278. Expressions within symbols of inclusion (parentheses, brackets, fraction bars, etc.) are simplified starting with the ________ (which?) inclusion symbol.
multiplication division	279. Then perform in order from left to right, as encountered, the operations of ____________ and ________.
addition; subtraction	280. Lastly, perform the ________ and __________ operations.
9; 12	281. $3 + 3 \cdot 3 = 3 + \_\_\_ = \_\_\_\_$.
3; 9; -1	282. $8 - 3(4 - 1) = 8 - 3(\_\_\_) = 8 - \_\_\_ = \_\_\_$.
-4; -20	283. $\frac{5(4-8)}{2} - \frac{12}{-3} = \frac{5(\_\_\_)}{2} - \frac{12}{-3} = \frac{\_\_\_}{2} - \frac{12}{-3}$
-4; -6	$= -10 - (\_\_\_) = \_\_\_$.
$\frac{-4}{2}$; 2; 10	284. $\left[\frac{8-(-2)}{5-3}\right]\left[\frac{5+(-9)}{3-5}\right] = \left[\frac{10}{(\_\_\_)}\right]\left[\frac{(\_\_\_)}{-2}\right] = [5][\_\_\_] = \_\_\_$.
-2	285. $2\left[9 + 3\left(\frac{6+(-16)}{2+3}\right) - 5\right] + 3 = 2[9 + 3(\_\_\_) - 5] + 3$
-6	$= 2[9 + (\_\_\_) - 5] + 3$
-2	$= 2[\_\_\_] + 3$
-4	$= \_\_\_ + 3$
-1	$= \_\_\_$.

In Frames 286-287, evaluate the expression $P + Prt$ when $P = 1{,}500$, $r = .05$, and $t = 3$.

.05; 3
.05; 3

286. In $P + Prt$, replace P by 1,500, r by _____ and t by _____ and obtain 1,500 + 1,500(_____)(___).

.15
225
1,725

287. Computing, we have

$$1{,}500 + 1{,}500(.05)(3) = 1{,}500 + 1{,}500(\_\_\_\_\_)$$
$$= 1{,}500 + \_\_\_\_\_$$
$$= \_\_\_\_\_\_\_.$$

TEST PROBLEMS

1. Let $A = \left\{-7, 6, -\frac{1}{7}, 0, \frac{6}{7}, 7\right\}$.

   a. List the members of $\{x \mid x \in A \text{ and } x \text{ is a natural number}\}$.

   b. List the members of $\{x \mid x \in A \text{ and } x \text{ is an even number}\}$.

2. Express each of the following relations using appropriate symbols.

   a. x is not less than y.

   b. x is between -2 and 1.

3. Graph each interval of real numbers.

   a. $\{x \mid -1 \le x < 3\}$.

   b. $[-2, +\infty)$.

In Problem 4, each of the statements is an application of one of the axioms R-1 through R-11 on page 15 of the text. Justify each statement by citing the appropriate axiom. Assume all variables represent real numbers.

4. a. $5 + (-5) = 0$.

   b. $x(y + z) = (y + z)x$.

   c. $0 + (x + y) = x + y$.

   d. $p + p(a + b) = p + pa + pb$.

In Problem 5, each of the statements is an application of one of the following four properties.

I. Addition law of equality.
II. Multiplication law of equality.
III. Zero-factor law.
IV. Double-negative law.

Justify each statement by citing the appropriate theorem. Assume all variables represent real numbers.

5. a. If $x + 2 = 4$, then $-3(x + 2) = -3(4)$.

   b. If $5 = n - 2$, then $2n + 5 = 3n - 2$.

   c. $-[-(x + 2)] = x + 2$.

   d. $0 \cdot (r + p) = 0$.

6. Rewrite without absolute-value notation.

   a. $|-8|$.

   b. $|x - 2|$ if $x - 2 < 0$.

7. Write each product or quotient using a basic numeral if possible.

   a. $(-2)(-2)(2)$.

   b. $\frac{5 \cdot (2)}{0}$.

   c. $\frac{-28}{4}$.

   d. $(-1)(-2)(-3)$.

8. Write each sum or difference using a basic numeral.

   a. $-4 + (4) - (-2)$.

   b. $(8 + 5 - 11) + (2 - 6)$.

   c. $(15 - 8) - (6 - 11)$.

   d. $8 - (-3 + 2 - 7)$.

9. a. Rewrite $\frac{5}{9}$ as a product.

   b. Rewrite $\frac{1}{11}(7)$ as a quotient.

10. a. Write $\frac{6 + 3 \cdot 4}{4 - 5 \cdot 2}$ as a basic numeral.

    b. Write $\frac{4 - \left(\frac{8 + 12}{2 - 7}\right) - 6}{4 \cdot 2 - 2}$ as a basic numeral.

2

POLYNOMIALS

2.1

DEFINITIONS

n; base
power

1. If $a \in R$ and $n \in N$, then a^n means that a appears in a product ___ times. The number a is called the _______ and the product a^n is called the _______ of the base.

In Frames 2-5, give the responses in exponent form.

y^3

2. $xxx + yyy = x^3 +$ ____.

$a^2b^3c^2$

3. $abb + aabbbcc = ab^2 +$ ________.

p^2; pr^3

4. $pp + prrr =$ ____ $+$ _____.

a^4b^2; a^2b^3

5. $3aaaabb + 4aabbb = 3$______ $+ 4$______.

In Frames 6-7, give the responses in factored form.

bb

6. $a^3 + 3b^2 = aaa + 3$____.

$4xxyyyz$

7. $4x^2y^3z =$ _________.

exponent; base
power

8. In a^7, 7 is the _________, a is the ______, and a^7 is the _______ of a.

expression; terms

9. Any collection of numerals, variables, and signs of operation of the form $A + B + C + \cdot \cdot \cdot$ is called a(n) ___________. The A, B, C, etc., are called _________.

expression

10. $6a^3b^2 - 3a^2b + 2ab^2$ is an ___________.

$6a^3b^2$; $-3a^2b$
$2ab^2$

11. The terms of $6a^3b^2 - 3a^2b + 2ab^2$ are _______, _______, and ______.

coefficient

12. Any factor or group of factors in a term is said to be the ___________ of the remaining factors of that term.

$3x^2$; 3
x^2; x^2y

13. In $3x^2y$, the coefficient of y is _____, of x^2y is ___, of $3y$ is ____, of 3 is _____.

Answer	Frame
$-7a^3b$ a^3bc^2; b	14. In $-7a^3bc^2$, the coefficient of c^2 is _______, of -7 is _______, of $-7a^3c^2$ is ___.
3; -7	15. The numerical coefficient (or just coefficient) in $3x^2y$ is ___, in $-7a^3bc^2$ is ____.
rational	16. An expression in which the operations involved consist solely of addition, subtraction, multiplication, and division is called a __________ expression.
rational	17. $\frac{x - y}{3z}$ is a __________ expression.
rational	18. Because the operation square root is in the expression, $3x - \sqrt{y}$ is not a __________ expression.
polynomial	19. Any rational expression in which no variable occurs in a denominator is called a ____________.
polynomial	20. $3x^2 - 4z + 6w$ is a ____________.
denominator	21. $\frac{5mn^2 - 4m}{5m - 4n}$ is not a polynomial because a variable occurs in the _____________.
monomial binomial trinomial	22. If the polynomial has only one term, it is called a _________. If it has two terms, it is called a _________, and if it has three terms, it is called a _________.
monomial	23. $7x^2yz$ is a __________.
binomial	24. $x^3 + 7x^2yz$ is a __________.
trinomial	25. $5m^2 - 3mn + 6n^2$ is a ___________.
exponent sum	26. The degree of a monomial is given by the __________ on the variable. If it has more than one variable, the degree is the _____ of the exponents on the variables.
second	27. $3x^2$ is of the ________ degree.
1; 3	28. $4xy^2 = 4x^1y^2$ is of degree ___ + 2 = ___.
2; 7	29. $-6m^2n^3p^2$ is of degree 2 + 3 + ___ = ___.
highest	30. The degree of a polynomial is the same as the degree of its term of _________ degree.
3	31. The degree of $x^2 + xy^2$ is ___.
5	32. The degree of $4m^2n^2 - 5mn^2p + 6m^2p^2n$ is ___.
	33. Because a power is a product, we should simplify any powers in an expression before performing other multiplication or division operations.
16; 9	34. $5^2 - (-4)^2 = 25 - (\_\_\_\_) = \_\_\_$.

4; 9; 36

4; 6

35. $\frac{4^2 \cdot 5}{8} - \frac{2^2 \cdot 3^2}{9} = \frac{16 \cdot 5}{8} - \frac{(\_\_)(\_\_)}{9} = \frac{80}{8} - \frac{\_\_}{9}$
$= 10 - (\_\_) = \_\_.$

In Frame 36, given $x = 4$, $y = -3$, evaluate $(3x - y)^2$.

4; -3; -3

15; 225

36. $(3x - y)^2 = [3(\_\_) - (\_\_)]^2 = (12 - \_\_)^2$
$= (\_\_)^2 = \_\_.$

In Frame 37, given $M = 32$, $V = 20$, $v = 4$, and $g = 16$, evaluate $\frac{M(V - v)^2}{g}$.

4; 16; 256

512

37. $\frac{M(V - v)^2}{g} = \frac{32[20 - (\_\_)]^2}{16} = \frac{32(\_\_)^2}{16} = \frac{\overset{2}{\cancel{32}}(\_\_)}{\underset{1}{\cancel{16}}}$
$= \_\_.$

variable

38. Symbols such as $P(x)$, $D(y)$, and $Q(w)$ are used to denote polynomials and the symbol in parentheses designates the ________ in the polynomial.

x

39. $P(x)$ can be used to denote the value of a polynomial for specific values of ___.

In Frames 40-45, take $P(x) = x^2 + 2x + 1$ and $D(z) = 2z - 1$.

3; 3; 16

40. $P(3) = (\_\_)^2 + 2(\_\_) + 1 = \_\_.$

1

41. $P(0) = \_\_.$

-1; -3

42. $D(-1) = 2(\_\_) - 1 = \_\_.$

-1

43. $D(0) = \_\_.$

16

16; 31

44. Using the results of Frame 40,
$D[P(3)] = D[\_\_]$. Then,
$D[P(3)] = D[16] = 2(\_\_) - 1 = \_\_.$

-3; 4

45. Using the results of Frame 42,
$P[D(-1)] = P[-3]$
$= (-3)^2 + 2(\_\_) + 1 = \_\_.$

2.2

SUMS AND DIFFERENCES

like

46. Terms that differ only in their numerical coefficients are called ______ terms.

$2x$

47. The like terms in $4x + 3y + 2x$ are $4x$ and ____.

$-7ab$
ab

48. The like terms in $5ab + 6a^2b - 7ab + ab$ are $5ab$, _______, and ____.

$3x^2yz$
$5x^2yz$

49. The like terms in $3x^2yz + 4xy^2z - 7xyz + 5x^2yz$ are ______ and _______.

Answer	Frame
equivalent simple	50. When all like terms in an expression have been combined, the resulting expression is said to be ________ to the original expression and to be in ______ form.

Express the given polynomial in simple form.

Answer	Frame
$6x$	51. $4x + 3y + 2x =$ ____ $+ 3y$.
$-ab$	52. $5ab + 6a^2b - 7ab + ab = 6a^2b +$ _____.
$8x^2yz$	53. $3x^2yz + 4xy^2 - 7xyz + 5x^2yz =$ _______ $+ 4xy^2 - 7xyz$.
addition	54. In any polynomial, the signs between the terms are regarded as indications of positive or negative coefficients, and the operation on these terms is understood to be ________.
$-5x$; x	55. $2x - 5x + 4x = (2x) + (\_\_\_\_) + (4x) =$ ___.
$-4m$; $-7m$; $-8m$	56. $3m - 4m - 7m = (3m) + (\_\_\_\_) + (\_\_\_\_) =$ _____.
$-3y^2$; x^2; $-y^2$ $6x^2 - 4y^2$	57. $5x^2 - 3y^2 + x^2 - y^2 = (5x^2) + (\_\_\_\_) + (\_\_\_) + (\_\_\_\_)$ $=$ __________.
b; c	58. $a - (b + c) = a - (\_\_) - (\_\_)$.
b; c	59. $a + (b + c) = a +$ ___ $+$ ___.

Insert appropriate signs between terms in Frames 60-63 where required and express in simple form:

Answer	Frame
$-$; $-$; $-2x - 3y$	60. $3x - (5x + 3y) = 3x$ ___ $5x$ ___ $3y =$ _________.
$+$; $+$; $8x + 3y$	61. $3x + (5x + 3y) = 3x$ ___ $5x$ ___ $3y =$ ________.
$-$; $-$; $+$ $4m^2 + m + 2$	62. $(6m^2 + 2m) - (2m^2 + m - 2) = 6m^2 + 2m$ ___ $2m^2$ ___ m ___ 2 $=$ ___________.
$+$; $-$; $+$ $7w^3 - 5w^2 + 3w$	63. $5w^3 - 3w^2 + w + (2w^3 - 2w^2 + 2w)$ $= 5w^3 - 3w^2 + w$ ___ $2w^3$ ___ $2w^2$ ___ $2w$ $=$ ______________.

In Frames 64-65, write each expression as a polynomial in simple form.

Answer	Frame
$-3x - 3$ 5	64. $(x^2 - 3x) + (x^2 - 2) - (3x + 3)$ $= x^2 - 3x + x^2 - 2$ ________ $= 2x^2 - 6x -$ ___.
$-2a^2 - 3b^2$ $a^2 - 3b^2$	65. $(4a^2 + b^2) - (a^2 + b^2) - (2a^2 + 3b^2)$ $= 4a^2 + b^2 - a^2 - b^2$ __________ $=$ _________.

In Frames 66-67, find the sum of the given polynomials by adding columns of like terms.

66.
$$\begin{array}{rrr} 4x^2 & -\ 5x & +\ 7 \\ x^2 & +\ 2x & +\ 5 \\ -3x^2 & -\ x & -\ 2 \\ \hline \_\_\_\_ & -\ 4x & +\ 10. \end{array}$$

67.
$$\begin{array}{rrrr} -3y^3 & -\ 2y^2 & +\ y & -\ 5 \\ y^3 & & -\ 5y & +\ 2 \\ y^3 & +\ 5y^2 & +\ 3y & -\ 1 \\ \hline \_\_\_\_\_\_ & & -\ y & -\ 4. \end{array}$$

$2x^2$; $-y^3 + 3y^2$

In Frames 68-70, subtract $5s^2 - 5s + 7$ from $2s^2 + 7s - 5$ using a vertical arrangement of like terms.

68.
$$\begin{array}{rr} 2s^2 + 7s & -\ 5 \\ \_\_\_\_\_\_ & +\ 7 \\ \hline \_\_\_\_\_\_\_\_. \end{array}$$

$5s^2 - 5s$

69. Replace each term in the bottom polynomial by its negative.
$$\begin{array}{rr} 2s^2 + 7s & -\ 5 \\ \_\_\_\_\_\_ & -\ 7 \\ \hline \_\_\_\_\_\_\_\_. \end{array}$$

$-5s^2 + 5s$

70. Add and obtain:
$$\begin{array}{rrr} 2s^2 & +\ 7s & -\ 5 \\ -5s^2 & +\ 5s & -\ 7 \\ \hline \_\_\_\_\_\_\_\_. \end{array}$$

$-3s^2 + 12s - 12$

In Frames 71-73, subtract $6x^2y - 3xy + xy^2$ from $4x^2y + xy - 5xy^2$ using a vertical arrangement of like terms.

71.
$$\begin{array}{c} 4x^2y + xy - 5xy^2 \\ \_\_\_\_\_\_\_\_\_\_ \\ \hline \_\_\_\_\_\_\_\_\_\_. \end{array}$$

$6x^2y - 3xy + xy^2$

72. Replace each term in the bottom polynomial with its negative.
$$\begin{array}{c} 4x^2y + xy - 5xy^2 \\ \_\_\_\_\_\_\_\_\_\_ \\ \hline \_\_\_\_\_\_\_\_\_\_. \end{array}$$

$-6x^2y + 3xy - xy^2$

73. Add and obtain:
$$\begin{array}{rrr} 4x^2y & +\ xy & -\ 5xy^2 \\ -6x^2y & +\ 3xy & -\ xy^2 \\ \hline \_\_\_\_\_\_\_\_. \end{array}$$

$-2x^2y + 4xy - 6xy^2$

74. **When grouping devices are nested in an expression, it is usually easier to remove the ________ devices and work outward.**

innermost

In Frames 75-79, simplify each given expression.

75. $2x - [3x + (2 - 2x)] = 2x - [\_\_\_\_\_\_\_\_]$
$= \_\_\_\_\_\_\_\_\_\_$
$= \_\_\_\_\_\_.$

$3x + 2 - 2x$
$2x - 3x - 2 + 2x$
$x - 2$

76. $3a + [4a - (a + 2)] = 3a + [\_\_\_\_\_\_\_\_]$
$= 3a + \_\_\_\_\_\_\_\_$
$= \_\_\_\_\_\_.$

$4a - a - 2$
$4a - a - 2$
$6a - 2$

77. $-(y - 2) + [3y - (4 + y) - 1]$
$= -y + 2 + [\_\_\_\_\_\_\_\_]$
$= -y + 2 + \_\_\_\_\_\_\_\_$
$= \_\_\_\_.$

Answers: $3y - 4 - y - 1$; $3y - 4 - y - 1$; $y - 3$

78. $m - \{m - [n - (2n + m)] - 2m\}$
$= m - \{m - [\_\_\_\_\_\_] - 2m\}$
$= m - \{\_\_\_\_\_\_\_\_\}$
$= \_\_\_\_\_\_\_\_$
$= \_\_\_\_.$

Answers: $n - 2n - m$; $m - n + 2n + m - 2m$; $m - m + n - 2n - m + 2m$; $m - n$

79. $[x - (2y + x)] - \{2x - [3x - (2x - y)] + y\}$
$= [x - 2y - x] - \{2x - [\_\_\_\_\_\_] + y\}$
$= -2y - \{\_\_\_\_\_\_\_\_\}$
$= \_\_\_\_\_\_\_\_$
$= \_\_\_\_.$

Answers: $3x - 2x + y$; $2x - 3x + 2x - y + y$; $-2y - 2x + 3x - 2x + y - y$; $-x - 2y$

In Frames 80-84, take $P(x) = x + 1$, $Q(x) = x^2 - 1$, and $R(x) = x^2 + x - 1$.

In Frames 80-81, find the value for $P(x) - [Q(x) - R(x)]$.

80. $P(x) - [Q(x) - R(x)] = x + 1 - [(x^2 - 1) - (\_\_\_\_\_\_)]$
$= x + 1 - [x^2 - 1 - x^2 \_\_\_\_\_\_].$

Answers: $x^2 + x - 1$; $-x + 1$

81. Within the brackets combine like terms to obtain:
$P(x) - [Q(x) - R(x)] = x + 1 - [\_\_\_] = \_\_\_\_.$

Answers: $-x$; $2x + 1$

In Frames 82-84, find the value for $R[Q(x)]$.

82. $R[Q(x)] = [Q(x)]^2 + \_\_\_\_ - 1.$

Answer: $Q(x)$

83. Replace each $Q(x)$ in the right-hand member of the equation with $x^2 - 1$ and obtain $R[Q(x)] = (x^2 - 1)^2 + \_\_\_\_ - 1.$

Answer: $x^2 - 1$

84. Simplify the right-hand member:
$R[Q(x)] = x^4 - 2x^2 + 1 + \_\_\_\_ - 1 = \_\_\_\_\_\_.$

Answers: $x^2 - 1$; $x^4 - x^2 - 1$

2.3

PRODUCTS AND QUOTIENTS OF MONOMIALS

85. For all natural numbers m and n the first law of exponents states that

$$a^m a^n = a^{\_\_\_}.$$

Answer: $m+n$

86. $x^3x^4 = x^{\_\_ + \_\_} = x^{\_\_}.$

Answers: 3; 4; 7

87. $xx^2x^3 = x^{\_\_ + \_\_ + \_\_} = x^{\_\_}.$

Answers: 1; 2; 3; 6

88. $(3x^2)(2x^3) = (3 \cdot 2)(\_\_\_ \cdot \_\_\_) = 6\_\_\_.$

Answers: x^2; x^3; x^5

x^2; y^3; $-6x^3y^4$

89. $(-2xy)(3x^2y^3) = (-2 \cdot 3)(x \cdot \_\_\_)(y \cdot \_\_\_) = \_\_\_\_\_\_\_\_.$

x; x; y^2; y; z; z^2
$30x^4y^4z^4$

90. $(-5x^2yz)(2xy^2z)(-3xyz^2)$
$= [(-5)(2)(-3)][(x^2)(\_\_)(\_\_)][y(\_\_)(\_\_)][z(\_\_)(\_\_)]$
$= \_\_\_\_\_\_\_\_.$

n; $(n+2)$; $2n+2$

91. $x^n \cdot x^{n+2} = x^{\_\_+\_\_} = x^{\_\_}.$

$-6a^{3n}$

92. $(2a^n)(-3a^{2n}) = [(2)(-3)][a^n \cdot a^{2n}] = \_\_\_\_\_.$

4; a^{n-2}
$n-2$
a^{3n-1}

93. $(-3a^{2n+1})(4a^{n-2}) = [(-3)(\_\_)][(a^{2n+1})(\_\_\_\_)]$
$= -12a^{2n+1+\_\_\_}$
$= -12(\_\_\_\_).$

a^{m-n}

94. For $m > n$, the second law of exponents states that
$\frac{a^m}{a^n} = \_\_\_\_\_\_ \quad (a \neq 0).$

$\frac{y^4}{y}$; $-2x^2y^3$

95. $\frac{8x^3y^4}{-4xy} = \left(\frac{8}{-4}\right)\left(\frac{x^3}{x}\right)\left(\_\_\_\right) = \_\_\_\_\_.$

$\frac{c^4}{c^2}$; $-3a^3b^2c^2$

96. $\frac{-15a^5b^3c^4}{5a^2bc^2} = \left(\frac{-15}{5}\right)\left(\frac{a^5}{a^2}\right)\left(\frac{b^3}{b}\right)\left(\_\_\_\right) = \_\_\_\_\_\_\_.$

$4(x - y)^2$

97. $\frac{12(x - y)^4}{3(x - y)^2} = \left(\frac{12}{3}\right)\left[\frac{(x - y)^4}{(x - y)^2}\right] = \_\_\_\_\_\_\_\_.$

z^2; $3xy^2z^2$

98. $\frac{-18x^2y^3z^2}{-6xy} = \left(\frac{-18}{-6}\right)\left(\frac{x^2}{x}\right)\left(\frac{y^3}{y}\right)(\_\_\_) = \_\_\_\_\_.$

x^{n+4}

99. $\frac{x^{2n+2}}{x^{n-2}} = x^{(2n+2)-(n-2)} = \_\_\_\_.$

$2n+1-n$
$a^{m+1}y^{n+1}$

100. $\frac{a^{2m}y^{2n+1}}{a^{m-1}y^n} = \left(\frac{a^{2m}}{a^{m-1}}\right)\left(\frac{y^{2n+1}}{y^n}\right) = a^{2m-(m-1)}y^{(\_\_\_\_)}$
$= \_\_\_\_\_\_.$

2.4

PRODUCTS OF POLYNOMIALS

a; a^2; $3a^3 - 3a^4$

101. $3a^2(a - a^2) = 3a^2(\_\_) - 3a^2(\_\_\_) = \_\_\_\_\_\_\_.$

$-2ab$; $-2ab$; $-2ab$
$-2a^2b^3 - 2a^2b^2 + 2a^3b^2$

102. $-2ab(ab^2 + ab - a^2b)$
$= (\_\_\_\_)(ab^2) + (\_\_\_\_)(ab) - (\_\_\_\_)(a^2b)$
$= \_\_\_\_\_\_\_\_\_\_\_\_.$

3; a^{n-1}
a^{3n}; a^{3n}
$6a^{3n} - 2a^{4n-1}$

103. $2a^{3n}(3 - a^{n-1}) = 2a^{3n}(\_\_) - 2a^{3n}(\_\_\_\_)$
$= (2 \cdot 3)(\_\_\_) - 2[(\_\_\_)(a^{n-1})]$
$= \_\_\_\_\_\_\_\_.$

104. **The following binomial products should be recognized on sight:**

$(x + a)(x + b) = x^2 + (\underline{\qquad})x + ab.$

$(x + a)^2 = x^2 + \underline{\quad} + a^2.$

$(x + a)(x - a) = \underline{\qquad}.$

$(ax + by)(cx + dy)$

$= (\underline{\quad})x^2 + (\underline{\qquad})xy + (\underline{\quad})y^2.$

$a + b$

$2ax$

$x^2 - a^2$

$ac;\quad ad + bc;\quad bd$

105. $(x - 3)(x + 1) = x^2 + [(\underline{\quad}) + (\underline{\quad})]x - 3$

$= x^2 + (\underline{\quad})x - 3.$

$-3;\quad 1$

-2

106. $(x - 4)(x - 6) = x^2 + [(\underline{\quad}) + (\underline{\quad})]x + \underline{\quad}$

$= \underline{\qquad\qquad}.$

$-4;\quad -6;\quad 24$

$x^2 - 10x + 24$

107. $(x - 5)^2 = x^2 + 2(\underline{\quad})x + \underline{\quad}.$

$-5;\quad 25$

108. $(x + 4)^2 = x^2 + 2(\underline{\quad}) + \underline{\quad}.$

$4x;\quad 16$

109. $(2a + 3)^2 = (2a)^2 + 2(\underline{\quad})(\underline{\quad}) + (3)^2 = 4a^2 + \underline{\quad} + 9.$

$2a;\quad 3;\quad 12a$

110. $(3a - 4)^2 = (3a)^2 + 2(3a)(\underline{\quad}) + (\underline{\quad})^2$

$= 9a^2 + \underline{\quad} + \underline{\quad}.$

$-4;\quad -4$

$-24a;\quad 16$

111. $(x - 3)(x + 3) = x^2 - (\underline{\quad})^2 = \underline{\qquad}.$

$3;\quad x^2 - 9$

112. $(2x + 5)(2x - 5) = (\underline{\quad})^2 - (\underline{\quad})^2 = \underline{\qquad}.$

$2x;\quad 5;\quad 4x^2 - 25$

113. $(a^{3n} + 2)(a^{3n} - 2) = (\underline{\quad})^2 - (\underline{\quad})^2 = \underline{\qquad}.$

$a^{3n};\quad 2;\quad a^{6n} - 4$

114. $(2x + 1)(3x^2 - x + 2)$

$= (2x + 1)(3x^2) + (2x + 1)(-x) + (2x + 1)(\underline{\quad})$

$= 6x^3 + 3x^2 + \underline{\qquad} + \underline{\qquad}$

$= \underline{\qquad\qquad}.$

2

$-2x^2 - x;\quad 4x + 2$

$6x^3 + x^2 + 3x + 2$

In Frame 115, use vertical form to obtain the product of the polynomials given in Frame 114.

115.
$$\begin{array}{l} 3x^2 - x + 2 \\ \underline{2x \quad + 1} \\ 6x^3 - 2x^2 + \underline{\quad} \\ \underline{\underline{\qquad} - x + \underline{\quad}} \\ 6x^3 + x^2 \ \underline{\qquad\qquad} . \end{array}$$

$4x$

$3x^2;\quad 2$

$+ 3x + 2$

116. $(z + 2)(z - 5)(z - 1) = (z + 2)[(\underline{\qquad})(\underline{\qquad})]$

$= (z + 2)[z^2 \underline{\qquad}]$

$= z(z^2 - 6z + 5) + \underline{\qquad\qquad}$

$= z^3 - 6z^2 + 5z + \underline{\qquad\qquad}$

$= z^3 - 4z^2 \ \underline{\qquad}.$

$z - 5;\quad z - 1$

$-6z + 5$

$2(z^2 - 6z + 5)$

$2z^2 - 12z + 10$

$-7z + 10$

In Frames 117-118, use vertical form to obtain the product of the polynomials given in Frame 116.

117. Commencing with step 2 of Frame 116, we write

$$\begin{array}{l} z^2 - 6z + 5 \\ \underline{\qquad\quad} \\ \underline{\qquad\qquad\quad}. \end{array}$$

Answer: $z + 2$

118.
$$\begin{array}{l} z^2 - 6z + 5 \\ \underline{z \quad + 2 \qquad\quad} \\ z^3 - 6z^2 + 5z \\ \qquad 2z^2 \; \underline{\qquad\qquad} \\ \underline{\qquad\qquad\qquad\qquad} \\ z^3 \; \underline{\qquad\qquad\qquad}. \end{array}$$

Answers: $-12z + 10$; $-4z^2 - 7z + 10$

119. $(a + 2)(a - 3)(a^2 - a + 2)$

$= [(\underline{\qquad})(\underline{\qquad})](a^2 - a + 2)$

$= [a^2 \underline{\qquad}](a^2 - a + 2)$

$= a^2(a^2 - a + 2) - a(a^2 - a + 2) \underline{\qquad\qquad}$

$= a^4 - a^3 + 2a^2 - a^3 + a^2 - 2a \underline{\qquad\qquad}$

$= a^4 \underline{\qquad\qquad\qquad}.$

Answers: $a + 2$; $a - 3$ / $-a - 6$ / $-6(a^2 - a + 2)$ / $-6a^2 + 6a - 12$ / $-2a^3 - 3a^2 + 4a - 12$

In Frame 120, use vertical form to obtain the product of the polynomials given in Frame 119.

120. Commencing with step 3 of Frame 119, we write

$$\begin{array}{l} a^2 - a - 6 \\ \underline{a^2 - a + 2 \quad} \\ a^4 - a^3 \; \underline{\quad} \\ \quad - a^3 + a^2 \; \underline{\quad} \\ \qquad\quad \underline{\qquad\qquad} \\ \underline{\qquad\qquad\qquad\qquad} \\ a^4 - 2a^3 \; \underline{\qquad\qquad}. \end{array}$$

Answers: $-6a^2$; $+6a$; $2a^2 - 2a - 12$; $-3a^2 + 4a - 12$

121. $(a^{2n} - 1)(a^{2n} + 2) = a^{2n+}\underline{\quad} - a^{2n} + \underline{\qquad} - 2$

$= a^{4n} \underline{\qquad\quad}.$

Answers: $2n$; $2a^{2n}$ / $+a^{2n} - 2$

In Frames 122-123, simplify each given polynomial.

122. $[(a + 2) - 3(2a - 1) + 2] = [a + 2 - 6a \underline{\quad} + 2]$

$= -5a + \underline{\quad}.$

Answers: $+3$; 7

123. $-2[2 -(1 - 2(a + 1) - a) + 2a]$

$= -2[2 - (1 \underline{\qquad} - a) + 2a]$

$= -2[2 - (-1 \underline{\quad}) + 2a]$

$= -2[2 + 1 \underline{\quad} + 2a]$

$= -2[3 \underline{\quad}]$

$= \underline{\qquad\quad}.$

Answers: $-2a - 2$ / $-3a$ / $+3a$ / $+5a$ / $-6 - 10a$

Answer	Frame
prime	124. If a natural number has no factors that are natural numbers other than itself or 1, it is said to be a ________ number.
composite	125. If a natural number greater than 1 is not a prime number, it is said to be ________.
2; composite	126. 8 = (2)(2)(____). Therefore 8 is a ________ number.
3; prime	127. The only factors of 3 are 1 and ___, so 3 is a ______ number.
2; composite	128. 12 = 2(___)(3). Therefore 12 is a ________ number.
13; composite	129. 26 = (2)(___). Therefore 26 is a ________ number.
11 prime	130. The only factors of 11 are 1 and (___). Therefore 11 is a ______ number.
completely	131. When a composite number is expressed as a product of prime factors, it is said to be ________ factored.

In Frames 132-139, express the given number in completely factored form.

Answer	Frame
2; 3	132. 8 = (2)(2)(___). 133. 12 = (2)(2)(___).
3; 3; 3	134. 15 = (5)(___). 135. 18 = (2)(___)(___).
5; 2; 7	136. -20 = -1(2)(2)(___). 137. -14 = -1(___)(___).
2; 2; 2; 3	138. 24 = (___)(___)(___)(___).
2; 3; 5	139. -30 = -1(___)(___)(___).

2.5

FACTORING MONOMIALS FROM POLYNOMIALS

A polynomial with integral coefficients is in completely factored form if:

Answer	Frame
integral	140. **It is written as a product with ________ (what kind?) coefficients.**
factor	141. **No polynomial (other than a monomial) in the factored form contains a ______ with integral coefficients.**

In Frames 142-154, express the given polynomial in completely factored form.

Answer	Frame
2	142. $3x + 6 = 3 \cdot x + 3 \cdot 2 = 3(x + \_\_\_)$.
2	143. $6x - 12 = 6 \cdot x - 6 \cdot 2 = 6(x - \_\_\_)$.

Answer	Frame
1	144. $xy^2 + x = x \cdot y^2 + x \cdot 1 = x(y^2 + \underline{\quad})$.
$4n$; $4n$	145. $15m^2n + 20mn^2 = (5mn)(3m) + (5mn)(\underline{\quad}) = 5mn(3m + \underline{\quad})$.
$2a^3$; $6a$; 3 $2a^3 + 6a - 3$	146. $6a^4 + 18a^2 - 9a = 3a(\underline{\quad}) + 3a(\underline{\quad}) - 3a(\underline{\quad})$ $= 3a(\underline{\qquad\qquad})$.
$x^2 + bx + b$	147. $ax^2 + abx + ab = a(\underline{\qquad\qquad})$.
$x^2yz + 3x - 2$	148. $x^2y^2z^2 + 3xyz - 2yz = yz(\underline{\qquad\qquad})$.
$4xy$; $5x - 6y$	149. $20x^2y - 24xy^2 = (\underline{\quad})(\underline{\qquad})$.
$3rs$; $4s + 5r - 1$	150. $12rs^2 + 15r^2s - 3rs = (\underline{\quad})(\underline{\qquad\qquad})$.
a	151. $b(a - 2) + a(a - 2) = (a - 2)(b + \underline{\quad})$.
$3y$	152. $4x(3a - b) - 3y(3a - b) = (3a - b)(4x - \underline{\quad})$.
$3x - 4y$	153. $3x(a - b) - 4y(a - b) = (a - b)(\underline{\qquad})$.
$r^2 - s^2$; $5p + 4q$	154. $5p(r^2 - s^2) + 4q(r^2 - s^2) = (\underline{\qquad})(\underline{\qquad})$.
$b - a$	155. $a - b$ and $b - a$ are negatives of each other. That is, $a - b = -(\underline{\qquad})$.
$2x$	156. $2x - y = -(y - \underline{\quad})$.
$3n - 5m$	157. $5m - 3n = -(\underline{\qquad})$.
$2 - m^2n^2$	158. $m^2n^2 - 2 = -(\underline{\qquad})$.
$b^2 - 2c^2$	159. $2a^2 - b^2 + 2c^2 = 2a^2 - (\underline{\qquad})$.
$-2x + 4y - 3z$	160. $2x - 4y + 3z = -(\underline{\qquad\qquad})$.
4	161. $-2x + 8 = -2(x - \underline{\quad})$.
-2	162. $-4x + 6 = \underline{\quad}(2x - 3)$.
$-3a$	163. $-6a + 15ab = (\underline{\quad})(2 - 5b)$.
$1 - y + y^2$	164. $-y + y^2 - y^3 = -y(\underline{\qquad\qquad})$.

In Frames 165-171, factor completely. Assume that the variables in the exponents denote natural numbers.

Answer	Frame
y^n; 1	165. $y^{2n} + y^n = y^n \cdot y^n + 1 \cdot y^n = y^n(\underline{\quad} + \underline{\quad})$.
x^n; 2	166. $x^{n+3} - 2x^3 = x^n \cdot x^3 - 2 \cdot x^3 = x^3(\underline{\quad} - \underline{\quad})$.

$1;\ x^n;\ 1$

167. $x^{4n} - x^{3n} = x^n x^{3n} - (\underline{\quad})x^{3n} = x^{3n}(\underline{\quad\quad} - \underline{\quad})$.

$a^{n+1};\ a^2 + 3$

168. $a^{n+3} + 3a^{n+1} = a^2 \cdot a^{n+1} + 3 \cdot a^{n+1} = (\underline{\quad\quad})(\underline{\quad\quad\quad})$.

$y^n;\ y^2 + 3y - 1$

169. $y^{n+2} + 3y^{n+1} - y^n = y^2 \cdot y^n + 3y \cdot y^n - 1 \cdot y^n$
$= (\underline{\quad})(\underline{\quad\quad\quad\quad})$.

$n;\ 1;\ x^n + 1$

170. $-x^{3n} - x^{2n} = -x^{2n} \cdot x^{(\underline{\quad})} - x^{2n}(\underline{\quad}) = -x^{2n}(\underline{\quad\quad\quad})$.

$y^2;\ -1;\ -y^a$

171. $-y^{a+2} + y^a = -y^a(\underline{\quad}) + (-y^a)(\underline{\quad}) = (\underline{\quad\quad})(y^2 - 1)$.

2.6

FACTORING QUADRATIC POLYNOMIALS

b

172. $x^2 + (a + b)x + ab = (x + a)(x + \underline{\quad})$.

2

173. $x^2 + 5x + 6 = x^2 + (3 + 2)x + 3 \cdot 2 = (x + 3)(x + \underline{\quad})$.

-3

$5;\ 3$

174. $x^2 + 2x - 15 = x^2 + (5 - 3)x + (5)(\underline{\quad})$
$= (x + \underline{\quad})(x - \underline{\quad})$.

$-4;\ 3$

$4;\ 3$

175. $x^2 - x - 12 = x + (-4 + 3) + (\underline{\quad})(\underline{\quad})$
$= (x - \underline{\quad})(x + \underline{\quad})$.

$5;\ 2$

$5;\ 2$

176. $x^2 + 7x + 10 = x^2 + (5 + 2)x + (\underline{\quad})(\underline{\quad})$
$= (x + \underline{\quad})(x + \underline{\quad})$.

$-6;\ -2$

$x - 6;\ x - 2$

177. $x^2 - 8x + 12 = x^2 + [(-6) + (-2)]x + (\underline{\quad})(\underline{\quad})$
$= (\underline{\quad\quad})(\underline{\quad\quad})$.

$m + 2p;\ m - 9p$

178. $m^2 - 7mp - 18p^2 = (\underline{\quad\quad})(\underline{\quad\quad})$.

$xy + 5;\ xy + 7$

179. $x^2y^2 + 12xy + 35 = (\underline{\quad\quad})(\underline{\quad\quad})$.

$y - 9;\ y - 4$

180. $y^2 - 13y + 36 = (\underline{\quad\quad})(\underline{\quad\quad})$.

$x \pm a$

181. $x^2 \pm 2ax + a^2 = (\underline{\quad\quad})^2$.

2

182. $x^2 + 4x + 4 = x^2 + 2 \cdot 2x + 2^2 = (x + \underline{\quad})^2$.

3

183. $x^2 + 6x + 9 = x^2 + 2 \cdot 3x + 3^2 = (x + \underline{\quad})^2$.

4

184. $x^2 - 8x + 16 = x^2 - 2 \cdot 4x + 4^2 = (x - \underline{\quad})^2$.

5; 5; 5	185. $x^2 - 10x + 25 = x^2 - 2(\underline{\quad})x + (\underline{\quad})^2 = (x - \underline{\quad})^2$.
8; 8; 8	186. $x^2 - 16x + 64 = x^2 - 2(\underline{\quad})x + (\underline{\quad})^2 = (x - \underline{\quad})^2$.
6; 6 $ab - 6$	187. $a^2b^2 - 12ab + 36 = (ab)^2 - 2(\underline{\quad})ab + (\underline{\quad})^2$ $= (\underline{\qquad\quad})^2$.
$7mn$; $7n$; $m + 7n$	188. $m^2 + 14mn + 49n^2 = m^2 + 2(\underline{\qquad}) + (\underline{\quad})^2 = (\underline{\qquad\quad})^2$.
$y - 9$	189. $y^2 - 18y + 81 = (\underline{\qquad\quad})^2$.
$xy + 10$	190. $x^2y^2 + 20xy + 100 = (\underline{\qquad\quad})^2$.
$m + 5p$	191. $m^2 + 10mp + 25p^2 = (\underline{\qquad\quad})^2$.
$x - a$	192. $x^2 - a^2 = (x + a)(\underline{\qquad\quad})$.
3; 3; 3	193. $x^2 - 9 = x^2 - (\underline{\quad})^2 = (x + \underline{\quad})(x - \underline{\quad})$.
1; 1; 1	194. $x^2 - 1 = x^2 - (\underline{\quad})^2 = (x + \underline{\quad})(x - \underline{\quad})$.
2; 2; 2	195. $x^2 - 4 = x^2 - (\underline{\quad})^2 = (x + \underline{\quad})(x - \underline{\quad})$.
y; 10; y; 10; y; 10	196. $y^2 - 100 = (\underline{\quad})^2 - (\underline{\quad})^2 = (\underline{\quad} + \underline{\quad})(\underline{\quad} - \underline{\quad})$.
ab; 4; ab; 4; ab; 4	197. $a^2b^2 - 16 = (\underline{\quad})^2 - (\underline{\quad})^2 = (\underline{\quad} + \underline{\quad})(\underline{\quad} - \underline{\quad})$.
$2x$; $3y$; $2x$; $3y$; $2x$; $3y$	198. $4x^2 - 9y^2 = (\underline{\quad})^2 - (\underline{\quad})^2 = (\underline{\quad} + \underline{\quad})(\underline{\quad} - \underline{\quad})$.
1; $3xy$; $1 + 3xy$; $1 - 3xy$	199. $1 - 9x^2y^2 = (\underline{\quad})^2 - (\underline{\quad})^2 = (\underline{\qquad\quad})(\underline{\qquad\quad})$.
$4x$; $5y$; $4x + 5y$; $4x - 5y$	200. $16x^2 - 25y^2 = (\underline{\quad})^2 - (\underline{\quad})^2 = (\underline{\qquad\quad})(\underline{\qquad\quad})$.
a^n; 3; $a^n - 3$; $a^n + 3$	201. $a^{2n} - 9 = (\underline{\quad})^2 - (\underline{\quad})^2 = (\underline{\qquad\quad})(\underline{\qquad\quad})$.
$x^n + y^{2n}$; $x^n - y^{2n}$	202. $x^{2n} - y^{4n} = (\underline{\qquad\quad})(\underline{\qquad\quad})$.
$cx + dy$	203. $acx^2 + (ad + bc)xy + bdy^2 = (ax + by)(\underline{\qquad\quad})$.
x	204. $3x^2 - 4x + 1 = (3x - 1)(\underline{\quad} - 1)$.
$8x$; x	205. $8x^2 + 21x - 9 = (\underline{\quad} - 3)(\underline{\quad} + 3)$.
$4a$; a	206. $4a^2 + 5a + 1 = (\underline{\quad} + 1)(\underline{\quad} + 1)$.
8; 1	207. $9x^2 + 21x - 8 = (3x + \underline{\quad})(3x - \underline{\quad})$.
$5x - 6$; $2x + 3$	208. $10x^2 + 3x - 18 = (\underline{\qquad\quad})(\underline{\qquad\quad})$.
$3x + a$; $x + 2a$	209. $3x^2 + 7ax + 2a^2 = (\underline{\qquad\quad})(\underline{\qquad\quad})$.
$3xy - 1$; $3xy - 1$	210. $9x^2y^2 - 6xy + 1 = (\underline{\qquad\quad})(\underline{\qquad\quad})$.
$2ab - 3$; $2ab - 3$	211. $4a^2b^2 - 12ab + 9 = (\underline{\qquad\quad})(\underline{\qquad\quad})$.
$3y - 5$; $3y + 2$	212. $9y^2 - 9y - 10 = (\underline{\qquad\quad})(\underline{\qquad\quad})$.
$3m - 7$; $2m + 3$	213. $6m^2 - 5m - 21 = (\underline{\qquad\quad})(\underline{\qquad\quad})$.
factored	214. If a polynomial of more than one term contains a monomial factor common to each of its terms, this monomial factor should be ________ from the polynomial before seeking other factors.

Answers	Frames
a	215. $4a^3 + 5a^2 + a = (\_\_)(4a^2 + 5a + 1)$
a; $4a + 1$; $a + 1$	$= \_\_(\_\_\_\_)(\_\_\_\_)$.
2; 2; $x - 5$; $x + 2$	216. $2x^2 - 6x - 20 = \_\_(x^2 - 3x - 10) = \_\_(\_\_\_\_)(\_\_\_\_)$.
x^2; $1 + 2y$; $1 - 2y$	217. $x^2 - 4x^2y^2 = \_\_(1 - 4y^2) = x^2(\_\_\_\_)(\_\_\_\_)$.
ab; $a^2 - b^2$; $a + b$; $a - b$	218. $a^3b - ab^3 = \_\_(\_\_\_\_) = ab(\_\_\_\_)(\_\_\_\_)$.
5; $2x - 3y$	219. $20x^2 - 60xy + 45y^2 = \_\_(4x^2 - 12xy + 9y^2) = 5(\_\_\_\_)^2$.
	220. **All polynomials should be factored completely.**
$x^2 - 1$	221. $x^4 + 3x^2 - 4 = (x^2 + 4)(\_\_\_\_)$
$x + 1$; $x - 1$	$= (x^2 + 4)(\_\_\_\_)(\_\_\_\_)$.
$a^2 - 4$; $a^2 - 1$	222. $a^4 - 5a^2 + 4 = (\_\_\_\_)(\_\_\_\_)$
$a - 1$; $a + 1$	$= (a + 2)(a - 2)(\_\_\_\_)(\_\_\_\_)$.
$x^2 + y^2$	223. $x^4 - y^4 = (\_\_\_\_)(x^2 - y^2)$
$x + y$; $x - y$	$= (x^2 + y^2)(\_\_\_\_)(\_\_\_\_)$.
$4x^2 + 3a^2$	224. $4x^4 - 33a^2x^2 - 27a^4 = (\_\_\_\_)(x^2 - 9a^2)$
$x - 3a$; $x + 3a$	$= (4x^2 + 3a^2)(\_\_\_\_)(\_\_\_\_)$.
3	225. $9x^4 - 30x^2 + 9 = \_\_(3x^4 - 10x^2 + 3)$
$3x^2 - 1$; $x^2 - 3$	$= 3(\_\_\_\_)(\_\_\_\_)$.
2	226. $8x^4 - 66a^2x^2 - 54a^4 = \_\_(4x^4 - 33a^2x^2 - 27a^4)$
$x^2 - 9a^2$	$= 2(4x^2 + 3a^2)(\_\_\_\_)$
$x + 3a$; $x - 3a$	$= 2(4x^2 + 3a^2)(\_\_\_\_)(\_\_\_\_)$.
$x^{2n} - 1$; $x^{2n} -1$	227. $x^{4n} - 2x^{2n} + 1 = (\_\_\_\_)(\_\_\_\_)$
$x^n - 1$; $x^n + 1$	$= (\_\_\_\_)(\_\_\_\_)(x^n - 1)(x^n + 1)$
$x^n - 1$; $x^n + 1$	$= (\_\_\_\_)^2(\_\_\_\_)^2$.

2.7

FACTORING OTHER POLYNOMIALS

Answers	Frames
b	228. $ax + ay + bx + by = a(x + y) + \_\_(x + y)$
b	$= (x + y)(a + \_\_)$.
b	229. $y^2 + ya - by - ab = y(y + a) - \_\_(y + a)$
b	$= (y + a)(y - \_\_)$.
x; y; x; y	230. $x^3 + x^2y + xy + y^2 = x^2(\_\_ + \_\_) + y(\_\_ + \_\_)$
y	$= (x + y)(x^2 + \_\_)$.
$1 - y$; $1 - x$	231. $1 - y - x + xy = 1(1 - y) - x(\_\_\_\_) = (1 - y)(\_\_\_\_)$.
n	232. $2m^2 + 3m - 2mn - 3n = m(2m + 3) - \_\_(2m + 3)$
$2m + 3$	$= (\_\_\_\_)(m - n)$.

$3 - x^3$; $3 - x^3$
$4 - y^2$
$2 - y$; $2 + y$

233. $12 - 4x^3 - 3y^2 + x^3y^2 = 4(\underline{\qquad\quad}) - y^2(\underline{\qquad\quad})$
$= (3 - x^3)(\underline{\qquad\quad})$
$= (3 - x^3)(\underline{\qquad\quad})(\underline{\qquad\quad})$.

$-$; $+$

234. $\mathbf{x^3 + a^3 = (x + a)(x^2 \underline{\quad} ax \underline{\quad} a^2)}$.

$+$; $+$

235. $\mathbf{x^3 - a^3 = (x - a)(x^2 \underline{\quad} ax \underline{\quad} a^2)}$.

$-$; $+$
$x^2 - 2x + 4$

236. $x^3 + 8 = x^3 + (2)^3 = (x + 2)(x^2 \underline{\quad} 2x \underline{\quad} 2^2)$
$= (x + 2)(\underline{\qquad\qquad})$.

2; $+$; $+$
$x^2 + 2x + 4$

237. $x^3 - 8 = x^3 - (\underline{\quad})^3 = (x - 2)(x^2 \underline{\quad} 2x \underline{\quad} 2^2)$
$= (x - 2)(\underline{\qquad\qquad})$.

1; $-$; $+$
$y^2 - y + 1$

238. $y^3 + 1 = y^3 + (\underline{\quad})^3 = (y + 1)(y^2 \underline{\quad} y \underline{\quad} 1^2)$
$= (y + 1)(\underline{\qquad\qquad})$.

$+$; $+$
$y^2 + y + 1$

239. $y^3 - 1 = y^3 - 1^3 = (y - 1)(y^2 \underline{\quad} y \underline{\quad} 1^2)$
$= (y - 1)(\underline{\qquad\qquad})$.

$3n$
$-$; $+$
$m^2 - 3mn + 9n^2$

240. $m^3 + 27n^3 = m^3 + (\underline{\quad})^3$
$= (m + 3n)[m^2 \underline{\quad} 3mn \underline{\quad} (3n)^2]$
$= (m + 3n)(\underline{\qquad\qquad})$.

$3n$
m; $3n$; $+$; $+$
$m^2 + 3mn + 9n^2$

241. $m^3 - 27n^3 = m^3 - (\underline{\quad})^3$
$= (\underline{\quad} - \underline{\quad})[m^2 \underline{\quad} 3mn \underline{\quad} (3n)^2]$
$= (m - 3n)(\underline{\qquad\qquad})$.

5; $2y$
5; $5 \cdot 2y$; $2y$
$25 - 10y + 4y^2$

242. $125 + 8y^3 = (\underline{\quad})^3 + (\underline{\quad})^3$
$= (5 + 2y)[(\underline{\quad})^2 - (\underline{\qquad}) + (\underline{\quad})^2]$
$= (5 + 2y)(\underline{\qquad\qquad})$.

$3x^2$; $2y$
$3x^2$; $2y$
$9x^4 + 6x^2y + 4y^2$

243. $27x^6 - 8y^3 = (\underline{\quad})^3 - (\underline{\quad})^3$
$= (3x^2 - 2y)[(\underline{\quad})^2 + (3x^2 \cdot 2y) + (\underline{\quad})^2]$
$= (3x^2 - 2y)(\underline{\qquad\qquad})$.

$4a$; $3b$
$4a$; $4a \cdot 3b$; $3b$
$16a^2 + 12ab + 9b^2$

244. $64a^3 - 27b^3 = (\underline{\quad})^3 - (\underline{\quad})^3$
$= (4a - 3b)[(\underline{\quad})^2 + (\underline{\qquad}) + (\underline{\quad})^2]$
$= (4a - 3b)(\underline{\qquad\qquad})$.

$x - 1$
x
$x^2 - 3x + 3$

245. $(x - 1)^3 + 1 = (x - 1)^3 + 1^3$
$= [(x - 1) + 1][(\underline{\qquad})^2 - 1 \cdot (x - 1) + 1^2]$
$= \underline{\quad}(x^2 - 2x + 1 - x + 1 + 1)$
$= x(\underline{\qquad\qquad})$.

TEST PROBLEMS

In Problem 1, identify each given polynomial as a monomial, binomial, or trinomial and give its degree.

1. a. $3y^2 + 5y - 7$. b. $m^2x^2 - 3mx^2y$.

2. If $P(x) = 3x^2 + 2x - 1$, find
   a. $P(0)$. b. $P(-1)$.

3. If $P(x) = x^2 - 3x$ and $Q(x) = x^4 - 2x^2$, find
   a. $P[Q(1)]$. b. $Q[P(1)]$.

In Problems 4-12, write each expression as a polynomial in simple form. Assume that all variables in exponents are natural numbers and no denominator equals zero.

4. $(2x - 5y) - (5x - 2y)$. 5. $3c - [2c - (3c - 1) + 1]$.

6. $(-3xy^2)(2x^2y)(xy)$. 7. $-3[(x - 3) - (x + 2)^2]$.

8. $\dfrac{-12x^3y^2z^3}{3xy^2z^2}$.

9. $y^{2n-1}y^{n+1}$.

10. $\dfrac{a^{3n}b^{4n+2}}{a^{2n}b^2}$. 11. $5p^3(p^2 - 2p + 1)$.

12. $(2x^n + 1)(2x^n - 1)$.

13. Write in completely factored form.
    a. 48 b. 112

In Problems 14-19, factor each polynomial completely. Assume that all variables in exponents are natural numbers.

14. $5m^3 + 2m^2 - m$. 15. $y^{n+3} + 4y^n$.

16. $3x^2 + 3xy - 18y^2$. 17. $8a^3 - b^3$.

18. $3m^2 - 5m + 3mn - 5n$. 19. $9x^4 - 16x^2y^2$.

3

FRACTIONS

3.1

BASIC PROPERTIES

a; b; $-a$

1. $\frac{-a}{b} = \frac{(\quad)}{-b} = -\frac{a}{(\quad)} = -\frac{(\quad)}{-b}$.

$\frac{a}{b} = -\frac{-a}{b} = -\frac{a}{-b} = \frac{-a}{-b}$ $(b \neq 0)$.

$\frac{-a}{b}$

2. If $a,b > 0$, the standard forms of a fraction are $\frac{a}{b}$ and $(\_\_)$.

$\frac{b}{b - a}$

3. If the denominator of a fraction is an expression containing more than one term, standard form will be either $\frac{-b}{a - b}$ or ____, $(a \neq b)$.

denominator

4. A fraction is undefined for any value of the variable that gives a value of zero to its ________.

In Frames 5-15, express the given fraction in standard form and specify any real values of the variables for which the fraction is undefined.

-6; y; y

5. $\frac{6}{-x} = \frac{(\quad)}{x}$ $(x \neq 0)$.

6. $-\frac{x}{-y} = \frac{x}{(\quad)}$ $(\_\_ \neq 0)$.

$\frac{5}{3}$; $\frac{5}{8}$

7. $\frac{-5}{-3} = \_\_$.

8. $-\frac{-5}{8} = \_\_$.

$\frac{5}{8}$; x,y

9. $-\frac{5}{-8} = \_\_$.

10. $\frac{-5}{-xy} = \frac{5}{xy}$ $(\_\_ \neq 0)$.

1; 3

11. $\frac{-1}{3 - x} = \frac{(\quad)}{x - 3}$ $(x \neq \_\_)$.

2; 5

12. $\frac{-2}{5 - x} = \frac{(\quad)}{x - 5}$ $(x \neq \_\_)$.

$2 - 3y$; 4

13. $\frac{3y - 2}{4 - x} = \frac{(\quad)}{x - 4}$ $(x \neq \_\_)$.

-2; $2y$

14. $-\dfrac{-2}{2y - x} = \dfrac{(\quad)}{x - 2y}$ $(x \neq \_\_\_)$.

$a - b$; $4b$

15. $-\dfrac{-a + b}{3a - 4b} = \dfrac{(\quad\quad)}{3a - 4b}$ $(3a \neq \_\_\_)$.

3.2

REDUCING AND BUILDING FRACTIONS

factor

16. A fraction is in lowest terms if the numerator and denominator have no common ______ other than one.

In Frames 17-29, reduce each fraction to lowest terms. Assume no denominator is zero.

a^2bc

17. $\dfrac{a^4b^2c}{a^2b} = \dfrac{(a^2b)\cdot(a^2bc)}{a^2b} = \_\_\_\_\_$.

$\dfrac{1}{xz}$

18. $\dfrac{x^3y^2}{x^4y^2z} = \dfrac{x^3y^2}{(x^3y^2)\cdot xz} = \dfrac{\quad}{\_\_\_}$.

$7rt^3$; $\dfrac{s^2}{2rt}$

19. $\dfrac{7r(-s)^2t^3}{14r^2t^4} = \dfrac{7rs^2t^3}{14r^2t^4} = \dfrac{(7rt^3)s^2}{(\quad\quad)2rt} = \dfrac{\quad}{\_\_\_\_\_}$.

$a + b$; 4

20. $\dfrac{4a + 4b}{a + b} = \dfrac{4(\quad\quad)}{1\cdot(a + b)} = \_\_\_$.

$m - 1$; m^3

21. $\dfrac{m^4 - m^3}{m - 1} = \dfrac{m^3(\quad\quad)}{1\cdot(m - 1)} = \_\_\_$.

$x - y$; -1; -1

22. $\dfrac{x - y}{y - x} = \dfrac{1\cdot(x - y)}{-1(\quad\quad)} = \dfrac{1}{(\quad)} = \_\_\_$.

$x + 2$; $x - 2$; $x - 2$

23. $\dfrac{x^2 - 4}{x + 2} = \dfrac{(\quad\quad)(\quad\quad)}{(x + 2)\cdot 1} = \_\_\_\_\_$.

$x - 2$; $-(x + 2)$ or $-x - 2$

24. $\dfrac{x^2 - 4}{2 - x} = \dfrac{(x + 2)(x - 2)}{-1(\quad\quad)} = \_\_\_\_\_\_$.

$2x - 3$

25. $\dfrac{8x - 12}{4} = \dfrac{4(2x - 3)}{4} = \_\_\_\_\_$.

ay; $y - 1$; $y - 1$

26. $\dfrac{ay^2 - ay}{ay} = \dfrac{(\quad)(\quad\quad)}{ay} = \_\_\_\_\_$.

$y - 3$; $y - 2$; $y - 3$

27. $\dfrac{y^2 - 5y + 6}{y - 2} = \dfrac{(\quad\quad)(\quad\quad)}{y - 2} = \_\_\_\_\_$.

$2x + 3$; $4x^2 - 6x + 9$;

$4x^2 - 6x + 9$

28. $\dfrac{8x^3 + 27}{2x + 3} = \dfrac{(\quad\quad)(\quad\quad\quad\quad)}{2x + 3}$

$= \_\_\_\_\_\_\_\_$.

$\dfrac{x + 4}{3x - 2}$; $\dfrac{x + 4}{3x - 2}$

29. $\dfrac{2x^2 + 5x - 12}{6x^2 - 13x + 6} = \dfrac{(2x - 3)(\quad\quad)}{(2x - 3)(\quad\quad)} = \dfrac{\quad}{\_\_\_\_\_}$.

c

30. The fundamental principle of fractions that is used to build equivalent fractions asserts that $\dfrac{a}{b} = \dfrac{ac}{b(\quad)}$.

31. **To obtain the building factor, divide the denominator (or numerator) of the given fraction into the equivalent fraction's ________.**

denominator (or numerator)

In Frames 32-40, express each given fraction as an equivalent fraction with the given denominator. Assume no denominator is zero.

32. $\frac{7}{9} = \frac{?}{27}$.

$27 \div 9 = 3$. Then $\frac{7}{9} = \frac{7(3)}{9(3)} = \frac{(\quad)}{27}$.

21

33. $\frac{4}{3y} = \frac{?}{21y}$.

$21y \div 3y =$ ____. Then $\frac{4}{3y} = \frac{4(\quad)}{3y(7)} =$ ____

7; 7; $\frac{28}{21y}$

34. $\frac{-x^2}{2y^2} = \frac{?}{8y^3}$. $8y^3 \div 2y^2 =$ ____.

4y

Then, $\frac{-x^2}{2y^2} = \frac{-x^2(\quad)}{2y^2(4y)} =$ ____.

$4y$; $\frac{-4x^2y}{8y^3}$

35. $\frac{a+2}{2} = \frac{?}{8(a-2)}$. $8(a-2) \div 2 =$ ____.

$4(a-2)$

Then, $\frac{a+2}{2} = \frac{(a+2)[\quad]}{2[4(a-2)]} =$ ____.

$4(a-2)$; $\frac{4a^2-16}{8(a-2)}$

36. $\frac{3}{2a-b} = \frac{?}{4a^2-b^2}$. $4a^2 - b^2 = (2a-b)(2a+b)$.

$(4a^2 - b^2) \div (2a - b) =$ ____.

$2a + b$

Then, $\frac{3}{2a-b} = \frac{3(\quad)}{(2a-b)(2a+b)} =$ ____.

$2a + b$; $\frac{6a+3b}{4a^2-b^2}$

37. $\frac{5}{x-y} = \frac{?}{y-x}$. $y - x = (-1)(x-y)$.

$(y-x) \div (x-y) = (-1)(x-y) \div (x-y) =$ ____.

-1;

Then, $\frac{5}{x-y} = \frac{5(\quad)}{(x-y)(-1)} =$ ____.

-1; $\frac{-5}{y-x}$

38. $\frac{5}{x-y} = \frac{?}{y^2-x^2}$.

$y^2 - x^2 = (-1)(x^2 - y^2) = (-1)(x+y)(\quad)$.

$x - y$

$(y^2 - x^2) \div (x - y) = (-1)(x^2 - y^2) \div (x - y)$

$= (-1)(\quad)(\quad) \div (x - y)$

$x + y$; $x - y$

$= (\quad)(\quad)$.

-1; $x + y$

Then, $\frac{5}{x-y} = \frac{5(\quad)(\quad)}{(x-y)(-1)(x+y)} =$ ____.

-1; $x + y$; $\frac{-5x-5y}{y^2-x^2}$

39. $\dfrac{3}{3x - 2y} = \dfrac{?}{27x^3 - 8y^3}$.

$27x^3 - 8y^3 = (3x - 2y)(9x^2 + 6xy + 4y^2)$.

$(27x^3 - 8y^3) \div (3x - 2y) =$ ______________.

Then, $\dfrac{3}{3x - 2y} = \dfrac{3(\_\_\_\_\_\_\_\_\_\_)}{(3x - 2y)(9x^2 + 6xy + 4y^2)}$

$=$ ______________.

$9x^2 + 6xy + 4y^2$

$9x^2 + 6xy + 4y^2$

$\dfrac{27x^2 + 18xy + 12y^2}{27x^3 - 8y^3}$

40. $\dfrac{y}{x - 1} = \dfrac{?}{xy + x - y - 1}$.

$xy + x - y - 1 = x(y + 1) - 1(y + 1) = (\_\_\_\_\_)(\_\_\_\_\_)$.

$(xy + x - y - 1) \div (x - 1) = (y + 1)(x - 1) \div (x - 1)$

$=$ ______.

Then, $\dfrac{y}{x - 1} = \dfrac{y(\_\_\_\_\_)}{(x - 1)(y + 1)} =$ ______________.

$y + 1$; $x - 1$

$y + 1$

$y + 1$; $\dfrac{y^2 + y}{xy + x - y - 1}$

3.3

QUOTIENTS OF POLYNOMIALS

In Frames 41-48, divide. Assume no denominator equals zero.

41. $\dfrac{3x^3 - 2x^2 + 5}{x} = \dfrac{3x^3}{x} - \dfrac{2x^2}{x} + \left(\_\_\right) = 3x^2 - \_\_\_\_ + \dfrac{5}{x}$.

$\dfrac{5}{x}$; $2x$

42. $\dfrac{12a^2y^2 + 8ay - 5}{4ay} = \dfrac{(\_\_\_\_\_)}{4ay} + \dfrac{(\_\_\_\_)}{4ay} - \dfrac{(\_\_)}{4ay}$

$=$ ______________.

$12a^2y^2$; $8ay$; 5

$3ay + 2 - \dfrac{5}{4ay}$

43. $\dfrac{x^2 + 4x - 20}{x + 7} =$ ______________.

$x - 3 + \dfrac{1}{x + 7}$

(detail below)

$$\begin{array}{r|l} & x - 3 \\ \hline x + 7 & x^2 + 4x - 20 \\ & \underline{x^2 + 7x} \\ & -3x - 20 \\ & \underline{-3x - 21} \\ & +1 \end{array}$$

44. $\dfrac{6a^2 + a - 3}{3a + 2} =$ ______________.

$2a - 1 + \dfrac{-1}{3a + 2}$

(detail below)

$$\begin{array}{r|l} & 2a - 1 \\ \hline 3a + 2 & 6a^2 + a - 3 \\ & \underline{6a^2 + 4a} \\ & -3a - 3 \\ & \underline{-3a - 2} \\ & -1 \end{array}$$

$3x^2 + x + 4 + \dfrac{-1}{x - 1}$

(detail below)

$$\begin{array}{r|l}
 & 3x^2 + x + 4 \\
\hline
x - 1 & 3x^3 - 2x^2 + 3x - 5 \\
 & \underline{3x^3 - 3x^2} \\
 & \qquad x^2 + 3x \\
 & \qquad \underline{x^2 - x} \\
 & \qquad\qquad 4x - 5 \\
 & \qquad\qquad \underline{4x - 4} \\
 & \qquad\qquad\quad - 1
\end{array}$$

45. $\dfrac{3x^3 - 2x^2 + 3x - 5}{x - 1} =$ ______________________.

$4y^2 - 4y + 5 + \dfrac{-7}{y + 1}$

(detail below)

$$\begin{array}{r|l}
 & 4y^2 - 4y + 5 \\
\hline
y + 1 & 4y^3 + 0y^2 + y - 2 \\
 & \underline{4y^3 + 4y^2} \\
 & \quad - 4y^2 + y \\
 & \quad \underline{- 4y^2 - 4y} \\
 & \qquad\qquad 5y - 2 \\
 & \qquad\qquad \underline{5y + 5} \\
 & \qquad\qquad\quad - 7
\end{array}$$

46. $\dfrac{4y^3 + y - 2}{y + 1} =$ ______________________.

$x^3 - x^2 + x - 1 + \dfrac{2}{x + 1}$

(detail below)

$$\begin{array}{r|l}
 & x^3 - x^2 + x - 1 \\
\hline
x + 1 & x^4 + 0x^3 + 0x^2 + 0x + 1 \\
 & \underline{x^4 + x^3} \\
 & \quad - x^3 + 0x^2 \\
 & \quad \underline{- x^3 - x^2} \\
 & \qquad\qquad x^2 + 0x \\
 & \qquad\qquad \underline{x^2 + x} \\
 & \qquad\qquad\quad - x + 1 \\
 & \qquad\qquad\quad \underline{- x - 1} \\
 & \qquad\qquad\qquad\quad 2
\end{array}$$

47. $\dfrac{x^4 + 1}{x + 1} =$ ______________________.

$2z^2 + 2z - 1 + \dfrac{2z - 3}{z^2 - z - 1}$

(detail below)

$$\begin{array}{r|l}
 & 2z^2 + 2z - 1 \\
\hline
z^2 - z - 1 & 2z^4 + 0z^3 - 5z^2 + z - 2 \\
 & \underline{2z^4 - 2z^3 - 2z^2} \\
 & \qquad 2z^3 - 3z^2 + z \\
 & \qquad \underline{2z^3 - 2z^2 - 2z} \\
 & \qquad\quad - z^2 + 3z - 2 \\
 & \qquad\quad \underline{- z^2 + z + 1} \\
 & \qquad\qquad\quad 2z - 3
\end{array}$$

48. $\dfrac{2z^4 - 5z^2 + z - 2}{z^2 - z - 1} =$ ______________________.

3.4

SUMS AND DIFFERENCES

divisible

49. The least common multiple (L.C.M.) of two or more natural numbers is the smallest natural number that is exactly ________ by each of the given numbers.

factored

greatest

50. To find the L.C.M. of a set of natural numbers, first express each number in completely ________ form. Then write as factors of a product each different prime factor occurring in any of the numbers, including each factor the ________ (greatest/least) number of times it occurs in any one of the given numbers.

In Frames 51-60, find the L.C.M. of each given set of numbers.

51. 4, 15, 16.

$4 = 2 \cdot 2$; $15 = 3 \cdot 5$; $16 = 2 \cdot 2 \cdot 2 \cdot 2$.

5

L.C.M. will have 2's, 3's, and ___'s occurring as factors.

4
1
1

2's occur as factors ___ times in 16;
3's occur as factors ___ time in 15;
5's occur as factors ___ time in 15;

240

L.C.M.: $2 \cdot 2 \cdot 2 \cdot 2 \cdot 3 \cdot 5 =$ _____.

52. 14, 21, 24.

2; 7; 3; 7

$14 =$ ___ $\cdot$ ___; $21 =$ ___ $\cdot$ ___; $24 = 2^3 \cdot 3$.

2; 3; 7

L.C.M. will have ___'s, ___'s, and ___'s occurring as factors.

3
1
1

2's occur as factors ___ times in 24;
3's occur as factors ___ time in 21 and 24;
7's occur as factors ___ time in 14 and 21;

168

L.C.M.: $2^3 \cdot 3 \cdot 7 =$ _____.

53. 8, 12, 15.

2^2; 3; 3; 5

$8 = 2^3$; $12 =$ ____ $\cdot$ ___; $15 =$ ___ $\cdot$ ___.

2^3; 120

L.C.M.: ____ $\cdot 3 \cdot 5 =$ _____.

54. $12xy$, $18x^2y^3$.

2^2; 3; 2; 3^2

$12xy =$ ____ $\cdot$ ____ $\cdot x \cdot y$; $18x^2y^3 =$ ___ $\cdot$ ____ $\cdot x^2 \cdot y^3$;

3^2; y^3; $36x^2y^3$

L.C.M.: $2^2 \cdot$ ____ $\cdot x^2 \cdot$ ____ $=$ ________.

55. $6ab$, $8b^2$, $3a^2b$.

$2^3 \cdot b^2$; $3 \cdot a^2 \cdot b$

$6ab = 2 \cdot 3 \cdot a \cdot b$; $8b^2 =$ ________; $3a^2b =$ __________.

$2^3 \cdot 3 \cdot a^2 \cdot b^2 = 24a^2b^2$

L.C.M.: ________________________.

56. $12xy^2z$, $9xyz^2$, $18x^2$.

L.C.M.: ______.

$36x^2y^2z^2$
(detail below)
$12xy^2z = 2^2 \cdot 3 \cdot x \cdot y^2 \cdot z$
$9xyz^2 = 3^2 \cdot x \cdot y \cdot z^2$
$18x^2 = 2 \cdot 3^2 \cdot x^2$

L.C.M.: $2^2 \cdot 3^2x^2y^2z^2$

57. $(a^2 - b^2)$, $(a + b)$.
$a^2 - b^2 = (\_\_\_\_)(\_\_\_\_)$; $a + b = a + b$

L.C.M.: $(\_\_\_\_)(\_\_\_\_)$.

$a + b$; $a - b$

$a + b$; $a - b$

58. $x^2 + 3x + 2$, $(x + 1)^2$.
$x^2 + 3x + 2 = (\_\_\_\_)(\_\_\_\_)$; $(x + 1)^2 = (x + 1)^2$.

L.C.M.: ______.

$x + 1$; $x + 2$

$(x + 1)^2(x + 2)$

59. $12x^2 - 12$, $x^2 - 3x + 2$, 8.

L.C.M.: ______.

$24(x + 1)(x - 1)(x - 2)$
(detail below)
$12x^2 - 12$
$= 2^2 \cdot 3 \cdot (x + 1)(x - 1)$
$x^2 - 3x + 2 = (x - 1)(x - 2)$
$8 = 2^3$

L.C.M.:
$2^3 \cdot 3(x + 1)(x - 1)(x - 2)$

60. x^2, $x^3 - x$, $(x - 1)^3$.

L.C.M.: ______.

$x^2(x + 1)(x - 1)^3$
(detail below)
$x^2 = x^2$;
$x^3 - x = x(x + 1)(x - 1)$;
$(x - 1)^3 = (x - 1)^3$;

L.C.M.: $x^2(x + 1)(x - 1)^3$.

61. To rewrite in simpler form sums involving fractions with the same denominator, we use the theorem:

$$\frac{a}{b} + \frac{b}{c} = \_\_\_\_ \quad (c \neq 0)$$

$\frac{a + b}{c}$

In Frames 62-72, write each sum or difference as a single fraction in lowest terms. Assume that no denominator equals zero.

62. $\frac{2}{3x} + \frac{5}{3x} = \_\_\_\_$.

$\frac{7}{3x}$

$2m - 1$; 1

$\frac{2}{b}$

63. $\frac{2m + 1}{b} - \frac{2m - 1}{b} = \frac{2m + 1}{b} + \frac{-(\qquad)}{b} = \frac{2m + 1 - 2m + \_\_\_}{b}$

$= \_\_\_$.

$2y - 3$; $\frac{3y + 1}{y^2 - y + 2}$

64. $\frac{y + 4}{y^2 - y + 2} + \frac{2y - 3}{y^2 - y + 2} = \frac{y + 4 + \_\_\_\_\_}{y^2 - y + 2} = \_\_\_\_\_\_\_$.

$\frac{6}{a - 3b}$

(detail below)

$\frac{3}{a - 3b} - \frac{b - 3}{a - 3b} + \frac{b}{a - 3b}$

$= \frac{3}{a - 3b} + \frac{-(b - 3)}{a - 3b} + \frac{b}{a - 3b}$

$= \frac{3 - b + 3 + b}{a - 3b} = \frac{6}{a - 3b}$

65. $\frac{3}{a - 3b} - \frac{b - 3}{a - 3b} + \frac{b}{a - 3b} = \_\_\_\_\_$.

denominator

66. To rewrite in simpler form sums of fractions with different denominators, we first rewrite each fraction equivalently with the L.C.M. of all of the denominators as the ________ of each new fraction. Then we apply the theorem of Frame 61.

$2x$; $\frac{3 - 2x}{ax}$

67. $\frac{3}{ax} - \frac{2}{a} = \frac{3}{ax} - \frac{\_\_\_}{ax} = \_\_\_\_\_$.

$5(5 - a)$

$25 - 5a$; $\frac{-3a + 19}{20}$

68. $\frac{a - 3}{10} + \frac{5 - a}{4} = \frac{2(a - 3)}{20} + \frac{\_\_\_\_\_}{20}$

$= \frac{2a - 6 + \_\_\_\_\_}{20} = \_\_\_\_\_$.

$(x + 2)(x + 2)$

$-x^2 - 4x - 4$

$\frac{-1}{(x + 3)(x + 2)}$

69. $\frac{x + 1}{x + 2} - \frac{x + 2}{x + 3} = \frac{(x + 1)(x + 3)}{(x + 2)(x + 3)} + \frac{-[\qquad\qquad]}{(x + 3)(x + 2)}$

$= \frac{x^2 + 4x + 3 \_\_\_\_\_\_\_}{(x + 3)(x + 2)}$

$= \_\_\_\_\_\_\_$.

$a - 1$

$-a + 1$; $\frac{2}{(a + 1)^2(a - 1)}$

70. $\frac{1}{a^2 - 1} - \frac{1}{a^2 + 2a + 1} = \frac{1}{(a + 1)(a - 1)} - \frac{1}{(a + 1)^2}$

$= \frac{a + 1}{(a + 1)^2(a - 1)} + \frac{-(\qquad)}{(a + 1)^2(a - 1)}$

$= \frac{a + 1 \_\_\_\_\_}{(a + 1)^2(a - 1)} = \_\_\_\_\_\_\_$.

-3; $\frac{-1}{x - 2}$

71. $\frac{3}{2 - x} + \frac{2}{x - 2} = \frac{\_\_\_}{x - 2} + \frac{2}{x - 2} = \_\_\_\_$.

$a - 3b$; $a - b$

$-3(a + 3b)$

$-3a - 9b$; $3a - 15b$

72. $\frac{6}{a^2 - 9b^2} - \frac{3}{a^2 - 4ab + 3b^2}$

$= \frac{6}{(a + 3b)(a - 3b)} + \frac{-3}{(\qquad)(\qquad)}$

$= \frac{6(a - b)}{(a + 3b)(a - 3b)(a - b)} + \frac{\_\_\_\_\_}{(a + 3b)(a - 3b)(a - b)}$

$= \frac{6a - 6b \_\_\_\_\_}{(a + 3b)(a - 3b)(a - b)} = \frac{\_\_\_\_\_}{(a + 3b)(a - 3b)(a - b)}$.

3.5

PRODUCTS AND QUOTIENTS

73. To rewrite products involving fractions, we use the theorem: For all $a,b,c,d \in R$ $(b,d \neq 0)$,

$$\frac{a}{b} \cdot \frac{c}{d} = \underline{\quad}$$

$\frac{ac}{bd}$

In Frames 74-84, write each product and each quotient as a single fraction in lowest terms. Assume no denominator is equal to zero.

74. $\frac{8}{15} \cdot \frac{3}{16} = \frac{8 \cdot 3}{15 \cdot 16} = \frac{8 \cdot 3}{5 \cdot 3 \cdot 8 \cdot 2} = \frac{(8 \cdot 3) \cdot 1}{(8 \cdot 3) \cdot 5 \cdot 2} = \underline{\quad}.$

$\frac{1}{10}$

75. $\frac{a^2}{xy} \cdot \frac{3x^3y}{4a} = \frac{a^2 \cdot \underline{\quad}}{xy \cdot 4a} = \frac{(axy)(\underline{\quad})}{(axy)(\underline{\quad})} = \underline{\quad}.$

$3x^3y$; $\frac{3ax^2}{4}$; $\frac{3ax^2}{4}$

76. $\frac{9x^2 - 25}{2x - 2} \cdot \frac{x^2 - 1}{6x - 10} = \frac{(9x^2 - 25)(\underline{\quad})}{(2x - 2)(\underline{\quad})}$

$= \frac{(3x - 5)(3x + 5)(\underline{\quad})(\underline{\quad})}{2(x - 1)(\underline{\quad})(\underline{\quad})}$

$= \frac{[(3x - 5)(x - 1)](3x + 5)(\underline{\quad})}{[(3x - 5)(x - 1)](\underline{\quad})(\underline{\quad})}$

$= \underline{\quad}.$

$x^2 - 1$; $6x - 10$

$x + 1$; $x - 1$
2; $3x - 5$

$x + 1$
2; 2

$\frac{(3x + 5)(x + 1)}{4}$

77. $\frac{3y}{4xy - 6y^2} \cdot \frac{2x - 3y}{12} = \underline{\quad}.$

$\frac{1}{8}$ (detail below)

$$\frac{3y}{4xy - 6y^2} \cdot \frac{2x - 3y}{12}$$
$$= \frac{(3y)(2x - 3y)}{(4xy - 6y^2) \cdot 12}$$
$$= \frac{3y(2x - 3y) \cdot 1}{2y(2x - 3y)(2^2 \cdot 3)} = \frac{1}{8}$$

78. $\frac{5y^2 - 5y}{4y - 40} \cdot \frac{y^2}{3} \cdot \frac{y^2 - 9y - 10}{2 - 2y} = \underline{\quad}.$

$\frac{-5y^3(y + 1)}{24}$

(detail below)

$$\frac{5y^2 - 5y}{4y - 40} \cdot \frac{y^2}{3} \cdot \frac{y^2 - 9y - 10}{2 - 2y}$$
$$= \frac{5y(y - 1) \cdot y^2 \cdot (y - 10)(y + 1)}{4(y - 10)(3)(-2)(y - 1)}$$
$$= \frac{[(y - 1)(y - 10)](5)(y^3)(y + 1)}{[(y - 1)(y - 10)](4)(3)(-2)}$$
$$= \frac{-5y^3(y + 1)}{24}$$

79. To rewrite quotients involving fractions we use the theorem: For $a,b,c,d \in R$ $(b,c,d \neq 0)$,

$$\frac{a}{b} \div \frac{c}{d} = \underline{\quad}.$$

$\frac{ad}{bc}$

80. $\frac{9ab^3}{x} \div \frac{3}{2x^3} = \frac{(9ab^3)(\underline{\quad})}{x(\underline{\quad})} = \frac{[3x](2 \cdot 3ab^3x^2)}{[3x]} = \underline{\quad}.$

$\frac{2x^3}{3}$; $6ab^3x^2$

b
$a + 2b$

b; b

$2a - b$
$a^2 - b^2$

$a + b$; $a - b$

$a + b$; $2a - b$

$$\frac{1}{(4a^2 + 2ab + b^2)(a - b)}$$

$$\frac{a - 5}{a + 5}$$

(detail below)

$$\frac{a^2 + a - 2}{a^2 + 2a - 3} \div \frac{a^2 + 7a + 10}{a^2 - 2a - 15}$$

$$= \frac{(a+2)(a-1)}{(a+3)(a-1)} \div \frac{(a+5)(a+2)}{(a-5)(a+3)}$$

$$= \frac{(a+2)(a-1)}{(a+3)(a-1)} \cdot \frac{(a-5)(a+3)}{(a+5)(a+2)}$$

$$= \frac{[(a+2)(a-1)(a+3)](a-5)}{[(a+2)(a-1)(a+3)](a+5)}$$

$$= \frac{(a - 5)}{(a + 5)}$$

$$\frac{(x - 1)(x + 2)}{x(x - 3)}$$

(detail below)

$$\frac{x^2-4x+3}{x^2} \cdot \frac{x^2 + x}{x^2-6x+9} \div \frac{x^2-2x-3}{x^2-x-6}$$

$$= \frac{(x-3)(x-1)}{x^2} \cdot \frac{x(x+1)}{(x-3)(x-3)}$$

$$\div \frac{(x-3)(x+1)}{(x-3)(x+2)}$$

$$= \frac{(x-3)(x-1)}{(x \cdot x)} \cdot \frac{x(x+1)}{(x-3)(x-3)}$$

$$\cdot \frac{(x-3)(x+2)}{(x-3)(x+1)}$$

$$= \frac{[x(x-3)^2(x+1)](x-1)(x+2)}{[x(x-3)^2(x+1)]x(x-3)}$$

$$= \frac{(x - 1)(x + 2)}{x(x - 3)}$$

81. $\frac{a^2 + 2ab}{a} \div \frac{a + 2b}{b} = \frac{(a^2 + 2ab)(\_\_\_)}{a(\_\_\_\_\_\_\_\_)}$

$= \frac{[a(a + 2b)](\_\_\_)}{[a(a + 2b)](\ 1\ )} = \_\_\_.$

82. $\frac{a + b}{8a^3 - b^3} \div \frac{a^2 - b^2}{2a - b} = \frac{(a + b)(\_\_\_\_\_\_\_\_)}{(8a^3 - b^3)(\_\_\_\_\_\_\_\_)}$

$$= \frac{(a + b)(2a - b)}{(2a - b)(4a^2 + 2ab + b^2)(\_\_\_\_\_\_\_)(\_\_\_\_\_\_\_)}$$

$$= \frac{[(a + b)(2a - b) \cdot 1}{[(\_\_\_\_\_\_\_)(\_\_\_\_\_\_\_)](4a^2 + 2ab + b^2)(a - b)}$$

$$= \frac{\qquad\qquad}{\_\_\_\_\_\_\_\_\_\_\_\_\_\_\_\_\_\_}.$$

83. $\frac{a^2 + a - 2}{a^2 + 2a - 3} \div \frac{a^2 + 7a + 10}{a^2 - 2a - 15} = \frac{\qquad}{\_\_\_\_\_\_\_}.$

84. $\frac{x^2 - 4x + 3}{x^2} \cdot \frac{x^2 + x}{x^2 - 6x + 9} \div \frac{x^2 - 2x - 3}{x^2 - x - 6} = \_\_\_\_\_\_\_\_\_\_\_\_\_\_\_.$

3.6

COMPLEX FRACTIONS

complex

85. A fraction that contains a fraction or fractions in either numerator or denominator or both is called a ________ fraction.

least common

86. The method often used when simplifying complex fractions requires that the numerator and denominator of the complex fraction be multiplied by the ________ ________ denominator of all the fractions in the complex fraction.

In Frames 87-93, simplify each complex fraction. Assume no denominator is equal to zero.

$\frac{16}{15}$

87. $$\frac{\frac{2}{3}}{\frac{5}{8}} = \frac{\frac{2}{3}(24)}{\frac{5}{8}(24)} = \underline{\quad}.$$

$30y$; $25x$; $\frac{24}{25}$

88. $$\frac{\frac{4x}{5y}}{\frac{5x}{6y}} = \frac{\frac{4x}{5y}(30y)}{\frac{5x}{6y}(\underline{\quad})} = \frac{24x}{\underline{\quad}} = \underline{\quad}.$$

y^2

y^2; $y^2 - 1$

$\frac{y^2}{y + 1}$

89. $$\frac{y - 1}{1 - \frac{1}{y^2}} = \frac{(y - 1)(\underline{\quad})}{\left(1 - \frac{1}{y^2}\right)(\underline{\quad})} = \frac{y^2(y - 1)}{\underline{\quad}}$$
$$= \frac{y^2(y - 1)}{(y + 1)(y - 1)} = \underline{\quad}$$

$\frac{y}{y}$; $\frac{y - 1}{y + 1}$

90. $$\frac{1 - \frac{1}{y}}{1 + \frac{1}{y}} = \frac{\left(1 - \frac{1}{y}\right)(\underline{\quad})}{\left(1 + \frac{1}{y}\right)(\underline{\quad})} = \underline{\quad}.$$

$\frac{y}{y}$; $\frac{y}{x + 2y}$

$\frac{2y}{2y}$; $\frac{6y}{x + 2y}$

$x + 2y$

$x + 2y$

$\frac{x + 2y - y}{x + 2y + 6y}$; $\frac{x + y}{x + 8y}$

91. $$\frac{1 - \frac{1}{\frac{x}{y} + 2}}{1 + \left(\frac{3}{\frac{x}{2y} + 1}\right)} = \frac{1 - \left(\frac{1}{\frac{x}{y} + 2}\right)\left(\frac{\underline{\quad}}{\underline{\quad}}\right)}{1 + \left(\frac{3}{\frac{x}{2y} + 1}\right)\left(\frac{\underline{\quad}}{\underline{\quad}}\right)} = \frac{1 - \frac{(\underline{\quad})}{(\underline{\quad})}}{1 + \frac{(\underline{\quad})}{(\underline{\quad})}}$$
$$= \frac{\left[1 - \left(\frac{y}{x + 2y}\right)\right](\underline{\quad})}{\left[1 + \left(\frac{6y}{x + 2y}\right)\right](\underline{\quad})}$$
$$= \underline{\quad} = \underline{\quad}.$$

$a - 2$

$a - 2$

$(a + 4)(a - 2) - 7$

$\frac{a^2 - 3a + 2 + 2}{a^2 + 2a - 8 - 7}$; $\frac{a^2 - 3a + 4}{a^2 + 2a - 15}$

92. $$\frac{a - 1 + \frac{2}{a - 2}}{a + 4 - \frac{7}{a - 2}} = \frac{\left(a - 1 + \frac{2}{a - 2}\right)(\underline{\quad})}{\left(a + 4 - \frac{7}{a - 2}\right)(\underline{\quad})}$$
$$= \frac{(a - 1)(a - 2) + 2}{\underline{\quad}}$$
$$= \underline{\quad} = \underline{\quad}.$$

$\dfrac{-2a}{a^2 + 1}$ (detail below)

$$\frac{\left(\frac{a-1}{a+1} - \frac{a+1}{a-1}\right)[(a+1)(a-1)]}{\left(\frac{a+1}{a-1} + \frac{a-1}{a+1}\right)[(a+1)(a-1)]}$$

$$= \frac{(a-1)(a-1) - (a+1)(a+1)}{(a+1)(a+1) + (a-1)(a-1)}$$

$$= \frac{a^2 - 2a + 1 - a^2 - 2a - 1}{a^2 + 2a + 1 + a^2 - 2a + 1}$$

$$= \frac{-4a}{2a^2 + 2} = \frac{2(-2a)}{2(a^2 + 1)} = \frac{-2a}{a^2 + 1}$$

93. $\dfrac{\frac{a-1}{a+1} - \frac{a+1}{a-1}}{\frac{a+1}{a-1} + \frac{a-1}{a+1}} =$ ________ .

TEST PROBLEMS

In all problems, assume no denominator is equal to zero.

In Problem 1, write each fraction in standard form.

1. a. $-\dfrac{-4}{7}$. b. $\dfrac{4 - 3x}{-x}$. c. $\dfrac{-5a^2b^2}{-2c^2}$. d. $\dfrac{m - 2n}{-n}$.

In Problems 2 and 3, reduce each fraction to lowest terms.

2. $\dfrac{8r^4 - 12r^3}{4r^2 - 6r}$.

3. $\dfrac{8a^3 - 27b^3}{4a^2 - 9b^2}$.

In Problems 4 and 5, divide.

4. $\dfrac{x^2 + 3x - 4}{x - 2}$

5. $\dfrac{6x^2 - 2xy + 9x - 3y}{2x + 3}$.

In Problems 6-13, write each expression as a single fraction in lowest terms.

6. $\dfrac{4}{3x} + \dfrac{5}{2y} - \dfrac{2}{xy}$.

7. $\dfrac{m}{m^2 - 1} - \dfrac{m}{m^2 - 2m + 1}$.

8. $\dfrac{3p}{4pq - 6q^2} \cdot \dfrac{2p - 3q}{12p}$.

9. $\dfrac{4y^2 + 8y + 3}{2y^2 - 5y + 3} \cdot \dfrac{6y^2 - 9y}{1 - 4y^2}$.

10. $\dfrac{m^2 + m - 2}{m^2 + 2m - 3} \div \dfrac{m^2 + 7m + 10}{m^2 - 2m - 15}$.

11. $\dfrac{8x^3 - 1}{x - 2} \div \dfrac{4x^2 + 2x + 1}{x^2 - 4x + 4}$.

12. $\dfrac{\frac{m-2}{m}}{\frac{m^2-4}{m^2}}$.

13. $\dfrac{\frac{a-1}{a+1} - \frac{a+1}{a-1}}{\frac{a-1}{a+1} + \frac{a+1}{a-1}}$.

4

FIRST-DEGREE EQUATIONS AND INEQUALITIES

4.1

SOLVING EQUATIONS

1. A sentence that can be labeled true or false is called a __________.

statement

2. A sentence that cannot be labeled true or false until replacements have been made for the variables of the sentence is called an ______ sentence.

open

3. Symbolic sentences involving only the equality relationship are called __________.

equations

4. If we replace the variable in an equation with a number, and the resulting statement is true, the number is called a ______ or __________ of the equation, and the equation is said to be __________.

root; solution
satisfied

5. The set of all roots or solutions of a given equation is called the _________ set of the equation.

solution

6. An equation that is not true for at least one member of the replacement set of the variable is called a _____________ equation.

conditional

7. An equation that is true for every member of the replacement set is called an _________.

identity

8. $2x - 6 = 0$ ____ (is/is not) satisfied by 3.

   Since $2x - 6 = 0$ ________ (is/is not) satisfied by 1, $2x - 6 = 0$ is a _____________ equation.

is

is not
conditional

9. $2x - 5 = 3x + 4$ ____ (is/is not) satisfied by -9.

   Since $2x - 5 = 3x + 4$ ________ (is/is not) satisfied by 0, $2x - 5 = 3x + 4$ is a _____________ equation.

is

is not
conditional

10. Equations that have identical solution sets are called ____________ equations.

equivalent

11. The addition of the same expression representing a real number to each member of an equation produces an ____________ equation.

equivalent

Answer	Frame
nonzero	12. **The multiplication of each member of an equation by the same expression representing a ________ real number produces an equivalent equation.**
variable	13. **The solutions should be checked when both members of an equation have been multiplied by an expression containing the ________.**

In Frames 14-18, solve the equation given in Frame 14.

Answer	Frame
	14. $3x + 2 = 4x - 6 + x$.
$5x$	15. $3x + 2 =$ ____ $- 6$.
8	16. $3x = 5x -$ ___.
$-2x$	17. _____ $= -8$.
4; $\{4\}$	18. $x =$ ___. The solution set is _____.

In Frames 19-23, solve the equation given in Frame 19.

Answer	Frame
	19. $2x - (3 - x) = 0$.
x	20. $2x - 3 +$ ___ $= 0$.
$3x$	21. ____ $- 3 = 0$.
3	22. $3x =$ ___.
1; $\{1\}$	23. $x =$ ___. The solution set is _____.

In Frames 24-30, solve the equation given in Frame 24.

Answer	Frame
	24. $3[2x - (x + 2)] = -3(3 - 2x)$.
2; $6x$	25. $3[2x - x -$ ___$] = -9 +$ ____.
x	26. $3[$___ $- 2] = -9 + 6x$.
6	27. $3x -$ ___ $= -9 + 6x$.
-3	28. $3x =$ ____ $+ 6x$.
$-3x$	29. ___ $= -3$.
1; $\{1\}$	30. $x =$ ___. The solution set is _____.

In Frames 31-35, solve the equation given in Frame 31.

Answer	Frame
	31. $0 = x^2 + (2 - x)(5 + x)$.
x^2	32. $0 = x^2 + 10 - 3x -$ ____.
10	33. $0 =$ ____ $- 3x$.
$3x$	34. ____ $= 10$.
$\frac{10}{3}$; $\left\{\frac{10}{3}\right\}$	35. $x =$ ____. The solution set is _____.

In Frames 36-42, solve the equation given in Frame 36.

Answer	Frame
	36. $\frac{2x}{3} - \frac{2x + 5}{6} = \frac{1}{2}.$
6	37. $6\left(\frac{2x}{3} - \frac{2x + 5}{6}\right) = (\_\_\_)\frac{1}{2}.$
6	38. $6\left(\frac{2x}{3}\right) - (\_\_\_)\left(\frac{2x + 5}{6}\right) = 3.$
$2x + 5$	39. $4x - (\_\_\_\_\_\_) = 3.$
$4x - 2x - 5$	40. $\_\_\_\_\_\_\_\_\_ = 3.$
8	41. $2x = \_\_\_.$
4; $\{4\}$	42. $x = \_\_\_.$ The solution set is ____.

In Frames 43-53, solve the equation given in Frame 43.

Answer	Frame
	43. $\frac{5}{x - 3} = \frac{x + 2}{x - 3} + 3.$
$x - 3$	44. $(x - 3)\left(\frac{5}{x - 3}\right) = (\_\_\_\_\_\_)\left(\frac{x + 2}{x - 3} + 3\right).$
$x + 2$	45. $5 = \_\_\_\_\_ + 3(x - 3).$
9	46. $5 = x + 2 + 3x - \_\_\_.$
$4x$	47. $5 = \_\_\_ - 7.$
12	48. $\_\_\_ = 4x.$
3	49. $\_\_\_ = x.$
variable	50. Check, because in Frame 44 above both members of the equation were multiplied by $(x - 3)$, an expression containing the _______.
	51. $\frac{5}{3 - 3} \overset{?}{=} \frac{3 + 2}{3 - 3} + 3.$
defined	52. $\frac{5}{0} \overset{?}{=} \frac{5}{0} + 3.$ No, because $\frac{5}{0}$ is not _______.
$\emptyset$	53. The solution set is ___.

4.2

USING FORMULAS

In Frames 54-56, solve $A = \frac{h}{2}(b + c)$ *for c, given that* $A = 120$, $h = 5$, *and* $b = 12$.

54. In the given formula replace A by 120, h by 5, and b by 12 and obtain

12 | $$120 = \frac{5}{2}(\_\_\_ + c).$$

55. $2(120) = 2\left(\frac{5}{2}\right)(12 + c)$

5 | $240 = \_\_\_\ (12 + c)$.

5c | 56. $240 = 60 + \_\_\_\_\_$

180 | $\_\_\_ = 5c$

36 | $\_\_\_ = c$.

If P dollars is invested at a simple interest rate r, then the amount A that is accumulated after t years is given by $A = P + Prt$. *In Frames 57-58, determine how many years it would take for \$2000 to accumulate to \$5780 if it is invested at 7% (0.07).*

57. In the given formula, replace P by 2000, A by 5780,

.07 | and r by _____ and obtain

.07 | $5780 = 2000 + 2000\ (\_\_\_\_)t$

140t | $5780 = 2000 + \_\_\_\_\_$.

3780 | 58. $\_\_\_\_\_ = 140t$

27 | $\_\_\_\_ = t$.

27 | It would take _____ years.

The monthly benefit B paid on a certain retirement plan is determined by

$$B = \frac{3}{7}w\left(1 + \frac{3n}{100}\right)$$

where w is the average monthly wage of the worker and n is the number of years worked. In Frames 59-60, determine what a worker's average monthly wage must be if his monthly benefit will be \$840 after working 25 years.

59. In the given formula, replace B by 840, and n by 25 and obtain

25 | $$840 = \frac{3}{7}w\left(1 + \frac{3(\_\_\_)}{100}\right)$$

3 | $$840 = \frac{3}{7}w\left(1 + \frac{\_\_\_}{4}\right)$$

7 | $$840 = \frac{3}{7}w\left(\frac{\_\_\_}{4}\right).$$

	60. Solving for w we have
$\frac{3}{4}$	$840 = \underline{\quad\quad} w$
3360	$\underline{\quad\quad} = 3w$
1120	$\underline{\quad\quad} = w.$
\$1120	His average monthly wage must be ______.

4.3

SOLVING EQUATIONS FOR SPECIFIED SYMBOLS

In all rational expressions, assume that no denominator equals zero.

In Frames 61-63, solve the equation given in Frame 61 for x.

	61. $ax = a^2b - ax.$
$2ax$	62. $\underline{\quad\quad} = a^2b.$
$\frac{ab}{2}$	63. $x = \underline{\quad\quad}.$

In Frames 64-68, solve the equation given in Frame 64 for x.

	64. $a^2x + b = ax.$
$ax - b$	65. $a^2x = \underline{\quad\quad}.$
$a^2x - ax$	66. $\underline{\quad\quad\quad} = -b.$
$a^2 - a$	67. $(\underline{\quad\quad\quad})x = -b.$
$\frac{-b}{a^2 - a}$ or $\frac{-b}{a(a - 1)}$	68. $x = \underline{\quad\quad\quad}.$

In Frames 69-76, solve the equation given in Frame 69 for x.

	69. $\frac{2x + 4a}{3a} - \frac{3x + 4a}{2a} = a.$
$6a^2$	70. $(2 \cdot 3a)\left(\frac{2x + 4a}{3a} - \frac{3x + 4a}{2a}\right) = \underline{\quad\quad}.$
3	71. $2(2x + 4a) - (\underline{\quad})(3x + 4a) = 6a^2.$
$4x + 8a - 9x - 12a$	72. $\underline{\quad\quad\quad\quad\quad\quad} = 6a^2.$
$4a$	73. $-5x - \underline{\quad\quad} = 6a^2.$
$4a$	74. $-5x = 6a^2 + \underline{\quad\quad}.$

$\frac{6a^2 + 4a}{-5}$

75. $x =$ ______.

$\frac{-(6a^2 + 4a)}{5}$

76. In standard form, $x =$ ______.

In Frames 77-79, solve the equation of Frame 77 for w.

77. $p = 21 + 2w$.

$2w$

78. $p - 21 =$ ____.

$\frac{p - 21}{2}$

79. $\frac{\text{______}}{} = w$.

In Frames 80-82, solve the equation given in Frame 80 for D.

80. $S = 3\pi d + 5\pi D$.

$S - 3\pi d$

81. ______ $= 5\pi D$.

$\frac{S - 3\pi d}{5\pi}$

82. ______ $= D$.

In Frames 83-88, solve the equation given in Frame 83 for r_2.

83. $\frac{1}{r} = \frac{1}{r_1} + \frac{2}{r_2}$.

rr_1r_2

84. $(rr_1r_2)\frac{1}{r} = (\text{______})\left(\frac{1}{r_1} + \frac{2}{r_2}\right)$.

$2rr_1$

85. $r_1r_2 = rr_2 +$ ______.

$r_1r_2 - rr_2$

86. ______ $= 2rr_1$.

$r_1 - r$

87. (______)$r_2 = 2rr_1$.

$\frac{2rr_1}{r_1 - r}$

88. $r_2 =$ ______.

4.4 APPLICATIONS

In Frames 89-92, set up an equation for the problem stated in Frame 89 and solve.

89. If the denominator of a certain fraction is 6 more than the numerator and the fraction is equal to $\frac{3}{4}$, find the numerator.

90. numerator: x
denominator: ______

$x + 6$

91. An equation is ______.

$\frac{x}{x + 6} = \frac{3}{4}$

92. Solving the equation,

$4x = 3($_______$)$ — $x + 6$

$4x = 3x +$ ____ — 18

$x =$ ____. — 18

Hence, the numerator is 18.

Check: Is $\frac{18}{18 + 6} = \frac{3}{4}$? _____ (yes/no). — yes

In Frames 93-100, set up an equation for the problem stated in Frame 93 and solve.

93. A man has $1000 more invested at 5% than he has invested at 4%. If his annual income from the two investments is $698, how much does he have invested at each rate?

94. amount invested at 4%: x

amount invested at 5%: __________ — $x + 1000$

95. ______________ = income from 4% investment. — $.04x$

96. ______________ = income from 5% investment. — $.05(x + 1000)$

97. {income from 4% investment} + {income from 5% investment} = total ________. — income

98. The equation is ____________________ = 698. — $.04x + .05(x + 1000)$

99. Solving the equation,

$100[.04x + .05(x + 1000)] =$ _____(698) — 100

$4x + 5x + 5000 =$ _______ — 69800

$9x =$ _______ — 64800

$x =$ _______. — 7200

100. Hence, the amount invested at 4% is $7200 and at 5% is x + $1000 or _______. Check: .04(7200) + .05(8200) = 698? _____(yes/no). — $8200; yes

In Frames 101-106, set up an equation for the problem stated in Frame 101 and solve.

101. The sum of the ages of Jane and Polly is 60. In 6 years Jane will be 3 times as old as Polly. How old is each now?

102. Jane's age now: x

Polly's age now: ____ $- x$ — 60

103. Set up a table.

	age now	age 6 years from now
Jane	x	$x + 6$
Polly	$60 - x$	__________

Answer: $60 - x + 6$ or $66 - x$

Answer	Frame
$3(66 - x)$	104. An equation relating the ages in 6 years is $x + 6 =$ __________.
$3x$ 192 48	105. Solving the equation, $x + 6 = 198 -$ ____ $4x =$ ____ $x =$ ____, which is Jane's age.
12 yes	106. $60 - x =$ ____, which is Polly's age Check: In 6 years will Jane be 3 times as old as Polly? Is $48 + 6 = 3(12 + 6)$? ____(yes/no).

In Frames 107-111, set up an equation for the problem stated in Frame 107 and solve.

107. The length of the base of an isosceles triangle is $\frac{2}{3}$ of the length of each of the equal sides. If its perimeter is 64 centimeters, find the length of each side.

108. A sketch of the triangle is

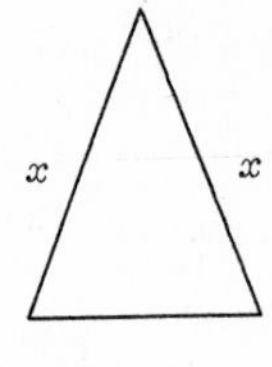

______

Answer	Frame
$\frac{2}{3}x$	(base of the triangle in Frame 108)
$x + x + \frac{2}{3}x = 64$	109. Since the perimeter is the sum of the lengths of the sides, an equation is ____________.
3; 3 $\frac{8}{3}$ 24 16	110. Solving the equation, $\frac{\_\_}{3}x + \frac{\_\_}{3}x + \frac{2}{3}x = 64$ $\_\_\,x = 64$ $x =$ ____. $\frac{2}{3}x =$ ____.
24 16 yes	111. Thus, the length of each of the equal sides is ____ centimeters and the length of the base is ____ centimeters. Check: Does $24 + 24 + 16 = 64$? ____ (yes/no).

In Frames 112-116, set up an equation for the problem stated in Frame 112 and solve.

112. Where should the fulcrum of an 8 foot crowbar be placed so that a 150 pound man can just balance a 450 pound weight?

113. A diagram of the problem is

8 - x

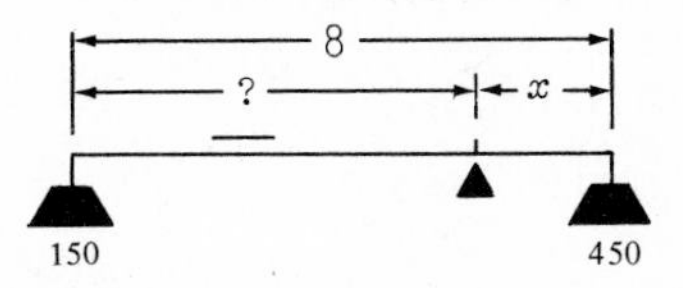

114. Using the law of the lever (page 116 of the text) with $F_1 = 450$, $d_1 = x$, $F_2 = 150$, and $d_2 = 8 - x$ we obtain the equation

$450x =$ ________.

150(8 - x)

115. Solving the equation,

$450x = 1200 -$ ______   150x

______ $= 1200$ 600x

$x =$ ___.   2

116. Hence, the fulcrum should be placed ___ feet from the 450 pound weight.   2

Check: Does 450(2) = 150(6)? _____ (yes/no).   yes

In Frames 117-123, set up an equation for the problem stated in Frame 117 and solve.

117. A collection of nickels, dimes, and quarters has a value of \$1.60. How many coins in each kind are there if the number of nickels is three less than twice the number of dimes and the number of quarters is two less than the number of dimes.

118. number of dimes: x

number of nickels: ________ $2x - 3$

number of quarters: _______. $x - 2$

119. A table for the problem is

Denomination	Value of 1 coin	Number of coins	Value of coins
dimes	10	x	$10x$
nickels	5	$2x - 3$	________
quarters	25	$x - 2$	________

5($2x$ - 3)

25(x - 2)

120. [value of dimes in cents] + [value of nickels in cents] + [value of quarters in cents] = [value of ________ in cents]

collection

121. Hence, an equation is

$10x + 5(2x - 3) +$ __________ $=$ _____.

25(x - 2); 160

122. Solving the equation,

$$10x + 10x - 15 + 25x - \_\_\_\_ = 160$$
$$45x - \_\_\_\_ = 160$$
$$45x = \_\_\_\_\_$$
$$x = 5.$$

Answers: 50; 65; 225

123. Then, $2x - 3 =$ ___ and $x - 2 =$ ___. Hence, there are 5 dimes, 7 _________, and 3 __________.

Check: Do 5 dimes, 7 nickels, and 3 quarters have a value of 160 cents? _____ (yes/no).

Answers: 7; 3; nickels; quarters; yes

In Frames 124-130, set up an equation for the problem of Frame 124 and solve.

124. How many grams of an alloy containing 45% silver must be melted with an alloy containing 60% silver to obtain 40 grams of an alloy containing 48% silver?

125. number of grams of 45% alloy: x

126. number of grams of silver in 45% alloy: ______

Answer: $45x$

127. A table for the problem is

Mixture	Part of silver in alloy	Amount of alloy	Amount of silver
45%	0.45	x	$0.45x$
60%	0.60	$40 - x$	____________
48%	0.48	40	________

Answers: $0.60(40 - x)$; $0.48(40)$

128. [no. of grams of silver in 45% alloy] + [no. of grams of silver in 60% alloy] = [no. of grams of silver in ____% alloy]

Answer: 48

129. An equation is

$$0.45x + 0.60(40 - x) = \_\_\_\_\_\_\_\_\_\_.$$

Answer: $0.48(40)$

130. Solving the equation,

$$100[0.45x + 0.060(40 - x)] = (\_\_\_\_\_)[0.48(40)]$$
$$45x + 60(40 - x) = \_\_\_\_\_\_\_\_$$
$$45x + 2400 - 60x = 1920$$
$$-15x = \_\_\_\_\_\_$$
$$x = \_\_\_\_.$$

Hence, 32 grams of 45% alloy are required.

Check: Does 32 grams of a 45% silver alloy when added to a 60% silver alloy form 40 grams of a 48% silver alloy? Does $0.45(32) + 0.60(40-32) = 0.48(40)$? _____ (yes/no).

Answers: 100; 48(40) or 1920; -480; 32; yes

In Frames 131-136, set up an equation for the problem stated in Frame 131 and solve.

131. Two ships traveling in opposite directions pass each other, one moving at 28 knots and the other at 36 knots. If they maintain these speeds, how long will it take for them to be 48 nautical miles apart?

132. Time it takes to be 48 nautical miles apart: t

133. A table for the problem is

	r	t	$d = rt$
ship A	28	t	$28t$
ship B	36	t	____

$36t$

134. $\begin{bmatrix}\text{distance ship } A\\ \text{travels}\end{bmatrix} + \begin{bmatrix}\text{distance ship } B\\ \text{travels}\end{bmatrix} =$ ____.

48

135. An equation is

$28t +$ ____ $= 48.$

$36t$

136. Solving the equation,

____$t = 48$

$t =$ ____.

64

$\frac{48}{64}$ or $\frac{3}{4}$

Hence, it would take $\frac{3}{4}$ of an hour.

Check: In $\frac{3}{4}$ of an hour, ship A would go $28\left(\frac{3}{4}\right) = 21$ miles while ship B would go $36\left(\frac{3}{4}\right) = 27$ miles. The sum of these distances is ____ nautical miles as required.

48

In Frames 137-141, set up an equation for the problem stated in Frame 137 and solve.

137. A pipe fills $\frac{3}{8}$ of a tank in 5 hours. At the end of 5 hours, a second pipe is turned on and with both pipes working together they finish filling the tank in another 3 hours. How long would it have taken the second pipe alone to fill the tank?

138. Time for second pipe alone to fill the tank: t

Portion of tank filled in 1 hour by first pipe alone:
$\frac{1}{5}$ of $\frac{3}{8} =$ ____

$\frac{3}{40}$

Portion of tank filled in 1 hour by second pipe alone:
____.

$\frac{1}{t}$

139. $\begin{bmatrix}\text{portion of}\\ \text{tank filled}\\ \text{by first pipe}\end{bmatrix} + \begin{bmatrix}\text{portion of}\\ \text{tank filled}\\ \text{by second pipe}\end{bmatrix} = \begin{bmatrix}\text{portion of}\\ \text{tank filled}\\ \text{by ____ pipes}\end{bmatrix}.$

both

140. Hence, an equation for the problem is

$$\frac{3}{40}\left(5 + 3\right) + \frac{1}{t}\left(\_\_\right) = 1.$$

3

141. Solving the equation,

$$\frac{24}{40} + \frac{3}{t} = 1$$

$$(5t) \cdot \left(\frac{3}{5} + \frac{3}{t}\right) = (\_\_) \cdot 1.$$

$$3t + \_\_ = 5t$$

$$\_\_ = 2t$$

$$\_\_ = t.$$

5t

15

15

$\frac{15}{2}$ or 7.5

Hence, it would take the second pipe alone 7.5 hours.

Check: Does $\frac{3}{40}\left(8\right) + \frac{3}{15/2} = \_\_$? Does $\frac{3}{5} + \frac{6}{15} = 1$? ____ (yes/no).

1

yes

4.5

INEQUALITIES

142. Open sentences involving the symbols <, ≤, >, ≥ are called __________.

inequalities

143. Any element of the replacement set of the variable for which an inequality is true is called a ________ of the inequality.

solution

144. The set of all solutions of an inequality is called the ________ set of the inequality.

solution

145. Inequalities that are true for every element of the replacement set are called ________ inequalities.

absolute

146. Inequalities that are not true for every element of the replacement set are called ________ inequalities.

conditional

147. **The addition of the same expression representing a real number to each member of an inequality produces an inequality of the _____ (same/opposite) sense.**

same

148. **If each member of an inequality is multiplied by the same expression representing a positive real number, the result is an equivalent inequality of the ______ (same/opposite) sense. If each member of an inequality is multiplied by the same expression representing a negative number, the result is an equivalent inequality of the ________ sense.**

same

opposite

In Frames 149-152, solve and graph the solution set of the inequality of Frame 149.

149. $2x + 3 \leq x - 1.$

150. $2x \leq x - \_\_.$

4

x

151. ___ ≤ -4.

$(-\infty, -4]$

152. In interval notation, the solution set is ________.

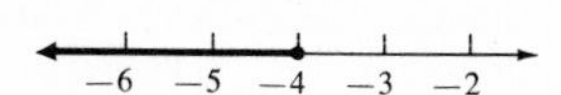

The graph is 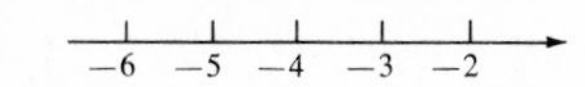

In Frames 153-158, solve and graph the solution set of the inequality of Frame 153.

153. $\dfrac{x - 6x}{2} < -20$.

2

154. $2\left(\dfrac{x - 6x}{2}\right) < (\_\_\_)(-20)$.

$x - 6x$

155. ________ < -40.

$-5x$

156. _____ < -40.

$>$

157. x ___ 8.

$(8, +\infty)$

158. In interval notation, the solution set is __________.

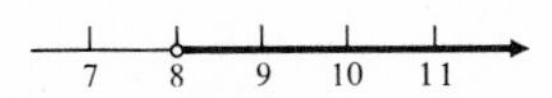

The graph is 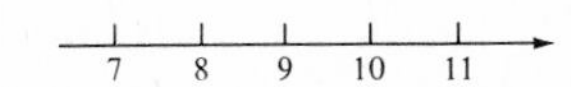

In Frames 159-166, solve and graph the inequality of Frame 159.

159. $\frac{3}{5}(3x + 2) - \frac{2}{3}(2x - 1) \leq 2$.

15

160. $15\left[\frac{3}{5}(3x + 2) - \frac{2}{3}(2x - 1)\right] \leq (\_\_\_\_)(2)$.

10

161. $9(3x + 2) - (\_\_\_\_)(2x - 1) \leq 30$.

10

162. $27x + 18 - 20x +$ ____ ≤ 30.

$7x$

163. ____ $+ 28 \leq 30$.

2

164. $7x \leq$ ___.

$\frac{2}{7}$

165. $x \leq$ ___.

$\left[-\infty, \frac{2}{7}\right]$

166. In interval notation, the solution set is ________.

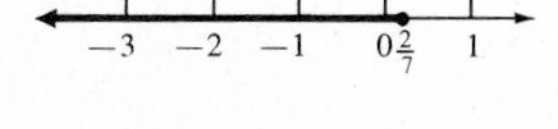

The graph is
−3 −2 −1 0 1

intersection ($\cap$)
disjoint

167. The operation on two sets A and B which forms a new set consisting of all the members that are common to A and B is called ______________. If $A \cap B = \emptyset$, A and B are called __________ sets.

In Frames 168-172, solve and graph the inequality of Frame 168.

168. $-3 < 3 - 2x < 9$

-3

169. $-3 - 3 < -3 + 3 - 2x <$ ____ $+ 9$.

6

170. $-6 < -2x <$ ___.

>; >

171. 3 ___ x ___ -3. (> or <)

(-3, 3)

172. In interval notation, the solution set is ___________.

The graph is

In Frames 173-179, graph the sets of Frame 173.

173. $\{x|2x - 3 < 5\} \cap \{x|-(2x - 3) < 5\} =$

8; >

174. $\{x|2x <$ ___$\} \cap \{x|2x - 3$ ___ $-5\} =$

4; -2

175. $\{x|x <$ ___$\} \cap \{x|2x >$ ____$\} =$

-1

176. $\{x|x < 4\} \cap \{x|x >$ ____$\}$.

177. The graph of $\{x|x < 4\}$ is

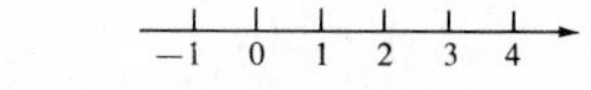

178. The graph of $\{x|x > -1\}$ is

179. The graph of $\{x|x < 4\} \cap \{x|x > -1\}$ is

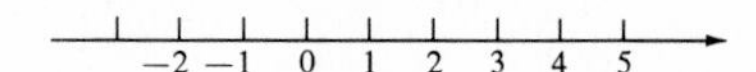

In Frames 180-184, set up an inequality for the problem of Frame 180 and solve.

180. The Fahrenheit and Celsius temperatures are related by the formula

$$C = \frac{5}{9}(F - 32).$$

Within what range must the temperature be in degrees Fahrenheit for the temperature to lie between 0° and 100°C?

0

181. We must have ____ < C < 100.

C

$\frac{5}{9}$(F - 32)

182. From the formula $C = \frac{5}{9}(F - 32)$, we substitute $\frac{5}{9}(F - 32)$ for ____ in Frame 181 and obtain

$0 < \_\_\_\_\_\_\_\_ < 100.$

$\frac{9}{5}$

180

183. Then

$\frac{9}{5} \cdot (0) < \frac{9}{5} \cdot \frac{5}{9}(F - 32) < \_\_ \cdot 100$

$0 < F - 32 < \_\_\_\_.$

212

184. Adding 32 to all members of the last inequality, we have the solution

$32 < F < \_\_\_\_.$

The Fahrenheit temperature must be between 32° and 212°.

4.6

EQUATIONS AND INEQUALITIES INVOLVING ABSOLUTE VALUE

$x - a$
$-(x - a)$

185. From the definition of the absolute value of a real number, we have

$$|x - a| = \begin{cases} \_\_\_\_\_\_ & \text{if } x \geq a. \\ \_\_\_\_\_\_ & \text{if } x < a. \end{cases}$$

union; ∪

186. The operation on two sets A and B, which forms a new set consisting of all the members of either A or B or both, is called ________ and is denoted by A _ B.

In Frames 187-188, solve $|x - 1| = 6$.

$-(x - 1)$

187. $x - 1 = 6$ or ________ $= 6$.

7; -5

188. Solve each equation and obtain

$x =$ ___ or $x =$ ____.

{7, -5}

189. The solution set is {7 -5} = ______.

In Frames 190- 193, solve the equation of Frame 190.

190. $\left|2x - \frac{1}{2}\right| = \frac{1}{4}$.

$-\left(2x - \frac{1}{2}\right) = \frac{1}{4}$

191. $2x - \frac{1}{2} = \frac{1}{4}$ or ______________.

$\frac{3}{8}$; $\frac{1}{8}$

192. Solve each equation and obtain

$x =$ ___ or $x =$ ___.

$\left\{\frac{3}{8}, \frac{1}{8}\right\}$

193. The solution set is $\left\{\frac{3}{8}\right\} \cup \left\{\frac{1}{8}\right\}$ = ____________.

In Frames 194 - 197 solve the equation of Frame 194.

194. $\left|2 - \frac{1}{3}x\right| = \frac{3}{4}$.

$-\left(2 - \frac{1}{3}x\right) = \frac{3}{4}$

195. $2 - \frac{1}{3}x = \frac{3}{4}$ or ______________.

$\frac{15}{4}$; $\frac{33}{4}$

196. Solve each equation and obtain

$x =$ ____ or $x =$ ____.

$\left\{\frac{15}{4}, \frac{33}{4}\right\}$

197. The solution set is $\left\{\frac{15}{4}\right\} \cup \left\{\frac{33}{4}\right\}$ = _____.

c

198. **An inequality of the form $|ax - b| < c$ can be equivalently written as $-c < ax - b <$ ___.**

In Frames 199- 202, solve and graph the solution set of the inequality of Frame 199 .

199. $|x| < 3$.

-3

200. ____ $< x < 3$.

$\{x \mid -3 < x < 3\}$;

201. The solution set is ___________ or

$(-3, 3)$

in interval notation _________.

202. The graph is

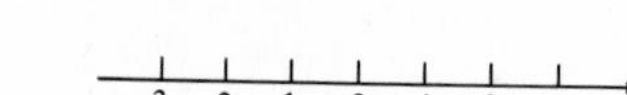

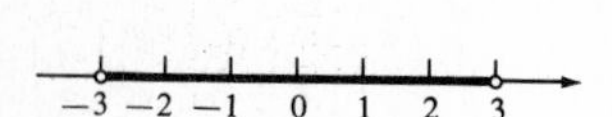

In Frames 203-207, solve and graph the solution set of the inequality of Frame 203.

203. $|x + 2| \leq 4$.

-4; 4

204. ____ $\leq x + 2 \leq$ ____.

2

205. $-6 \leq x \leq$ ____.

$\{x \mid -6 \leq x \leq 2\}$;

$[-6, 2]$

206. The solution set is ____________ or in interval notation __________.

207. The graph is

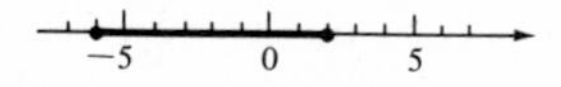

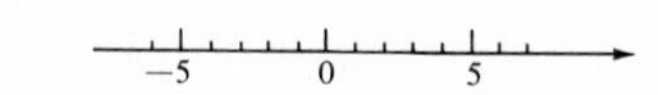

In Frames 208-213, solve and graph the solution set of Frame 208.

208. $|5 - 2x| \leq 15$.

-15; 15

209. _____ $\leq 5 - 2x \leq$ ____.

-20; 10

210. _____ $\leq -2x \leq$ ____.

$\geq$; $\geq$

211. 10 ___ x ___ -5.

$\{x \mid -5 < x < 10\}$;

$[-5, 10]$

212. The solution set is _________________ or in interval notation __________.

213. The graph is

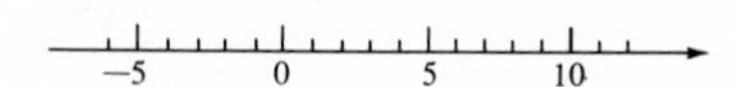

214. An inequality of the form $|ax + b| > c$ can be equivalently written as $ax + b > c$ or $-(ax + b) > c$.

In Frames 215-219, solve and graph the solution set of the inequality of Frame 215.

215. $|x| > 2$.

$-x$

216. $x > 2$ or ____ > 2.

$<$

217. $x > 2$ or x ___ -2.

$(2, +\infty)$

218. The solution set is $(-\infty, -2) \cup$ __________.

219. The graph is

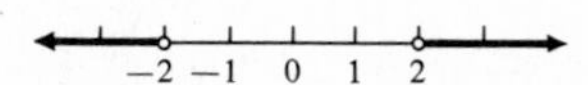

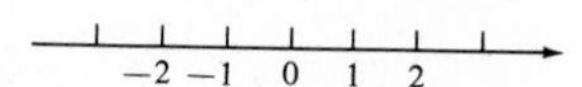

In Frames 220-224, solve and graph the solution set of the inequality $|4 - 3x| \geq 10$ *of Frame 219.*

10; 10

220. $4 - 3x \geq$ ____ or $-(4 - 3x) \geq$ ____.

$\leq$

221. $4 - 3x \geq 10$ or $4 - 3x$ ____ -10.

-14

222. $-3x \geq 6$ or $-3x \leq$ ____.

$\leq$; $\geq$

223. x ____ -2 or x ____ $\frac{14}{3}$.

$\left[\frac{14}{3}, +\infty\right]$

224. The solution set is $(-\infty, -2]$ ____.

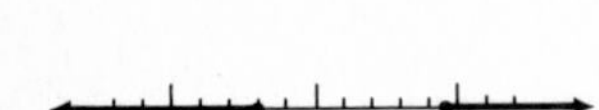

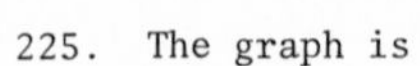

225. The graph is

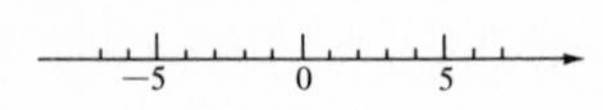

0
2

226. $|2x - 4| = 2x - 4$ is true if and only if $2x - 4 \geq$ ___. Solving this inequality, we have $x \geq$ ___.

0
$x < \frac{1}{3}$

227. $|3x - 1| = -(3x - 1)$ is true if and only if $3x - 1 <$ ___. Solving this inequality, we have ______.

TEST PROBLEMS

In Problems 1-5, solve for x.

1. $5[2 + 3(x - 2)] + 20 = 0$.

2. $(2x + 1)(x - 3) = (x - 2) + 2x^2$.

3. $\frac{x}{5} - 9 = \frac{x}{2}$.

4. $\frac{x}{x + 2} - \frac{x^2 + 8}{x^2 - 4} = \frac{3}{x - 2}$.

5. $\frac{x}{a} - \frac{a}{b} = \frac{b}{c}$.

6. Solve the equation

$$s = \frac{a}{1 - r} \text{ for } r.$$

7. Solve $\frac{1}{C_n} = \frac{1}{C_1} + \frac{1}{C_2}$ for C if $C_2 = 15$ and $C_n = 5$.

8. Solve $\frac{2x + y}{2} = \frac{4x - y}{3}$ for x in terms of y.

In Problems 9-12, solve and graph each solution set on a number line.

9. $\frac{2x - 3}{2} \leq 5$.

10. $\frac{1}{2}(x + 2) \geq \frac{2x}{3}$.

11. $2 \leq 3x - 4 \leq 8$.

12. $\{x | 2x - 3 < 5\} \cap \{x | 2x - 3 > -5\}$.

13. Solve $|2x + 5| = 2$.

14. Solve $\left|2x + \frac{1}{2}\right| = \frac{1}{4}$.

In Problems 15 and 16, solve and graph each solution set.

15. $|x + 5| > 2$.

16. $|2x + 4| < 6$.

17. When a number is divided by 8, it has 4 as a quotient and 5 as a remainder. Find the number.

18. The width of a rectangle is $\frac{3}{4}$ of its length. When the length is increased by 8 feet and the width by 6 feet, the area is increased by 96 square feet. Find the length and width of the original rectangle.

19. How much water should be added to 2 gallons of pure acid to obtain a 10% solution?

20. A collection of coins containing nickels, dimes, and quarters has a value of \$4.55. If there are 5 more dimes than nickels, and the number of quarters is 25 less than twice the number of dimes, how many of each kind of coin is in the collection?

21. An airplane left an airport and traveled at an average rate of 300 miles per hour. One hour later a second airplane left the airport traveling in the same direction and caught up with the first plane in 4 hours. What was the rate of the second plane?

5

EXPONENTS, ROOTS, AND RADICALS

5.1

POSITIVE INTEGRAL EXPONENTS

In this section, a and b are real numbers and m and n are positive integers.

Answers	Frames
a^{m+n} a^{m-n}, $(a \neq 0)$	1. $a^m \cdot a^n =$ ______ and $\frac{a^m}{a^n} =$ ______.
a^{n-m}	2. For $m < n$, $a \neq 0$, $\frac{a^m}{a^n} = \frac{1}{\_\_\_\_}$.
6; x^3	3. $\frac{x^6}{x^9} = \frac{1}{x^{9-(\_\_)}} = \frac{1}{(\_\_)}$.
3; $\frac{1}{5^4}$	4. $\frac{5^3}{5^7} = \frac{1}{5^{7-(\_\_)}} =$ ______.
a^{mn}	5. $(a^m)^n =$ ______.
2; 4^6	6. $(4^3)^2 = 4^{3\cdot(\_\_)} =$ ____.
3; x^{12}	7. $(x^4)^3 = x^{4\cdot(\_\_)} =$ ____.
$a^n b^n$	8. $(ab)^n =$ ______.
x^4y^4	9. $(xy)^4 =$ ______.
$9x^4y^6$	10. $(3x^2y^3)^2 = 3^2 \cdot (x^2)^2 \cdot (y^3)^2 =$ ______.
$\frac{a^n}{b^n}$	11. $b \neq 0$, $\left(\frac{a}{b}\right)^n =$ ______.

$\frac{3^4}{7^4}$

12. $\left(\frac{3}{7}\right)^4 = \underline{\quad}$.

$\frac{4x^6}{y^4}$

13. $\left(\frac{2x^3}{y^2}\right)^2 = \frac{2^2 \cdot (x^3)^2}{(y^2)^2} = \underline{\quad\quad}$ $(y \neq 0)$.

In Frames 14-24, write each given expression as a product or quotient in which each variable occurs at most once in each term and involves positive exponents only. Assume no denominator equals zero.

a^5

14. $a^4 \cdot a = \underline{\quad}$.

c^8

15. $c^4 \cdot c^4 = \underline{\quad}$.

$\frac{y}{y^4}$; $\frac{1}{a^2y^3}$

16. $\frac{ay}{a^3y^4} = \frac{a}{a^3} \cdot \underline{\quad} = \underline{\quad\quad}$.

a^6

17. $(-a^3)^2 = (-1 \cdot a^3)^2 = (-1)^2(a^3)^2 = \underline{\quad}$.

$(t^3)^4$; $r^8s^4t^{12}$

18. $(r^2st^3)^4 = (r^2)^4(s)^4(\underline{\quad\quad}) = \underline{\quad\quad\quad}$.

$\frac{a^6}{b^6}$

19. $\left(\frac{a^2}{b^2}\right)^3 = \frac{(a^2)^3}{(b^2)^3} = \underline{\quad}$.

$\frac{16}{27x^9}$; $\frac{16}{27x^7}$

20. $\frac{(4x)^2}{(3x^3)^3} = \frac{(4^2)(x^2)}{(3^3)(x^3)^3} = \frac{16x^2}{\underline{\quad\quad}} = \underline{\quad\quad}$.

$\frac{-27}{y^6z^3}$; $\frac{-3}{y^2z}$

21. $\left(\frac{y^2z}{3}\right)^2\left(\frac{-3}{y^2z}\right)^3 = \left(\frac{y^4z^2}{9}\right)\left(\underline{\quad\quad}\right) = \underline{\quad\quad}$.

$\frac{-27}{64ab^5}$

(detail below)

$\left(\frac{a}{b}\right)^2\left(-\frac{3}{4ab}\right)^3$

$= \left(\frac{a^2}{b^2}\right)\left(-\frac{27}{64a^3b^3}\right) = \frac{-27}{64ab^5}$

22. $\left(\frac{a}{b}\right)^2\left(-\frac{3}{4ab}\right)^3 = \underline{\quad\quad}$.

x^{4n}; x^{3n-2}

23. $\frac{x^{3n}x^{n}}{x^{n+2}} = \frac{x^{4n}}{x^{n+2}} = \underline{\quad\quad}$.

y^{n^2-n}; y^{n^2-2n}

24. $\frac{(y^{n-1})^n}{y^n} = \frac{y^{n^2-n+n}}{y^n} = \underline{\quad\quad}$.

5.2

ZERO AND NEGATIVE INTEGRAL EXPONENTS

In this section, a is any real number.

1

25. $a^0 = \_\_\_ \ (a \neq 0).$

1

26. $7^0 = \_\_\_.$

1

27. $\left(\frac{2}{3}\right)^0 = \_\_\_.$

$\frac{1}{a^n}$

28. $a^{-n} = \frac{}{\_\_\_} \ (a \neq 0).$

In Frames 29-36, write each expression as a basic numeral or fraction in lowest terms.

1; $\frac{1}{16}$

29. $4^{-2} = \frac{\_\_\_}{4^2} = \_\_\_.$

3^2; 3^2; 18

30. $\frac{2}{3^{-2}} = \frac{2}{\frac{1}{(\_\_\_)}} = 2 \cdot (\_\_\_) = \_\_\_.$

5^2; $\frac{1}{25}$; $\frac{626}{25}$

31. $5^2 + 5^{-2} = 5^2 + \frac{1}{(\_\_\_)} = 25 + \_\_\_ = \frac{625}{25} + \frac{1}{25} = \_\_\_.$

1; $\frac{1}{4}$

32. $\frac{4^{-1}}{5^0} = \frac{\frac{1}{4}}{(\_\_\_)} = \_\_\_.$

$\frac{1}{2^2}$; $\frac{1}{4}$; $4 \cdot 9$; $\frac{9}{4}$

33. $\frac{2^{-2}}{3^{-2}} = \frac{\_\_\_}{\frac{1}{3^2}} = \frac{\_\_\_}{\frac{1}{9}} = \frac{\frac{1}{4}(\_\_\_)}{\frac{1}{9}(4 \cdot 9)} = \_\_\_.$

4^2; 16

34. $\frac{1}{4^{-2}} = \frac{1}{\frac{1}{(\_\_\_)}} = \frac{1}{\frac{1}{16}} = \_\_\_.$

$\frac{3}{4}$; $\frac{4}{3}$

35. $\left(\frac{3}{4}\right)^{-1} = \frac{1}{(\_\_\_)} = \_\_\_.$

$\frac{1}{3^2}$; 3; $\frac{2}{9}$

36. $3^{-1} - 3^{-2} = \frac{1}{3^1} - (\_\_\_) = \frac{1}{3} - \frac{1}{9} = \frac{\_\_\_}{9} - \frac{1}{9} = \_\_\_.$

In Frames 37-42, write each expression as a product or quotient of powers in which each variable occurs but once and all exponents are positive. Assume all variable bases represent positive real numbers.

x^{-2}; $\frac{1}{x^2}$

37. $x^2 \cdot x^{-4} = x^{2+(-4)} = \_\_\_ = \frac{}{\_\_\_}.$

$\frac{1}{x^7}$

38. $\frac{x^{-3}}{x^4} = \frac{1}{x^{4-(-3)}} = \underline{\qquad}$.

y^{-4}; y^{-12}

y^{12}; $\frac{x^6}{y^{12}}$

39. $(x^2y^{-4})^3 = (x^2)^3(\underline{\qquad})^3 = x^6(\underline{\qquad})$

$= x^6 \cdot \frac{1}{(\underline{\qquad})} = \underline{\qquad}$.

$\frac{a^2y^4}{b^2x^3}$

40. $\frac{a^2x^{-3}}{b^2y^{-4}} = \frac{a^2 \cdot \frac{1}{x^3}}{b^2 \cdot \frac{1}{y^4}} = \frac{\left(a^2\frac{1}{x^3}\right) \cdot (x^3y^4)}{\left(b^2\frac{1}{y^4}\right) \cdot (x^3y^4)} = \underline{\qquad}$.

$a^{-4}b^{-2}$; a^4b^2

41. $\left(\frac{a^{-2}b^{-1}}{5^0a^2b}\right)^{-1} = \left(\frac{\qquad}{1}\right)^{-1} = \underline{\qquad}$.

-2

42. $\left[\left(\frac{a^2b^{-3}c^{-2}}{a^{-3}bc}\right)^2\right]^{-1} = \left(\frac{a^2b^{-3}c^{-2}}{a^{-3}bc}\right)^{(\underline{\qquad})}$

$= \left(\frac{\qquad}{a^{-5}}\right)^{-2} = \underline{\qquad}$.

$b^{-4}c^{-3}$; $\frac{b^8c^6}{a^{10}}$

In Frames 43-46, write each expression as a single fraction involving positive exponents only. Assume that each variable base represents a positive real number.

a; $\frac{2b}{a}$

43. $\frac{a^{-1}}{b^{-1}} + \frac{b}{a} = \frac{b}{\underline{\qquad}} + \frac{b}{a} = \underline{\qquad}$.

$\frac{1}{(x + y)^2}$

44. $(x + y)^{-2} = \underline{\qquad}$.

ab; $b + a$

45. $\frac{a^{-1} + b^{-1}}{(ab)^{-1}} = (\underline{\qquad})\left(\frac{1}{a} + \frac{1}{b}\right) = \underline{\qquad}$.

$\frac{\frac{1}{y}}{\frac{1}{y}}$; $\frac{xy}{xy}$; $\frac{y - x}{y + x}$

46. $\frac{x^{-1} - y^{-1}}{x^{-1} + y^{-1}} = \frac{\frac{1}{x} - \underline{\qquad}}{\frac{1}{x} + \underline{\qquad}} = \frac{\left(\frac{1}{x} - \frac{1}{y}\right)(\underline{\qquad})}{\left(\frac{1}{x} + \frac{1}{y}\right)(\underline{\qquad})} = \underline{\qquad}$.

In Frames 47-49, write each expression as a product free of fractions in which each variable occurs but once. Assume that each variable base represents a positive real number.

x

47. $x^{-2n}x^{2n+1} = x^{-2n+2n+1} = \underline{\qquad}$.

y^{n-1-n}; $x^{-n+1}y^{-1}$

48. $\frac{x^{2n}y^{n-1}}{x^{3n-1}y^{n}} = x^{2n-(3n-1)}(\underline{\qquad}) = \underline{\qquad}$.

$y^{2n-(-n)}$

49. $\left(\frac{x^{2n-1}y^{2n}}{x^{-2}y^{-n}}\right)^2 = \left[x^{2n-1-(-2)}\underline{\qquad}\right]^2$

$= \left[x^{2n+1}\underline{\qquad}\right]^2 = \underline{\qquad}$.

y^{3n}; $x^{4n+2}y^{6n}$

5.3

SCIENTIFIC NOTATION

1; 10

50. Scientific notation for a real number requires that the number be expressed as the product of a number between ___ and ___ and an appropriate power of ten.

right

51. A number may be written in scientific notation by moving the decimal point to change the number to a number between one and ten and then indicating multiplication by a power of ten where the exponent on the ten counts the number of places the decimal point was moved and is positive or negative as the decimal point was moved to the left or _______, respectively.

In Frames 52-58, express the given number in scientific notation.

Answer	Frame
10^2	52. $265 = 2.65 \times$ _______.
10^4	53. $26{,}500 = 2.65 \times$ _____.
10^{-2}	54. $.0265 = 2.65 \times$ ________.
10^6	55. $2{,}650{,}000 = 2.65 \times$ ______.
2.65; 10^{-6}	56. $.00000265 =$ ______ $\times$ ______.
3.48×10^5	57. $348{,}000 =$ ____________.
3.48×10^{-3}	58. $.00348 =$ ______________.

positive

59. A number written in scientific notation may be expressed in standard form by moving the decimal point to the left or right, as the exponent on the ten is negative or ________, a number of places equal to the absolute value of that exponent.

In Frames 60-63, express the scientific notation in standard form.

Answer	Frame
3750	60. $3.75 \times 10^3 =$ ____________.
.00375	61. $3.75 \times 10^{-3} =$ __________.
180,000	62. $1.8 \times 10^5 =$ _________.
.000018	63. $1.8 \times 10^{-5} =$ _________.

In Frames 64-67, compute.

10^{-3}; 10^{-1}

64. $\dfrac{10^4 \times 10^{-2}}{10^3} = 10^4 \times 10^{-2} \times \_\_\_\_\_\_ = 10^{4-2-3} = \_\_\_\_\_\_.$

10^{-9}; 10^{11}

65. $\dfrac{(4 \times 10^4) \times (6 \times 10^{-2})}{3 \times 10^{-9}} = \left(\dfrac{4 \times 6}{3}\right)\left(\dfrac{10^4 \times 10^{-2}}{\_\_\_\_\_\_}\right) = 8 \times \_\_\_\_\_\_.$

9.3×10^{-3}

3.1×10^{-1}

10^{-5}

10^{-2}; 10^{-1}

66. $\dfrac{.06 \times .084 \times .0093}{.00021 \times .31}$

$$= \frac{(6 \times 10^{-2}) \times (8.4 \times 10^{-2}) \times (\underline{\qquad\qquad})}{(2.1 \times 10^{-4}) \times (\underline{\qquad\qquad})}$$

$$= \frac{6 \times 8.4 \times 9.3}{2.1 \times 3.1} \times \frac{10^{-7}}{\underline{\qquad}} = \left(\frac{6}{1} \times \frac{8.4}{2.1} \times \frac{9.3}{3.1}\right) \times \frac{10^{-7}}{10^{-5}}$$

$$= 72 \times \underline{\qquad} = 7.2 \times \underline{\qquad}.$$

8×10^{3}

2×10^{2}

10^{4}; 10^{4}

67. $$\frac{4200 \times .016 \times 8000}{.00028 \times 120 \times 200} = \frac{(4.2 \times 10^{3})(1.6 \times 10^{-2})(\underline{\qquad\qquad})}{(2.8 \times 10^{-4})(1.2 \times 10^{2})(\underline{\qquad\qquad})}$$

$$= \frac{4.2 \times 1.6 \times 8}{2.8 \times 1.2 \times 2} \times \frac{\underline{\qquad}}{10^{0}} = 8 \times \underline{\qquad}.$$

5.4

RATIONAL EXPONENTS

In this section, n is a natural number and a is a real number.

a

root

68. For $a > 0$, $(a^{1/n})^{n} = a^{n/n} = \underline{\qquad}$. The number $a^{1/n}$ is called an nth ________ of a.

positive

$-a^{1/n}$

69. When n is even and $a > 0$, $a^{1/n}$ denotes the (positive/negative) __________ nth root of a. The negative nth root of a is written as ________.

8

70. $64^{1/2} = (8^{2})^{1/2} = \underline{\quad}$.

-8

71. $-64^{1/2} = -(8^{2})^{1/2} = \underline{\quad}$.

positive

72. If $a < 0$ and n is even, the number $a^{1/n}$ is not a real number, because an even power of either a positive or negative number is __________.

-16

73. $(-16)^{1/4}$ is not a real number, because there is no real number whose 4th power is equal to _____.

negative

74. If $a < 0$ and n is odd, there is a real number $a^{1/n}$, because an odd power of a negative real number is __________.

-27

75. $(-27)^{1/3} = -3$, since $(-3)^{3} = \underline{\quad}$.

a^{m}

76. For $a \geq 0$, m, n positive integers, $a^{m/n} = (a^{1/n})^{m} = (\underline{\quad})^{1/n}$.

Answer	Frame
$9^{1/2}$; 27	77. $9^{3/2} = (\_\_\_\_)^3 = (3)^3 = \_\_\_\_$.
9^3; 27	78. $9^{3/2} = (\_\_\_\_)^{1/2} = (729)^{1/2} = \_\_\_\_$.
$\lvert a\rvert^{m/n}$ $\lvert a\rvert$	79. When m and n are even natural numbers, $(a^m)^{1/n} = \_\_\_\_\_\_$, and for the special case when $m = n$, $(a^n)^{1/n} = \_\_\_\_\_\_$.
$\lvert -3\rvert = 3$	80. $(-3^2)^{1/2} = \_\_\_\_\_\_$.
$\frac{1}{a^{m/n}}$	81. For m and n positive integers, $a \neq 0$, and $a^{1/n}$, a real number, $a^{-(m/n)} = \_\_\_\_\_\_$.
$3^{1/2}$; $\frac{1}{4^{2/3}}$	82. $3^{-(1/2)} = \frac{1}{(\_\_\_\_)}$ and $4^{-(2/3)} = \left(\frac{}{\_\_\_\_}\right)$.
do	83. We assume that powers with rational exponents, positive, negative, or zero ____ (do/do not) obey all the laws of exponents set forth earlier.

In Frames 84-88, write each of the following using a basic numeral or fraction in lowest terms. Assume that all bases are positive unless otherwise specified.

Answer	Frame
6; 4	84. $36^{1/2} = \_\_\_$ and $64^{1/3} = \_\_\_$.
4; 2; 16	85. $32^{4/5} = (32^{1/5})^{(\_\_\_)} = (\_\_\_)^4 = \_\_\_\_$.
-2; -3; $\frac{1}{(-3)^2}$ or $\frac{1}{9}$	86. $(-27)^{-2/3} = [(-27)^{1/3}]^{(\_\_\_)} = (\_\_\_)^{-2} = \_\_\_\_\_\_$.
-3; 3; $\frac{1}{3^3}$ or $\frac{1}{27}$	87. $81^{-3/4} = (81^{1/4})^{(\_\_\_)} = (\_\_\_)^{-3} = \_\_\_\_\_\_$.
4; $\frac{1}{2}$; $\frac{1}{16}$	88. $\left[\frac{1}{8}\right]^{4/3} = \left[\left(\frac{1}{8}\right)^{1/3}\right]^{(\_\_\_)} = [\_\_\_]^4 = \_\_\_\_$.

In Frames 89-96, write each of the following as a product or quotient of powers in which each variable occurs but once and all exponents are positive.

Answer	Frame
1/4; $x^{5/8}$	89. $x^{3/8} \cdot x^{1/4} = x^{(3/8)+\_\_\_} = x^{(3/8)+(2/8)} = \_\_\_\_\_\_$.
1/6; $x^{4/6}$ or $x^{2/3}$	90. $\frac{x^{5/6}}{x^{1/6}} = x^{5/6-(\_\_\_)} = \_\_\_\_\_\_$.

2; $y^{2/3}$

91. $(y^{1/3})^2 = y^{(1/3)(\_\_\_)} = \_\_\_\_\_\_.$

8; $3/4$; a^6

92. $(a^8)^{3/4} = a^{(\_\_\_)(\_\_\_\_\_)} = \_\_\_\_.$

$1/2$

93. $x^{-(3/4)} \cdot x^{1/2} = x^{-(3/4)+(\_\_\_\_\_)}$

$x^{-1/4}$; $\dfrac{1}{x^{1/4}}$

$= x^{-(3/4)+(2/4)} = \_\_\_\_\_\_ = \dfrac{}{\_\_\_\_\_\_}.$

$1/2$

94. $(a^{2/5}b)^{1/2} = (a^{2/5})^{1/2} \cdot b^{(\_\_\_\_\_)}$

$1/2$; $a^{1/5}b^{1/2}$

$= a^{2/5(\_\_\_\_\_)}b^{1/2} = \_\_\_\_\_\_\_\_.$

$(x^3)^{2/3}$; $x^{3(2/3)}$; $\dfrac{x^2}{y^4}$

95. $\left(\dfrac{x^3}{y^6}\right)^{2/3} = \dfrac{\_\_\_\_\_\_}{(y^6)^{2/3}} = \dfrac{\_\_\_\_\_\_}{y^{6(2/3)}} = \dfrac{}{\_\_\_\_}.$

$-3/4$

96. $\left(\dfrac{x^{3/4}y^{1/4}z^{-2}}{x^{1/4}y^{-3/4}z^{0}}\right)^{-1/2} = \left(x^{(3/4)-(1/4)}y^{1/4-\_\_\_\_\_}z^{-2-0}\right)^{-1/2}$

1

$= (x^{1/2}y^{(\_\_\_)}z^{-2})^{-1/2}$

-2; $-1/2$

$= x^{(1/2)(-1/2)}y^{-1/2}z^{(\_\_\_\_)(\_\_\_\_)}$

z

$= x^{-1/4}y^{-1/2}(\_\_)$

$x^{1/4}y^{1/2}$

$= \dfrac{z}{\_\_\_\_\_\_\_\_}.$

In Frames 97-98, write the product in a form in which each base of a power occurs only once in each term.

$y^{(2/5)+(3/5)}$; $y^{3/5} + y$

97. $y^{2/5}(y^{1/5} + y^{3/5}) = y^{(2/5)+(1/5)} + \_\_\_\_\_\_\_\_ = \_\_\_\_\_\_\_\_.$

$x^{(-3/4)+(7/4)}$

98. $x^{-3/4}(x^{1/4} + x^{7/4}) = x^{(-3/4)+(1/4)} + \_\_\_\_\_\_\_\_$

$x^{-1/2} + x$

$= \_\_\_\_\_\_\_\_.$

In Frames 99-102, factor as indicated.

$x^{3/4}$ (see below)

$\dfrac{7}{8} = \dfrac{1}{8} + ?$

99. $x^{7/8} = x^{1/8}(\_\_\_\_\_\_).$

$y^{-3/5}$ (see below)

$\dfrac{2}{5} = 1 + ?$

100. $y^{2/5} = y(\_\_\_\_\_\_).$

$x^{1/3} - x^{2/3}$
(see below)
$x^{2/3} - x^{1} = x^{1/3}(x^{?} - x^{?})$

101. $x^{2/3} - x = x^{1/3}(\_\_\_\_\_\_\_\_\_\_).$

$y - 1$
(see below)
$y^{1/2} - y^{-1/2}$
$= y^{-1/2}(y^{?} - y^{?})$

102. $y^{1/2} - y^{-1/2} = y^{-1/2}(\_\_\_\_\_\_).$

In Frames 103-106, simplify. Assume $m,n > 0$ and all bases are positive.

b^9

103. $(a^3)^{n/3} \cdot (b^{3n})^{3/n} = a^n(\_\_\_)$.

$\frac{b^{1/2}}{a^n}$; $\frac{a^{n/2}}{b}$

104. $\left(\frac{a^n}{b}\right)^{3/2}\left(\frac{b}{a^{2n}}\right)^{1/2} = \left(\frac{a^{3n/2}}{b^{3/2}}\right)\left(\_\_\_\right) = \frac{a^{(3n/2)-n}}{b^{(3/2)-1/2}} = \_\_\_$.

y^n; xy

105. $\left[\frac{x^{n+2}y^{n+1}}{x^2y}\right]^{1/n} = [(x^n)(\_\_\_)]^{1/n} = \_\_\_$.

$\frac{x^{4n}}{y^{2n}}$; $\frac{x^{16n}}{y^{8n}}$

106. $\left[\left(\frac{x^{4n}}{y^{2n}}\right)^{1/n}\right]^{4n} = \left(\_\_\_\right)^4 = \_\_\_$.

In Frames 107-110, consider variable bases to be any member of the set of real numbers (positive or negative) and state all necessary restrictions.

4 (See Frame 79.)

107. $[(-4)^2]^{1/2} = |-4| = \_\_\_$.

$|3x|$ (See Frame 79.)

108. $[9x^2]^{1/2} = [(3x)^2]^{1/2} = \_\_\_$.

$4x^3$; $|4x^3|$

109. $[16x^6]^{1/2} = [(\_\_\_)^2]^{1/2} = \_\_\_$.

$\frac{2}{|y|(y+1)^{1/2}}$; 0; -1

(detail below)

$y \neq 0$, because no factor in a denominator can equal zero. $y > -1$ because $(y + 1)$ must be greater than zero in order for $(y + 1)^{1/2}$ to be real and not zero.

110. $\frac{2}{[y^2(y+1)]^{1/2}} = \_\_\_$ ($y \neq \_\_\_$; $y > \_\_\_$).

irrational

rational

111. Any real number that cannot be expressed as the quotient of two integers is called an ___ number; any expression such as $a^{1/n}$ represents a rational number if and only if a is the nth power of a ___ (what kind?) number.

rational

112. Because $9 = 3^2$, $9^{1/2}$ is a ___ (what kind?) number.

rational

5

113. Because $-8 = (-2)^3$, $(-8)^{1/3}$ is a ___ number and $5^{1/3}$ is an irrational number because there is no rational number whose third power is equal to ___.

5.5

RADICALS

In this section n is a natural number, a is a real number, and m is an integer.

Answer	Frame
$\sqrt[n]{a}$	114. **An alternative form for the nonnegative nth root of a, where $n \geq 2$, is defined by $a^{1/n} =$ _______.**
$\sqrt[3]{2}$; $\sqrt[5]{5}$	115. $2^{1/3} =$ ____. 116. $5^{1/5} =$ ___.
a^m	117. **Since $a^{m/n} = (a^m)^{1/n} = (a^{1/n})^m$, $a^{m/n} = \sqrt[n]{\_\_\_}$.**
4	118. $8^{2/3} = \sqrt[3]{8^2} = \sqrt[3]{64} =$ ___.
^{2}; 4	119. $8^{2/3} = (\sqrt[3]{8})^{—} = (2)^2 =$ ___.
$\|a\|$	120. **Since, for n even, $a^{n/n} = \|a\|$, $\sqrt[n]{a^n} =$ _____ for n even.**
-2; 2; -3; 3	121. $\sqrt{(-2)^2} = \|\_\_\_\_\| =$ ___. 122. $\sqrt[4]{(-3)^4} = \|\_\_\_\_\| =$ ___.
a	123. **If n is odd, $\sqrt[n]{a^n} =$ ___.**
-2; 2	124. $\sqrt[3]{(-2)^3} =$ ____. 125. $\sqrt[3]{2^3} =$ ____.

In Frames 126-144, assume that each variable and each radicand represents a positive real number and that no denominator equals zero.

In Frames 126-133, write in radical form.

Answer	Frame
$\sqrt[3]{6}$; $6\sqrt[3]{x^2}$	126. $6^{1/3} =$ ____. 127. $6x^{2/3} = 6(x)^{2/3} =$ _______.
$\sqrt[3]{36x^2}$; $2\sqrt[6]{a^2b}$	128. $(6x)^{2/3} =$ ______. 129. $2(a^2b)^{1/6} =$ _______.
$\sqrt[5]{(3r - 2s)^2}$	130. $(3r - 2s)^{2/5} =$ _______.
2^3; $\dfrac{1}{\sqrt[4]{8}}$	131. $2^{-3/4} = \dfrac{1}{2^{3/4}} = \dfrac{1}{\sqrt[4]{\_\_\_}} = \_\_\_$.
$\sqrt[5]{a}$; $\sqrt[5]{b^3}$	132. $a^{1/5} - b^{3/5} =$ ____ - _____.
$\dfrac{1}{\sqrt[3]{x^2 - y^2}}$	133. $(x^2 - y^2)^{-1/3} = \dfrac{1}{(x^2 - y^2)^{1/3}} = \_\_\_\_\_\_\_\_\_\_$.

In Frames 134-139, represent each of the given expressions with positive fractional exponents.

$9^{1/3}$

134. $\sqrt[3]{9} =$ ______.

$12a^4$; $12^{1/5}a^{4/5}$

135. $\sqrt[5]{12a^4} = (\_\_\_\_)^{1/5} = 12^{1/5}(a^4)^{1/5} =$ __________.

ab^3; $-4a^{1/4}\,b^{3/4}$

136. $-4\sqrt[4]{ab^3} = -4(\_\_\_\_)^{1/4} =$ __________.

$(a - 4b)^{1/3}$

137. $\sqrt[3]{a - 4b} =$ __________.

$x^{1/4} - 3y^{1/4}$

138. $\sqrt[4]{x} - 3\sqrt[4]{y} =$ __________.

$\dfrac{3}{x^{3/4}}$

139. $\dfrac{3}{\sqrt[4]{x^3}} = \dfrac{\quad}{\_\_\_\_}$.

In Frames 140-146, find each root indicated.

-1; 1; -3

140. $\sqrt{(-1)^2} = |\_\_\_| =$ ___. 141. $-\sqrt[4]{81} = -\sqrt[4]{3^4} =$ ___.

y^2

142. $\sqrt[4]{y^8} = \sqrt[4]{(y^2)^4} =$ ___.

$\frac{2}{5}ab^2$; $\frac{2}{5}ab^2$

143. $\sqrt{\frac{4}{25}a^2b^4} = \sqrt{(\_\_\_)^2} =$ ______.

$-\frac{2}{3}xy^2$; $-\frac{2}{3}xy^2$

144. $\sqrt[3]{-\frac{8}{27}x^3y^6} = \sqrt[3]{(\_\_\_\_)^3} =$ ______.

In Frame 145, assume that variables and radicands represent any elements of the set of real numbers. Use absolute value notation as needed.

$2a - b$; $\dfrac{1}{|2a - b|}$

(See Frame 120.)

145. $\dfrac{1}{\sqrt{4a^2 - 4ab + b^2}} = \dfrac{1}{\sqrt{(\_\_\_\_\_\_)^2}} =$ ________ $(2a \neq b)$.

rational

146. Any radical expression $\sqrt[n]{a}$ represents a rational number if and only if a is the nth power of a ________ number.

5.6

CHANGING FORMS OF RADICALS

In this section n is a natural number, and assume that all variables in radicands denote positive real numbers only, and no denominator equals zero.

$\sqrt[n]{b}$

147. $\sqrt[n]{ab} = \sqrt[n]{a} \cdot$ _____.

$\sqrt{2}$; $3\sqrt{2}$

148. $\sqrt{18} = \sqrt{9} \cdot \sqrt{2} = 3\sqrt{2}$

$\sqrt[4]{2x}$; $2xy\sqrt[4]{2x}$

149. $\sqrt[4]{32x^5y^4} = \sqrt[4]{2^4x^4y^4} \cdot \sqrt[4]{2x} = 2xy\sqrt[4]{2x}$

$\sqrt[n]{b}$

150. $\sqrt[n]{\frac{a}{b}} = \frac{\sqrt[n]{a}}{\_\_\_\_}$.

36 ; $\frac{5}{6}$

151. $\sqrt{\frac{25}{36}} = \frac{\sqrt{25}}{\sqrt{\_\_\_}} = \_\_\_$.

$\frac{\sqrt{5}}{3}$

152. $\sqrt{\frac{5}{9}} = \frac{\sqrt{5}}{\sqrt{9}} = \_\_\_\_$.

$4x$; $\frac{\sqrt[3]{4x}}{2}$

153. $\sqrt[3]{\frac{2x}{4}} = \sqrt[3]{\frac{4x}{8}} = \frac{\sqrt[3]{\_\_\_\_}}{\sqrt[3]{8}} = \_\_\_\_$.

$\sqrt[n]{a^m}$

154. If m is an integer and c is a natural number with $n \geq 2$, then $\sqrt[cn]{a^{cm}} = \_\_\_\_\_\_$.

$\sqrt[3]{2^2}$ or $\sqrt[3]{4}$

155. $\sqrt[6]{16} = \sqrt[2\cdot 3]{2^{2\cdot 2}} = \_\_\_\_$.

$\sqrt{y}$

156. $\sqrt[4]{y^2} = \sqrt[2\cdot 2]{y^{2\cdot 1}} = \_\_\_$.

157. A radical expression is in simplest form if the following conditions exist:

index

a. the radicand contains no polynomial factor raised to a power equal to or greater than the ______ of the radical.

fractions

b. the radicand contains no ______.

denominators

c. no radical expressions are contained in the ______ of fractions.

factors

d. the index of the radical and exponents on factors in the radicand have no common ______.

In Frames 158–165, change the given expression to simplest form.

10; 10^2; $10\sqrt{5}$

158. $\sqrt{500} = \sqrt{(\_\_\_)^2(5)} = \sqrt{\_\_\_\_}\sqrt{5} = \_\_\_\_$.

$(2z^3)^2$; $-2z^3\sqrt{6}$

159. $-\sqrt{24z^6} = -\sqrt{(2z^3)^2(6)} = -\sqrt{\_\_\_\_}\sqrt{6} = \_\_\_\_$.

5; 5^3; $5\sqrt[3]{2}$

160. $\sqrt[3]{250} = \sqrt[3]{(\_\_\_)^3(2)} = \sqrt[3]{\_\_\_}\sqrt[3]{2} = \_\_\_\_$.

$2y^2\sqrt[4]{4x^2y}$

(see below)

$$\sqrt[4]{64x^2y^9} = \sqrt[4]{2^4(y^2)^4}\sqrt[4]{4x^2y}$$

$$= 2y^2\sqrt[4]{4x^2y}.$$

161. $\sqrt[4]{64x^2y^9} = \_\_\_\_\_\_$.

$3\sqrt[4]{2}$

162. $\sqrt[4]{6}\,\sqrt[4]{27} = \sqrt[4]{2 \cdot 3}\sqrt[4]{3^3} = \sqrt[4]{3^4 \cdot 2} =$ ______.

$2\sqrt[3]{15}$

163. $\sqrt[3]{10}\sqrt[3]{12} =$ ______.

(see below)

$\sqrt[3]{10}\sqrt[3]{12} = \sqrt[3]{2 \cdot 5}\sqrt[3]{2^2 \cdot 3}$

$= \sqrt[3]{2^3 \cdot 5 \cdot 3}$

$= 2\sqrt[3]{15}.$

$a\sqrt[5]{a}$ (see below)

164. $\sqrt[5]{a^2}\sqrt[5]{a^4} =$ ______.

$\sqrt[5]{a^2}\sqrt[5]{a^4} = \sqrt[5]{a^6}$

$= \sqrt[5]{a^5}\sqrt[5]{a} = a\sqrt[5]{a}$

165. $\sqrt{80000} = \sqrt{8(100)^2} = \sqrt{2(2)^2(100)^2}$

$= \sqrt{(2)^2(100)^2}\sqrt{} =$ ______.

$2;\ 200\sqrt{2}$

166. $\sqrt{\dfrac{1}{6}} = \dfrac{\sqrt{1}(\_\_\_)}{\sqrt{6}(\_\_\_)} = \dfrac{}{\_\_\_}.$

$\dfrac{\sqrt{6}}{\sqrt{6}};\ \dfrac{\sqrt{6}}{6}$

167. $\sqrt{\dfrac{2}{5}} = \dfrac{\sqrt{2}(\_\_\_)}{\sqrt{5}\sqrt{5}} = \dfrac{}{\_\_\_}.$

$\sqrt{5};\ \dfrac{\sqrt{10}}{5}$

168. $\dfrac{\sqrt{y}}{\sqrt{8}} = \dfrac{\sqrt{y}(\sqrt{2})}{\sqrt{8}(\sqrt{2})} = \dfrac{}{\_\_\_}.$

$\dfrac{\sqrt{2y}}{4}$

169. $\dfrac{\sqrt{a}}{\sqrt{2b}} = \dfrac{\sqrt{a}(\sqrt{2b})}{\sqrt{2b}(\sqrt{2b})} = \dfrac{}{\_\_\_}.$

$\dfrac{\sqrt{2ab}}{2b}$

170. $\dfrac{3}{\sqrt[4]{27}} = \dfrac{3}{\sqrt[4]{3^3}} = \dfrac{3(\sqrt[4]{3})}{\sqrt[4]{3^3}(\sqrt[4]{3})} = \dfrac{}{\_\_\_} = \sqrt[4]{3}.$

$\dfrac{3\sqrt[4]{3}}{3}$

171. $\sqrt[3]{\dfrac{1}{5xy}} = \dfrac{\sqrt[3]{1}(\sqrt[3]{5^2x^2y^2})}{\sqrt[3]{5xy}(\sqrt[3]{5^2x^2y^2})} = \dfrac{}{\_\_\_}.$

$\dfrac{\sqrt[3]{25x^2y^2}}{5xy}$

172. $\dfrac{\sqrt{xy}\sqrt{xy^3}}{\sqrt{y}} = \sqrt{\dfrac{}{}} = \sqrt{x^2(\_\_\_)} = \sqrt{\_\_\_}\sqrt{y} =$ ______.

$\dfrac{x^2y^4}{y};\ y^3;\ x^2y^2;\ xy\sqrt{y}$

173. $\dfrac{\sqrt[5]{a^2}\sqrt[5]{a^2b^2}}{\sqrt[5]{a^4b^4}} = \sqrt[5]{\dfrac{}{a^4b^4}} = \sqrt[5]{\dfrac{}{b^2}}$

$= \sqrt[5]{\dfrac{1 \cdot (\_\_\_)}{b^2 \cdot b^3}} = \dfrac{}{\sqrt[5]{b^5}} = \dfrac{}{\_\_\_}.$

$a^4b^2;\ 1$

$b^3;\ \sqrt[5]{b^3};\ \dfrac{\sqrt[5]{b^3}}{b}$

In Frames 174-175, rationalize numerators.

174. $\dfrac{\sqrt{5}}{5} = \dfrac{\sqrt{5}(\_\_\_)}{5(\sqrt{5})} = \dfrac{}{5\sqrt{5}} = \dfrac{}{\_\_\_}.$

$\sqrt{5};\ 5;\ \dfrac{1}{\sqrt{5}}$

$\sqrt{xy}$; xy; $\dfrac{x}{\sqrt{xy}}$

175. $\dfrac{\sqrt{xy}}{y} = \dfrac{\sqrt{xy}(\_\_\_\_)}{y(\sqrt{xy})} = \dfrac{\_\_\_\_}{y\sqrt{xy}} = \dfrac{\quad}{\_\_\_\_}$.

In Frames 176-178, reduce the order of each radical.

$\sqrt{27} = 3\sqrt{3}$

176. $\sqrt[4]{3^6} = \sqrt[(2)\cdot 2]{3^{(2)\cdot 3}} = \_\_\_\_$.

$\sqrt[5]{8}$

177. $\sqrt[10]{64} = \sqrt[10]{2^6} = \_\_\_\_$.

$9xy$; $\sqrt{9xy} = 3\sqrt{xy}$

178. $\sqrt[4]{81x^2y^2} = \sqrt[4]{(\_\_\_\_)^2} = \_\_\_\_$.

In Frames 179-181, express as radicals of the same order and multiply.

^{4}; ^{3}; 2^3; $\sqrt[12]{648}$

179. $\sqrt[3]{3}\sqrt[4]{2} = \sqrt[12]{3^{(\_\_)}}\sqrt[12]{2^{(\_\_)}} = \sqrt[12]{3^4(\_\_\_\_)} = \_\_\_\_$.

$\sqrt[4]{x^2}$; $\sqrt[4]{x^3}$

180. $\sqrt[4]{x}\sqrt{x} = \sqrt[4]{x}(\_\_\_\_) = \_\_\_\_$.

$\sqrt[10]{(3\cdot y)^5}$; $243y^5$

$\sqrt[10]{972x^4y^5}$

181. $\sqrt[5]{2x^2}\sqrt{3y} = \sqrt[10]{(2x^2)^2}\ (\_\_\_\_\_\_) = \sqrt[10]{4x^4 \cdot \_\_\_\_\_\_}$

$= \_\_\_\_\_\_$.

5.7

EXPRESSIONS CONTAINING RADICALS

In this section, assume that all radicands and variables are positive real numbers.

In Frames 182-187, write each sum or difference as a single term.

$-4\sqrt{5}$

182. $2\sqrt{5} - 6\sqrt{5} = (2 - 6)\sqrt{5} = \_\_\_\_$.

$\sqrt{3^2}$

$15\sqrt{3}$; $5\sqrt{3}$

183. $4\sqrt{75} - 5\sqrt{27} = 4\sqrt{5^2}\ \sqrt{3} - 5(\_\_\_\_)\sqrt{3}$

$= 20\sqrt{3} - \_\_\_\_ = \_\_\_\_$.

184. $2\sqrt{18r^3} - \sqrt{8r^3} + 3\sqrt{2r^3}$

$\sqrt{4r^2}$; $\sqrt{r^2}$

$= 2\sqrt{(3r)^2}\sqrt{2r} - (\_\_\_\_)\sqrt{2r} + 3(\_\_\_\_)\sqrt{2r}$

$2r$; $3r$; $7r\sqrt{2r}$

$= 6r\sqrt{2r} - (\_\_\_)\sqrt{2r} + (\_\_\_)\sqrt{2r} = \_\_\_\_$.

$\sqrt[3]{3^3}$

2^2

$9\sqrt[3]{2}$; $29\sqrt[3]{2}$

185. $5\sqrt[3]{128} + 3\sqrt[3]{54} = 5\sqrt[3]{2^6}\ \sqrt[3]{2} + 3(\_\_\_\_)\sqrt[3]{2}$

$= 5\cdot(\_)\sqrt[3]{2} + 3\cdot 3\sqrt[3]{2}$

$= 20\sqrt[3]{2} + \_\_\_\_ = \_\_\_\_$.

$\dfrac{3 - \sqrt{2}}{7}$

186. $\dfrac{3}{7} - \dfrac{\sqrt{2}}{7} = \dfrac{\quad}{\_\_\_\_}$.

$5\sqrt{3}$; $\dfrac{9 - 5\sqrt{3}}{15}$

187. $\dfrac{3}{5} - \dfrac{\sqrt{3}}{3} = \dfrac{9}{15} - \dfrac{\_\_\_\_}{15} = \dfrac{\quad}{\_\_\_\_}$.

In Frames 188-194, write each expression without parentheses and then write all radicals in simple form.

$10 - 2\sqrt{3}$

188. $2(5 - \sqrt{3}) = 2 \cdot 5 - 2 \cdot \sqrt{3} =$ ______.

$\sqrt{5}$; $3\sqrt{2} - \sqrt{10}$

189. $\sqrt{2}(3 - \sqrt{5}) = 3 \cdot \sqrt{2} - \sqrt{2} \cdot$ (___) = ______.

2; $\sqrt{2}$

190. $(1 + \sqrt{2})(2 - \sqrt{2}) = 1 \cdot 2 + 2\sqrt{2} - \sqrt{2} -$ ___ = ___.

$x - 4$

191. $(\sqrt{x} + 2)(\sqrt{x} - 2) =$ ______.

16; $y - 8\sqrt{y} + 16$

192. $(\sqrt{y} - 4)^2 = (\sqrt{y})^2 - 2 \cdot 4\sqrt{y} +$ ___ = ______.

2; 3; $12 - 2\sqrt{15}$

193. $(\sqrt{5} - \sqrt{3})(3\sqrt{5} + \sqrt{3}) = 3\sqrt{25} -$ (___)$\sqrt{15}$ - ___ = ______.

$4\sqrt[3]{16}$

$\sqrt[3]{2}$

$9 - 8\sqrt[3]{2}$

194. $(3 - 2\sqrt[3]{4})(3 + 2\sqrt[3]{4}) = 3^2 - [2\sqrt[3]{4}]^2 = 9 -$ ______

$= 9 - 4\sqrt[3]{8}$ (___)

= ______.

In Frames 195-198, change each expression to the form indicated. Assume all radicals are positive real numbers.

$3 + 2\sqrt{2}$ (detail below)

$15 + 10\sqrt{2} = 5 \cdot 3 + 5 \cdot 2\sqrt{2}$

$= 5(3 + 2\sqrt{2})$.

195. $15 + 10\sqrt{2} = 5($______$)$.

$1 + \sqrt{2x}$ (detail below)

$6 + \sqrt{72x} = 6 + \sqrt{36}\sqrt{2x}$

$= 6 + 6\sqrt{2x}$

$= 6(1 + \sqrt{2x})$.

196. $6 + \sqrt{72x} = 6($______$)$.

$1 - 2\sqrt{2}$ (detail below)

$\sqrt{12} - 4\sqrt{6} = (2\sqrt{3}) - 2\sqrt{3}(2\sqrt{2})$

$= 2\sqrt{3}(1 - 2\sqrt{2})$.

197. $\sqrt{12} - 4\sqrt{6} = 2\sqrt{3}($______$)$.

$\sqrt{3} - 1$ (detail below)

$\sqrt{3y} - \sqrt{y} = \sqrt{3}\,\sqrt{y} - \sqrt{y}$

$= \sqrt{y}(\sqrt{3} - 1)$

198. $\sqrt{3y} - \sqrt{y} = \sqrt{y}($______$)$.

In Frames 199-201, reduce each fraction to lowest terms.

$4 + 2\sqrt{5}$; $4 + 2\sqrt{5}$

199. $\dfrac{8 + 4\sqrt{5}}{2} = \dfrac{2(\text{______})}{2} =$ ______.

4; $3 - \sqrt{3}$; $3 - \sqrt{3}$

200. $\dfrac{12 - 2\sqrt{12}}{4} = \dfrac{12 - (\text{___})\sqrt{3}}{4} = \dfrac{4(\text{______})}{4} =$ ______.

x; $1 - xy$; $1 - xy$

201. $\dfrac{\sqrt{2x} - y\sqrt{2x^3}}{\sqrt{2x}} = \dfrac{\sqrt{2x} - y(\text{___})\sqrt{2x}}{\sqrt{2x}} = \dfrac{\sqrt{2x}(\text{______})}{\sqrt{2x}} =$ ______.

$a - b$

202. $a + b$ and ______ are conjugates of each other.

$2 - \sqrt{3}$

$3x + 4y$

203. The conjugate of $2 + \sqrt{3}$ is ______.

The conjugate of $3x - 4y$ is ______.

conjugate

204. **To rationalize denominators of fractions in which radicals occur in one or both of two terms of a binomial, multiply the numerator and denominator of the fraction by the ________ of the denominator.**

In Frames 205-208, rationalize denominators.

$2 - \sqrt{2}$; $2 - \sqrt{2}$; $\frac{2 - \sqrt{2}}{2}$

205. $\frac{1}{2 + \sqrt{2}} = \frac{1(\_\_\_\_)}{(2 + \sqrt{2})(2 - \sqrt{2})} = \frac{\_\_\_\_}{4 - 2} = \_\_\_\_.$

$2 + \sqrt{5}$; -1

$-1 - \sqrt{5}$; $1 + \sqrt{5}$

206. $\frac{\sqrt{5} - 3}{2 - \sqrt{5}} = \frac{(-3 + \sqrt{5})(\_\_\_\_)}{(2 - \sqrt{5})(2 + \sqrt{5})} = \frac{-6 - \sqrt{5} + 5}{\_\_\_\_}$

$= \frac{\_\_\_\_}{-1} = \_\_\_\_.$

$\sqrt{2} + y$; $\frac{y\sqrt{2} + y^2}{2 - y^2}$

207. $\frac{y}{\sqrt{2} - y} = \frac{y(\_\_\_\_)}{(\sqrt{2} - y)(\sqrt{2} + y)} = \frac{}{\_\_\_\_}.$

$\sqrt{6} + 2$; 2; $\frac{\sqrt{6}}{2}$

208. $\frac{3 - \sqrt{6}}{\sqrt{6} - 2} = \frac{(-\sqrt{6} + 3)(\_\_\_\_)}{(\sqrt{6} - 2)(\sqrt{6} + 2)} = \frac{-6 + \sqrt{6} + 6}{\_\_\_\_} = \frac{}{\_\_\_\_}.$

In Frames 209-211, write each expression as a single fraction in which the denominator is rationalized.

$3\sqrt{3}$; $6\sqrt{3}$; $\frac{5\sqrt{6} - 6\sqrt{3}}{6}$

209. $\frac{5}{\sqrt{6}} - \frac{3}{\sqrt{3}} = \frac{5\sqrt{6}}{6} - \frac{\_\_\_\_}{3} = \frac{5\sqrt{6}}{6} - \frac{\_\_\_\_}{6} = \_\_\_\_.$

$2\sqrt{3x}$; $\sqrt{x}$; $3\sqrt{x}$; $\frac{2\sqrt{3x} - 3\sqrt{x}}{3x}$

210. $\frac{2}{\sqrt{3x}} - \frac{1}{\sqrt{x}} = \frac{\_\_\_\_}{3x} - \frac{\_\_\_\_}{x} = \frac{2\sqrt{3x}}{3x} - \frac{\_\_\_\_}{3x} = \_\_\_\_.$

$\sqrt{x - 1}$

$x - 1$

$\frac{-\sqrt{x - 1}}{x - 1}$

211. $\sqrt{x - 1} - \frac{x}{\sqrt{x - 1}} = \sqrt{x - 1} - \frac{x(\_\_\_\_)}{(\sqrt{x - 1})^2}$

$= \frac{\sqrt{x - 1}(\_\_\_\_)}{x - 1} - \frac{x\sqrt{x - 1}}{x - 1}$

$= \frac{x\sqrt{x - 1} - \sqrt{x - 1} - x\sqrt{x - 1}}{x - 1} = \_\_\_\_$

In Frames 212-213, rationalize numerators.

$\sqrt{3} + \sqrt{2}$; $\sqrt{3} + \sqrt{2}$; $\frac{1}{3 + \sqrt{6}}$

212. $\frac{\sqrt{3} - \sqrt{2}}{\sqrt{3}} = \frac{(\sqrt{3} - \sqrt{2})(\_\_\_\_)}{\sqrt{3}(\_\_\_\_)} = \_\_\_\_.$

$\sqrt{x} - \sqrt{y}$; $\frac{x - y}{y(\sqrt{x} - \sqrt{y})}$

213. $\frac{\sqrt{x} + \sqrt{y}}{y} = \frac{(\sqrt{x} + \sqrt{y})(\_\_\_\_)}{y(\sqrt{x} - \sqrt{y})} = \_\_\_\_.$

TEST PROBLEMS

In all problems, assume that all radicands and variables are positive real numbers and no denominator is equal to zero.

In Problems 1-4, write each of the statements as a product or quotient in which each variable occurs once at most in each term and involves positive exponents only.

1. a. $(x^3y^2)^3$. b. $(-2x^3)^2(-2y^2)^2$.

2. a. $\left(\dfrac{x^3}{y^4}\right)^3$. b. $\left(\dfrac{x^2}{y^2}\right)^2\left(-\dfrac{2xy}{5}\right)^3$.

3. a. $\dfrac{(-x^3)^2(-x)^3}{(x^2)^3}$. b. $(x^{2n+1} \cdot x^{n-2})^2$.

4. a. $\left(\dfrac{x^{2n}x^{3n}}{x^{5n-2}}\right)^3$. b. $\left(\dfrac{x^{2n-2}}{x^{2n}}\right)^{2n}$.

In Problems 5 and 6, simplify. Assume m,n > 0.

5. $\left(\dfrac{x^{3n}y^{4n}}{x^{2n}y^{n}}\right)^{1/3}$. 6. $\left[\left(\dfrac{x^{4n}}{y^{3n}}\right)^{1/4n}\right]^{1/3}$.

In Problems 7 and 8, write each of the statements as a product or quotient of powers in which each variable occurs but once, and all exponents are positive.

7. a. $x^{-5}x^2$. b. $(x^4y^{-2})^{-1/3}$.

8. a. $\dfrac{x^{-3}y^{-2}z^{-1}}{x^0y^{-4}z^2}$. b. $\left(\dfrac{3^0x^{-2}y^{-3}}{x^2y^{-2}}\right)^{-2}$.

9. Simplify $\dfrac{(2 \times 10^{-4})(4 \times 10^6)(5 \times 10^2)}{8 \times 10^2}$.

10. Simplify $\dfrac{.009 \times .0008}{.0036}$.

In Problems 11-13, factor as indicated.

11. $x^{1/3} - x = x(? - ?)$.

12. $y^{1/3} - y^{-1/3} = y^{-1/3}(? - ?)$.

13. $x^{2n+2} - y^n = y^n(? - ?)$.

14. a. Express $(2 - y^2)^{-3/4}$ in radical notation.

    b. Express $\sqrt[3]{2m^2n}$ in exponential notation.

15. Simplify.

    a. $\sqrt[4]{81x^6y^3}$. b. $\sqrt{\frac{5}{6}}$. c. $\frac{2m}{\sqrt{2mn}}$. d. $\frac{\sqrt[3]{x}\sqrt[3]{y^2}}{\sqrt[3]{x^2y}}$.

16. Write each sum or difference as a single term.

    a. $\sqrt{48} + \sqrt{75}$. b. $\sqrt{9m} + 3\sqrt{36m} - 2\sqrt{25m}$.

    c. $\frac{2\sqrt{8}}{3} + \frac{3\sqrt{18}}{2}$. d. $\sqrt[3]{8x} - \sqrt[3]{-64x^4}$.

17. Simplify.

    a. $\sqrt{5}(\sqrt{6} - \sqrt{5})$. b. $(\sqrt{2} - \sqrt{2x})(\sqrt{2} + \sqrt{2x})$.

    c. $\sqrt[4]{4}\,(\sqrt[4]{4} + \sqrt[4]{64})$.

18. Reduce to lowest terms.

    a. $\frac{\sqrt{15} + \sqrt{21}}{\sqrt{3}}$. b. $\frac{m\sqrt{mn^3} - \sqrt{mn}}{\sqrt{mn}}$.

19. Simplify.

    a. $\frac{2\sqrt{3} - 2}{2\sqrt{3} + 2}$. b. $\frac{\sqrt{x} + \sqrt{y}}{\sqrt{x} - \sqrt{y}}$.

6

SECOND-DEGREE EQUATIONS AND INEQUALITIES

6.1

SOLUTION OF QUADRATIC EQUATIONS BY FACTORING

$ax^2 + bx + c = 0$

1. The standard form of a quadratic equation is ________ where a, b, and c are constants representing real numbers.

0; 0

2. $ab = 0$ if and only if $a =$ ___ or $b =$ ___.

In Frames 3-4, solve the equation given in Frame 3 for x.

0; 0

3. $3x(x + 1) = 0$ if and only if $3x =$ ___ or $x + 1 =$ ___.

0; -1
{0,-1}

4. If $3x = 0$, $x =$ ___ or if $x + 1 = 0$, $x =$ ____. The solution set is ________.

In Frames 5-10, solve the equation given in Frame 5.

5. $2x(2x + 1)(3x - 1) = 0$.

0

6. By inspection, $2x = 0$ if $x =$ ___.

0

7. Set $2x + 1 =$ ___ and solve.

$-\frac{1}{2}$

8. $x =$ ____.

0

9. Set $3x - 1 =$ ___ and solve:

$\frac{1}{3}$

$x =$ ___.

$\left\{0, -\frac{1}{2}, \frac{1}{3}\right\}$

10. The solution set is __________.

In Frames 11-13, solve the equation in Frame 11 for x.

11. $(x + 4)(x - 3) = 0$.

-4
3

12. By inspection, $x + 4 = 0$ for $x =$ ____ and $x - 3 = 0$ for $x =$ ___.

Answer	Frame
$\{-4, 3\}$	13. The solution set is ____.
	In Frames 14-19, solve the equation given in Frame 14 for x.
	14. $(4x + 3)(2x - 1) = 0$.
0	15. Set $4x + 3 =$ ____ and solve.
$-\frac{3}{4}$	16. $x =$ ____.
0	17. Set $2x - 1 =$ ____ and solve.
$\frac{1}{2}$	18. $x =$ ____.
$\left\{-\frac{3}{4}, \frac{1}{2}\right\}$	19. The solution set is ____.
	In Frames 20-24, solve the equation given in Frame 20 for x.
	20. $3bx(2ax + c) = 0$.
0	21. By inspection $3bx = 0$ if $x =$ ____.
0	22. Set $2ax + c =$ ____ and solve for x.
$\frac{-c}{2a}$	23. $x =$ ____.
$\left\{0, -\frac{c}{2a}\right\}$	24. The solution set is ____.
	In Frames 25-30, solve the equation given in Frame 25 for x.
	25. $3x^2 = 5x$.
0	26. $3x^2 - 5x =$ ____.
$3x - 5$	27. $x($____$) = 0$.
0; 0	28. Set $x =$ ____ and $3x - 5 =$ ____.
$\frac{5}{3}$	29. Solve $3x - 5 = 0$ and obtain $x =$ ____.
$\left\{0, \frac{5}{3}\right\}$	30. The solution set is ____.
	In Frames 31-33, solve the equation given in Frame 31 for x.
	31. $x^2 - \frac{9}{16} = 0$.
$x + \frac{3}{4}$; $x - \frac{3}{4}$	32. $($____$)($____$) = 0$.
$\left\{\frac{3}{4}, \frac{3}{4}\right\}$	33. By inspection, the solution set is ____.

In Frames 34-38, solve the equation given in Frame 34 for x.

34. $4c^2 - 9x^2 = 0$.

$2c - 3x$; $2c + 3x$

35. (________)(________) = 0.

0; 0

36. $2c - 3x =$ ___ or $2c + 3x =$ ___.

$\frac{2c}{3}$; $-\frac{2c}{3}$

37. Solve each equation and obtain $x =$ ____ or $x =$ ____.

$\left\{\frac{2c}{3}, -\frac{2c}{3}\right\}$

38. The solution set is ________.

In Frames 39-41, solve the equation given in Frame 39 for x.

39. $x^2 - 6x - 27 = 0$.

$x - 9$; $x + 3$

40. (______)(______) = 0.

$\{9, -3\}$

41. By inspection, the solution set is _______.

In Frames 42-47, solve the equation given in Frame 42 for x.

42. $12x^2 = 8x + 15$.

$12x^2 - 8x - 15$

43. ______________ = 0.

$6x + 5$; $2x - 3$

44. (_______)(_______) = 0.

0; 0

45. $6x + 5 =$ ___ or $2x - 3 =$ ___.

$-\frac{5}{6}$; $\frac{3}{2}$

46. $x =$ ___ or $x =$ ___.

$\left\{-\frac{5}{6}, \frac{3}{2}\right\}$

47. The solution set is _______.

In Frames 48-53, solve the equation given in Frame 48 for x.

48. $2x(x - 2) = x + 3$.

$2x^2 - 5x - 3$

49. Write Frame 48 in standard form and obtain ____________ = 0.

$2x + 1$; $x - 3$

50. (_______)(_______) = 0.

0; 0

51. $2x + 1 =$ ___ or $x - 3 =$ ___.

$-\frac{1}{2}$; 3

52. $x =$ ___ or $x =$ ___.

$\left\{-\frac{1}{2}, 3\right\}$

53. The solution set is _______.

In Frames 54-61, solve the equation given in Frame 54 for x.

54. $\frac{x}{4} - \frac{3}{4} = \frac{1}{x}$.

$4x$

55. Multiply both members of Frame 54 by the L.C.D., ____.

Answer	Frame
$(4x)\frac{1}{x}$	56. $(4x)\left(\frac{x}{4} - \frac{3}{4}\right) =$ ______.
$x^2 - 3x$; 4	57. ______ = ___.
$x^2 - 3x - 4$	58. ______ = 0.
$x - 4$; $x + 1$	59. (______)(______) = 0.
$\{4,-1\}$	60. By inspection, the solution set is ______.
variable	61. Check, because in Frame 56 both members were multiplied by an expression containing the ______. By inspection, the solutions check.

In Frames 62-73, solve the equation given in Frame 62 for x.

Answer	Frame
	62. $\frac{x}{x - 1} - \frac{x}{x + 1} = \frac{4}{3}$.
$3(x - 1)(x + 1)$	63. Multiply both members of Frame 62 by the L.C.D., ______.
$3(x - 1)(x + 1)\left(\frac{4}{3}\right)$	64. $3(x - 1)(x + 1)\left(\frac{x}{x - 1} - \frac{x}{x + 1}\right) =$ ______.
$4(x - 1)(x + 1)$	65. $3x(x + 1) - 3x(x - 1) =$ ______.
$- 3x^2 + 3x$	66. $3x^2 + 3x$ ______ $= 4x^2 - 4$.
$6x$	67. ___ $= 4x^2 - 4$.
0	68. ___ $= 4x^2 - 6x - 4$.
$2x + 1$; $x - 2$	69. $0 = 2($______$)($______$)$.
$\left\{-\frac{1}{2}, 2\right\}$	70. By inspection, the solution set is ______.
	71. Check, because in Frame 64 both members were multiplied by an expression containing the variable.
$\frac{4}{3}$	72. For $x = 2$; $\frac{2}{2 - 1} - \frac{2}{2 + 1} =$ ___. Check.
$\frac{4}{3}$	For $x = -\frac{1}{2}$; $\frac{-\frac{1}{2}}{-\frac{1}{2} - 1} - \frac{-\frac{1}{2}}{-\frac{1}{2} + 1} =$ ___. Check.
$\left\{2, -\frac{1}{2}\right\}$	73. Therefore, the solution set is ______.

In Frames 74-79, solve the equation given in Frame 74 for x.

Answer	Frame
	74. $x^3 - 6x^2 - x + 30 = 0$, where one solution is 3.
$x - 3$	75. If 3 is a solution, ______ is a factor of $x^3 - 6x^2 - x + 30$.
$x^2 - 3x - 10$	76. By division of $x^3 - 6x^2 - x + 30$ by $x - 3$, find the other factor to be ______.
$x - 3$; $x^2 - 3x - 10$	77. Therefore, the given equation can be written as (______)(______) = 0.

$x - 5$; $x + 2$

78. Factoring completely, $(x - 3)$(_______)(_______) = 0.

{3,5,-2}

79. By inspection, the solution set is __________.

In Frames 80-82, write in standard form with integral coefficients the quadratic equation whose roots are 4 and 2.

$x - 2$

80. Factors of the required equation are $x - 4$ and _______.

$(x - 4)(x - 2) = 0$

81. Then, the equation in factored form is ____________________.

$x^2 - 6x + 8$

82. Multiplying, _____________ = 0.

In Frames 83-87, write in standard form with integral coefficients the quadratic equation whose roots are $\frac{2}{3}$ *and* $-\frac{1}{2}$.

$x - \frac{2}{3}$; $x + \frac{1}{2}$

83. Factors of the required equation are _______ and _______.

$\left(x - \frac{2}{3}\right)\left(x + \frac{1}{2}\right) = 0$

84. Then, the equation in factored form is ____________________.

$x^2 - \frac{1}{6}x - \frac{1}{3}$

85. Multiplying, _____________ = 0.

6

86. Multiply both members by L.C.D. = ___.

$6x^2 - x - 2$

87. _____________ = 0.

6.2

SOLUTION OF EQUATIONS OF THE FORM $x^2 = a$; COMPLETING THE SQUARE

$\sqrt{a}$; $-\sqrt{a}$
roots

88. **If a is any real number, the equation $x^2 = a$ has as its solutions x = ____, x = _____. This method is called the extraction of _______.**

In Frames 89-91 solve the equation given in Frame 89 for x by the extraction of roots.

89. $5x^2 = 125$.

25

90. x^2 = ____.

5; -5

91. x = ___; x = ____. The solution set is {5,-5}.

In Frames 92-96, solve the equation given in Frame 92 by the extraction of roots.

92. $\frac{2x^2}{7} = 8$.

56

93. $2x^2$ = ____.

28

94. x^2 = ____.

Answer	Frame
$\sqrt{28}$; $-\sqrt{28}$	95. $x =$ ____; $x =$ ____.
$2\sqrt{7}$; $-2\sqrt{7}$	96. Simplify and obtain the solution set {____, ____}.

In Frames 97-101, solve the equation given in Frame 97 by extraction of roots.

Answer	Frame
	97. $\frac{bx^2}{a} = c$.
ac	98. $bx^2 =$ ____.
$\frac{ac}{b}$	99. $x^2 =$ ____.
$\sqrt{\frac{ac}{b}}$; $-\sqrt{\frac{ac}{b}}$	100. $x =$ ____, $x =$ ____.
$\frac{\sqrt{abc}}{b}$; $-\frac{\sqrt{abc}}{b}$	101. Simplify and obtain the solution set {____, ____}.

In Frames 102-104, solve the equation given in Frame 102 by the extraction of roots.

Answer	Frame
	102. $(x - 2)^2 = 16$.
4; -4	103. $x - 2 =$ ___, $x - 2 =$ ___.
6; -2	104. Hence, the solution set is {___, ___}.

In Frames 105-108, solve the equation given in Frame 105 by the extraction of roots.

Answer	Frame
	105. $(x - 4)^2 = 18$.
$\sqrt{18}$; $-\sqrt{18}$	106. $x - 4 =$ ____, $x - 4 =$ ____.
$4 + \sqrt{18}$; $4 - \sqrt{18}$	107. $x =$ ____, $x =$ ____.
$4 + 3\sqrt{2}$; $4 - 3\sqrt{2}$	108. Simplify and obtain the solution set {____, ____}.

In Frames 109-112, solve the equation given in Frame 109 by the extraction of roots.

Answer	Frame
	109. $(ax - b)^2 = 16$.
4; -4	110. $ax - b =$ ___, $ax - b =$ ___.
$b + 4$; $b - 4$	111. $ax =$ ____, $ax =$ ____.
$\frac{b+4}{a}$; $\frac{b-4}{a}$	112. Hence, the solution set is {____, ____}.
binomial	113. **When an expression is a perfect square trinomial, it can be written as the square of a ______.**
$\left(\frac{b}{2}\right)^2$	114. **To complete the square of $x^2 + bx$, $b \in R$, add ______ to it.**

In Frames 115-118, complete the square of the binomial in Frame 115 and express the results as the square of a binomial.

Answer	Frame
	115. $x^2 + 5x$.
$\frac{5}{2}$	116. Add $\left(\_\_\right)^2$ to $x^2 + 5x$.
$x^2 + 5x + \frac{25}{4}$	117. The resulting perfect square trinomial is __________.
$x + \frac{5}{2}$	118. $x^2 + 5x + \frac{25}{4} = \left(\_\_\_\_\right)^2$.

In Frames 119-122, complete the square of the binomial given in Frame 119 and express the results as the square of a binomial.

Answer	Frame
	119. $x^2 - \frac{5}{3}x$.
$-\frac{5}{6}$	120. Add $\left(\_\_\right)^2$ to $x^2 - \frac{5}{3}x$.
$x^2 - \frac{5}{3}x + \frac{25}{36}$	121. The resulting perfect square trinomial is __________.
$x - \frac{5}{6}$	122. $x^2 - \frac{5}{3}x + \frac{25}{36} = \left(\_\_\_\_\right)^2$.

In Frames 123-128, solve the equation given in Frame 123 by completing the square.

Answer	Frame
	123. $x^2 + x - 6 = 0$.
6	124. $x^2 + x =$ ___.
$\frac{1}{4}$; $\frac{1}{4}$	125. Completing the square, $x^2 + x +$ ___ $= 6 +$ ___.
$x + \frac{1}{2}$; $\frac{25}{4}$	126. $\left(\_\_\_\_\right)^2 =$ ___.
$\frac{5}{2}$; $-\frac{5}{2}$	127. $x + \frac{1}{2} =$ ___, $x + \frac{1}{2} =$ ___.
2; -3	128. Solve each equation for x and obtain the solution set {___,___}.

In Frames 129-134, solve the equation given in Frame 129 by completing the square.

Answer	Frame
	129. $x^2 + 3x - 1 = 0$
1	130. $x^2 + 3x =$ ___.
$\frac{9}{4}$; $\frac{9}{4}$	131. Completing the square, $x^2 + 3x +$ ___ $= 1 +$ ___.
$x + \frac{3}{2}$; $\frac{13}{4}$	132. $\left(\_\_\_\_\right)^2 =$ ___.

$\frac{\sqrt{13}}{2}$; $-\frac{\sqrt{13}}{2}$

133. $x + \frac{3}{2} =$ ____, $x + \frac{3}{2} =$ ____.

$\frac{-3 - \sqrt{13}}{2}$

134. Solve each equation for x and obtain the solution set $\left\{\frac{-3 + \sqrt{13}}{2}, \_\_\_\_\right\}$.

6.3

COMPLEX NUMBERS

-1; i

135. $\sqrt{-1}$ is defined by $(\sqrt{-1})(\sqrt{-1}) =$ ____ and is denoted by ____.

-1

136. Equivalent to Frame 135 is the statement $i^2 =$ ____.

$i\sqrt{b}$

137. For any positive real number b, $\sqrt{-b} =$ ____.

pure

138. The number represented by the symbol $i\sqrt{b}$ $(b > 0)$ is called a ____ imaginary number.

complex

139. The set of expressions of the form $a + bi$, $a,b \in R$ is called the set of ____ numbers.

complex

140. The set of real numbers, R, is a subset of the set of ____ numbers, C.

imaginary

141. If $b \neq 0$, then each number in the set C is called an ____ number.

In Frames 142-146, write each expression in the form $a + bi$ or $a + ib$.

i; $7i$

142. $\sqrt{-49} = (\_\_\_)\sqrt{49} =$ ____.

$7i$; $3 - 14i$

143. $3 - 2\sqrt{-49} = 3 - 2(\_\_\_) =$ ____.

$5i$; $4i$

144. $\frac{4\sqrt{-25}}{5} = \frac{4(\_\_\_)}{5} =$ ____.

$3i\sqrt{2}$; $1 + i\sqrt{2}$

145. $\frac{-3 - \sqrt{-18}}{-3} = \frac{-3}{-3} - \frac{\sqrt{-18}}{-3} = 1 - \frac{(\_\_\_)}{-3} =$ ____.

$2i\sqrt{2}$; $-\frac{1}{2} + \frac{\sqrt{2}}{2}i$

146. $\frac{-2 + \sqrt{-8}}{4} = \frac{-2}{4} + \frac{\sqrt{-8}}{4} = -\frac{1}{2} + \frac{(\_\_\_)}{4} =$ ____.

$(a + c) + (b + d)i$

147. The sum of two complex numbers is defined by $(a + bi) + (c + di) =$ ____.

$(a - c) + (b - d)i$

148. The difference of two complex numbers is defined by $(a + bi) - (c + di) =$ ____.

-1; -2
$-1 - 3i$

149. $(2 - i) + (-3 - 2i) = [2 + (-3)] + (\underline{\quad} + \underline{\quad})i$
$= \underline{\qquad}$.

2; -3; -1; -2
$5 + i$

150. $(2 - i) - (-3 - 2i) = (\underline{\quad} - \underline{\quad}) + (\underline{\quad} - \underline{\quad})i$
$= \underline{\qquad}$.

$6 - i$

151. $(3 - 3\sqrt{-1}) + (3 + 2\sqrt{-1}) = (3 - 3i) + (3 + 2i) = \underline{\qquad}$.

$3i$; $2 + i$

152. $\sqrt{-16} + (2 - \sqrt{-9}) = 4i + (2 - \underline{\quad}) = \underline{\qquad}$.

$i\sqrt{27}$
$3i\sqrt{3}$
$-2\sqrt{3} - 3\sqrt{3}$
$2 - 5\sqrt{3}i$

153. $(4 - \sqrt{-12}) - (2 + \sqrt{-27}) = (4 - i\sqrt{12}) - (2 + \underline{\qquad})$
$= (4 - 2i\sqrt{3}) - (2 + \underline{\qquad})$
$= (4 - 2) + (\underline{\qquad\qquad})i$
$= \underline{\qquad}$.

binomials

154. **The product of two complex numbers is defined in such a way that it can be obtained by using the same pattern of multiplication used with two real ________.**

-1; 5

155. $(2 + i)(2 - i) = 4 - i^2 = 4 - (\underline{\quad}) = \underline{\quad}$.

$9i^2$; -1; 13

156. $(2 - 3i)(2 + 3i) = 4 - (\underline{\quad}) = 4 - 9(\underline{\quad}) = \underline{\quad}$.

4; i^2; $13 + 4i$

157. $(6 - i)(2 + i) = 12 + (\underline{\quad})i - (\underline{\quad}) = \underline{\qquad}$.

$-4\sqrt{2}$; $2i^2$
$23 - 4i\sqrt{2}$

158. $(3 - i\sqrt{2})(7 + i\sqrt{2}) = 21 - (\underline{\qquad})i - (\underline{\qquad})$
$= \underline{\qquad\qquad}$.

$\sqrt{2}$; $\sqrt{2}$
$5\sqrt{2}$; 2
4

159. $(3 + \sqrt{-2})(2 + \sqrt{-2}) = (3 + i\underline{\quad})(2 + i\underline{\quad})$
$= 6 + (\underline{\qquad})i + (\underline{\quad})i^2$
$= \underline{\quad} + 5i\sqrt{2}$

$8\sqrt{3}$; 48
$-47 - 8i\sqrt{3}$

160. $(1 - 4\sqrt{-3})^2 = (1 - 4i\sqrt{3})^2 = 1 - (\underline{\qquad})i + (\underline{\quad})i^2$
$= \underline{\qquad\qquad}$.

161. **When one of the factors in a product is either a real number or a pure imaginary number, the product can be found by a direct application of the distributive law.**

$2i^2$; $2 + 3i$

162. $i(3 - 2i) = 3i - (\underline{\quad}) = \underline{\qquad}$.

$12i^2$; $-12 - 6i$

163. $-3i(2 - 4i) = -6i + \underline{\qquad} = \underline{\qquad}$.

$2i$; $4i$
$-2 + 6i$

164. $2(3 + \sqrt{-1}) - 4(2 - \sqrt{-1}) = 2(3 + i) - 4(2 - i)$
$= 6 + \underline{\quad} - 8 + \underline{\quad}$
$= \underline{\qquad}$.

$c - di$

165. **The quotient of two complex numbers is defined by $\frac{a + bi}{c + di} = \frac{(a + bi)(\underline{\qquad})}{(c + di)(c - di)}$, which can then be written in standard form.**

$1 - i$; 2; $1 - i$

166. $\frac{2}{1 + i} = \frac{2(\underline{\qquad})}{(1 + i)(1 - i)} = \frac{2 - 2i}{(\underline{\quad})} = \frac{2}{2} - \frac{2i}{2} = \underline{\qquad}$.

$3 + 2i$; $-9 - 6i$

167. $\dfrac{-3}{3 - 2i} = \dfrac{-3(\_\_\_\_\_\_)}{(3 - 2i)(3 + 2i)} = \dfrac{(\_\_\_\_\_\_)}{9 - 4i^2}$

$= \dfrac{-9 - 6i}{(\_\_\_)} = (\_\_\_) - (\_\_\_)i.$

13; $\dfrac{-9}{13}$; $\dfrac{6}{13}$

$3 - 2i$; $9 - 4i^2$

168. $\dfrac{2i}{3 + 2i} = \dfrac{2i(\_\_\_\_\_\_)}{(3 + 2i)(3 - 2i)} = \dfrac{6i - 4i^2}{\_\_\_\_\_\_}$

$= \dfrac{\_\_\_ + 6i}{13} = (\_\_\_) + (\_\_\_)i.$

4; $\dfrac{4}{13}$; $\dfrac{6}{13}$

$3 - \sqrt{3}i$

$\sqrt{3}$; $3 - \sqrt{3}i$

169. $\dfrac{4}{3 + \sqrt{-3}} = \dfrac{4}{3 + \_\_\_ i} = \dfrac{4 \cdot (\_\_\_\_\_\_)}{(3 + \sqrt{3}i) \cdot (\_\_\_\_\_\_)}$

$= \dfrac{12 - 4\sqrt{3}i}{9 - \_\_\_\_} = \dfrac{12 - 4\sqrt{3}i}{(\_\_\_)}$

$= \dfrac{12}{12} - \_\_\_\_ = 1 - \_\_\_\_ .$

$3i^2$; 12

$\dfrac{4\sqrt{3}}{12}i$; $\dfrac{\sqrt{3}}{3}i$

In Frames 170-174, solve the equation given in Frame 170.

170. $x^2 + 28 = 0.$

-28

171. $x^2 = \_\_\_\_.$

$-\sqrt{-28}$

172. $x = \sqrt{-28}$ or $x = \_\_\_\_.$

$\sqrt{28}$; $\sqrt{28}$

173. $x = i\ \_\_\_\_$ or $x = -i\ \_\_\_\_.$

$2\sqrt{7}$

$2i\sqrt{7}$; $-2i\sqrt{7}$

174. Since $\sqrt{28}$ simplifies to ____, the solution set is {____, ____}.

In Frames 175-180, solve the equation given in Frame 175.

175. $x^2 + 4x + 5 = 0.$

-5

176. $x\ \ + 4x = \_\_\_\_\_\_.$

4

177. $x^2 + 4x + 4 = -5 + \_\_\_\_.$

-1

178. $(x + 2)^2 = \_\_\_\_.$

Answer: $-\sqrt{-1}$

179. $x + 2 = \sqrt{-1}$ or $x + 2 =$ ______.

Answer: $-2 - i$

180. $x = -2 + i$ or $x =$ ______
$\{-2 + i, -2 - i\}$.

6.4

THE QUADRATIC FORMULA

Answer: $\dfrac{-b \pm \sqrt{b^2 - 4ac}}{2a}$

181. **The solutions of the general quadratic equation $ax^2 + bx + c = 0$ are given by $x =$ __________.**

In Frames 182-188, solve the equation given in Frame 182 for x, using the quadratic formula.

182. $x^2 - 5x = 6$.

Answer: $x^2 - 5x - 6$

183. Write in standard form: __________ = 0.

Answer: 1; -5; -6

184. $a =$ ___, $b =$ ___, $c =$ ___.

Answer: -5; -6

185. $x = \dfrac{-(\_\_\_) \pm \sqrt{(-5)^2 - 4(1)(\_\_\_)}}{2(1)}$.

Answer: 49; 7

186. $x = \dfrac{5 \pm \sqrt{\_\_\_}}{2} = \dfrac{5 \pm \_\_\_}{2}$.

Answer: 7; 7

187. $x = \dfrac{5 + \_\_\_}{2}$; $x = \dfrac{5 - \_\_\_}{2}$.

Answer: {6,-1}

188. The solution set is ______.

In Frames 189-195, solve the equation given in Frame 189 for z, using the quadratic formula.

189. $6z^2 = 5z + 6$.

Answer: $6z^2 - 5z - 6$

190. Write in standard form: __________ = 0.

Answer: 6; -5; -6

191. $a =$ ___, $b =$ ___, $c =$ ___.

Answer: -5; -5; 6; -6

192. $x = \dfrac{-(\_\_\_) \pm \sqrt{(\_\_\_)^2 - 4(\_\_\_)(\_\_\_)}}{2(6)}$.

Answer: 169

193. $x = \dfrac{5 \pm \sqrt{\_\_\_}}{12}$.

Answer: 13; 13

194. $x = \dfrac{5 + \_\_\_}{12}$; $x = \dfrac{5 - \_\_\_}{12}$.

$\left\{\frac{3}{2}, -\frac{2}{3}\right\}$

195. The solution set is ________.

In Frames 196-203, solve the equation given in Frame 196 for y, using the quadratic formula.

196. $\frac{y^2 + 3}{2} + \frac{y}{4} = 1.$

4

197. $(4)\left(\frac{y^2 + 3}{2} + \frac{y}{4}\right) = (\_\_)(1).$

y

198. $2(y^2 + 3) + \_\_ = 4.$

$2y^2 + y + 2$

199. ____________ = 0.

2; 1; 2

200. $a = \_\_, b = \_\_, c = \_\_.$

1; 1; 2; 2

201. $y = \frac{-(\_\_) \pm \sqrt{(\_\_)^2 - 4(\_\_)(\_\_)}}{2(2)}$

-15

202. $y = \frac{-1 \pm \sqrt{\_\_}}{4}.$

$\left\{-\frac{1}{4} + i\frac{\sqrt{15}}{4}, -\frac{1}{4} - i\frac{\sqrt{15}}{4}\right\}$

203. The solution set in $a + bi$ form is

______________________.

In Frames 204-208, solve the equation given in Frame 204 for x.

204. $3x^2 + xy + 5y = 1.$

$5y - 1$

205. Write in standard form: $(3)x^2 + (y)x + (\_\_\_\_\_\_) = 0.$

3; y; $5y - 1$

206. $a = \_\_, b = \_\_, c = \_\_\_\_.$

y; y; 3; $5y - 1$

207. $x = \frac{-(\_\_) \pm \sqrt{(\_\_)^2 - 4(\_\_)(\_\_\_\_)}}{2(3)}$

$\frac{-y \pm \sqrt{y^2 - 60y + 12}}{6}$

208. $x =$ ____________________.

In Frames 209-213, state the discriminant test for the nature of the solutions of a quadratic equation.

real

209. If $b^2 - 4ac = 0$, there is one ______ solution.

imaginary

210. If $b^2 - 4ac < 0$, there are two __________ solutions.

real

211. If $b^2 - 4ac > 0$, there are two unequal _____ solutions.

rational

212. If $b^2 - 4ac$ is the square of a rational number and if $a,b,c \in Q$, the solutions are _________.

irrational

213. If $b^2 - 4ac$ is positive and not the square of a rational number, the solutions are _________.

In Frames 214-216, find only the discriminant for the given equation and determine whether the solutions are: (a) one real, (b) real and unequal, (c) imaginary and unequal.

-2; 1; -3; 16

real and unequal

214. $y^2 - 2y - 3 = 0$.

$b^2 - 4ac = (\_\_)^2 - 4(\_\_)(\_\_) = \_\_$.

Since 16 is a positive number, the roots are _______________.

$y^2 + 4y + 4$

16; 1; 4; 0

one

215. $y^2 + 4y = -4$,

$\_\_\_\_\_\_\_\_ = 0$.

$b^2 - 4ac = \_\_\_ - 4(\_\_)(\_\_) = \_\_$.

Since $b^2 - 4ac = 0$, the number of real roots is _____.

$y^2 + \frac{1}{3}y + \frac{2}{3}$

3

$3y^2 + y + 2$

1; 3; 2; -23

imaginary

216. $y^2 = -\frac{1}{3}y - \frac{2}{3}$.

$\_\_\_\_\_\_\_\_ = 0$.

$3\left(y^2 + \frac{1}{3}y + \frac{2}{3}\right) = (\_\_)(0)$.

$\_\_\_\_\_\_\_\_ = 0$.

$b^2 - 4ac = \_\_\_ - 4(\_\_)(\_\_) = \_\_\_$.

Since $b^2 - 4ac$ is negative, the roots are _________.

In Frames 217-224, determine k so that the solutions of $x^2 - x + k = 2$ are imaginary numbers.

$x^2 - x + (k - 2)$

1; -1; $k - 2$

<

217. $x^2 - x + k = 2$.

$\_\_\_\_\_\_\_\_ = 0$.

218. a = ___, b = ___, c = _____.

219. For the roots to be imaginary, $b^2 - 4ac$ ___ 0.

Answers	Frames
-1; 1; $k-2$	220. Therefore, $(\_\_\_)^2 - 4(\_\_\_)(\_\_\_\_\_) < 0$.
8	221. $1 - 4k + \_\_\_ < 0$.
9	222. $-4k + \_\_\_ < 0$.
-9	223. $-4k < \_\_\_$.
$>$	224. $k \_\_\_ \frac{9}{4}$.

6.5

APPLICATIONS

In Frames 225-230, solve the problem stated in Frame 225.

Answers	Frames
	225. Find two consecutive odd integers whose product is 63.
	226. the smaller odd integer: x
$x + 2$	the next odd integer: _____
x; $x + 2$	227. $(\_\_\_)(\_\_\_\_\_) = 63$.
$x^2 + 2x - 63$	228. ________ $= 0$.
	Solving the above equation by factoring:
7; -9	229. $x = \_\_\_$ or $x = \_\_\_$.
	230. Hence, there are two pairs of consecutive odd integers that solve the problem
7; 9; -9; -7	$x = \_\_\_$ and $x + 2 = \_\_\_$; or $x = \_\_\_$ and $x + 2 = \_\_\_$.
	By inspection, both pairs check.

In Frames 231-238, solve the problem stated in Frame 231.

Answers	Frames
	231. The difference of a number and twice its reciprocal is $\frac{31}{4}$. Find the number.
	232. The number: x
$\frac{1}{x}$	the reciprocal: ___
$x - \frac{2}{x}$	233. _____ $= \frac{31}{4}$.
$4x^2 - 31x - 8$	234. ________ $= 0$.
	235. Solve the above equation by factoring:
$-\frac{1}{4}$; 8	$x = \_\_\_$; or $x = \_\_\_$.

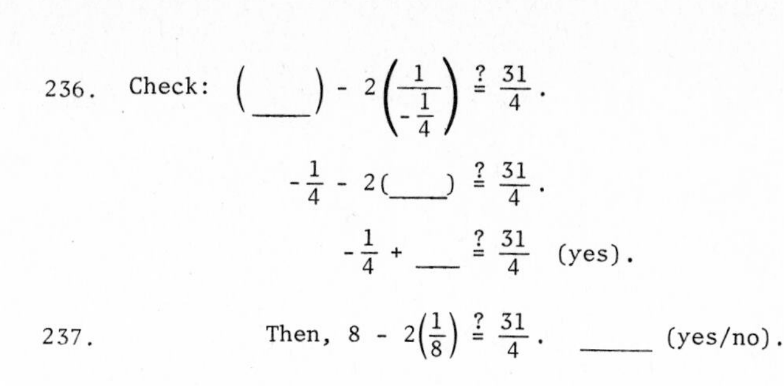

Answers: $-\frac{1}{4}$; -4; 8

236. Check: $\left(\underline{\quad}\right) - 2\left(\frac{1}{-\frac{1}{4}}\right) \overset{?}{=} \frac{31}{4}$.

$-\frac{1}{4} - 2(\underline{\quad}) \overset{?}{=} \frac{31}{4}$.

$-\frac{1}{4} + \underline{\quad} \overset{?}{=} \frac{31}{4}$ (yes).

Answer: yes

237. Then, $8 - 2\left(\frac{1}{8}\right) \overset{?}{=} \frac{31}{4}$. _____ (yes/no).

Answer: $-\frac{1}{4}$; 8

238. Then the problem is solved by ____ or ___.

In Frames 239-247, solve the problem stated in Frame 239.

239. If increasing the length of the sides of a square by 12 feet results in a square with nine times the area of the original square, what was the length of a side of the original square?

240. Draw a diagram and on it represent the given data and what is to be found.

Answer: $x + 12$

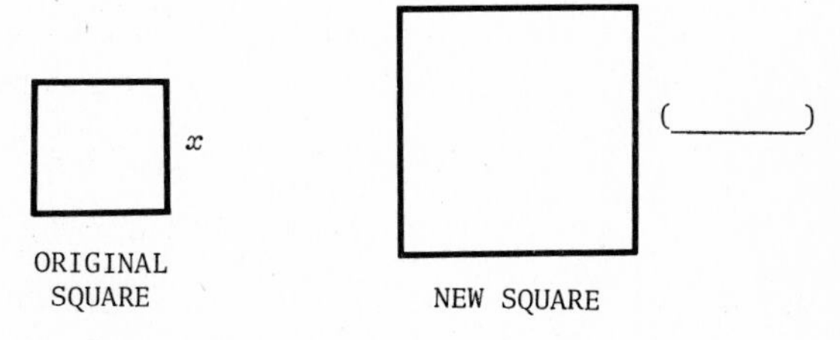

Answer: s^2

241. The area of any square of side (s) is given by $A =$ ______.

Answer: $9x^2$

242. Now use information from Frame 241 to obtain:

$(x + 12)^2 =$ _____.

Answer: $x^2 - 3x - 18$ (detail below) $-8x^2 + 24x + 144 = 0$

243. ______________ = 0.

Answer: 6; -3

244. Solving, obtain:

$x =$ ___ or $x =$ ____.

Answer: -3

245. It makes no sense for the side of a square to be taken as ____; so it is rejected as a solution to the problem.

Answers: 36; 324; yes

246. Check $x = 6$: Original square's area = ____.

New square's area = _____.

$324 \overset{?}{=} 9(36)$. _____ (yes/no).

Answer: 6 feet

247. The solution to the problem is _______.

In Frames 248-256, solve the problem stated in Frame 248.

248. A hitchhiker walked 3 miles. He then obtained a ride for 12 miles. The total time for the trip was $1\frac{1}{4}$ hours. If the rate of the automobile was 30 miles per hour greater than his rate walking, what was each rate?

249. the rate walking: x
the rate of the car: ______

Answer: $x + 30$

250. A chart for the problem is

	d	r	$t = d/r$
walking	3	x	$\frac{3}{x}$
riding	12	$x + 30$	______

Answer: $\frac{12}{x + 30}$

251. $\left[\text{time walking}\right] + \left[\text{time riding}\right] =$ ______.

Answer: $1\frac{1}{4}$ or $\frac{5}{4}$

252. $\frac{3}{x} + \frac{12}{x + 30} =$ ______.

$4x(x + 30) \cdot \left[\frac{3}{x} + \frac{12}{x + 30}\right] =$ ____________ $\cdot \left(\frac{5}{4}\right)$

$12(x + 30) + 48x =$ ____________

____________ $= 0.$

Answers: $\frac{5}{4}$; $4x(x + 30)$; $5x(x + 30)$; $x^2 + 18x - 72$

253. Using the quadratic formula,

$x = \frac{-18 \pm \sqrt{18^2 - 4(1)(\_\_\_\_)}}{2(1)} = \frac{-18 \pm \sqrt{\_\_\_\_}}{2}$

$= \frac{-18 \pm \sqrt{36 \cdot 17}}{2} = \frac{-18 \pm (\_\_\_)\sqrt{17}}{2}$

$= -9 \pm 3\sqrt{17}.$

Answers: -72; 612; 6

254. $x = -9 - 3\sqrt{17}$ cannot be a solution of the problem because it makes no sense for a rate to be ________ (positive/negative).

Answer: negative

255. Hence, the rate walking is $-9 + 3\sqrt{17}$ mph or approximately 3.4 mph and the rate riding is $[-9 + 3\sqrt{17} + \_\_\_\_]$ $= (21 + 3\sqrt{17})$ mph or approximately 33.4 mph.

Answer: 30

256. Check: Does $\frac{3}{3.4} + \frac{12}{33.4} \approx$ ____ ?

Does $0.88 + 0.36 \approx 1.25$? _____ (yes/no).

Answers: $1\frac{1}{4}$ or 1.25; yes

6.6

EQUATIONS INVOLVING RADICALS

257. In order to solve equations involving radicals we shall assume that if each member of an equation is raised to the same power, the solution set of the resulting equation will contain all of the solutions of the ________ equation.

original

258. Because raising each member of an equation to the same power may result in an equation with extraneous roots, each root of the resulting equation must be checked in the ________ equation.

original

In Frames 259-263, solve the equation of Frame 259 and check.

259. $\sqrt{x - 3} = 8.$

260. $(\sqrt{x - 3})^2 = (\_\_)^2.$

8

261. ______ = 64.

$x - 3$

262. $x = 67.$

263. Check: $\sqrt{67 - 3} \stackrel{?}{=} 8.$ _____ (yes/no).

yes

In Frames 264-267, solve the equation of Frame 264 and check.

264. $\sqrt[3]{x + 2} = 4.$

265. ______ $= 4^3.$

$x + 2$

266. $x =$ ____.

62

267. Check: $\sqrt[3]{62 + 2} \stackrel{?}{=} 4.$ _____ (yes/no).

yes

In Frames 268-278, solve the equation of Frame 268 and check.

268. $4\sqrt{y} + \sqrt{1 + 16y} = 5.$

269. $4\sqrt{y} =$ ____________.

$5 - \sqrt{1 + 16y}$

270. $(4\sqrt{y})^2 = (\_\_\_\_\_\_\_\_\_\_\_\_)^2.$

$5 - \sqrt{1 + 16y}$

271. $16y =$ ________________________.

$25 - 10\sqrt{1 + 16y} + 1 + 16y$

272. _____ $= -10\sqrt{1 + 16y}$.

-26

273. $13 =$ __________.

$5\sqrt{1 + 16y}$

274. $169 =$ ____________.

$25(1 + 16y)$

Answer	Frame
$400y$	275. $169 = 25 + \_\_\_\_\_\_$.
144	276. $\_\_\_\_\_ = 400y$.
$\frac{144}{400}$ or $\frac{9}{25}$	277. $\frac{\quad}{\_\_\_\_\_} = y$.
	278. Check: $4\sqrt{\frac{9}{25}} + \sqrt{1 + 16\left(\frac{9}{25}\right)} \overset{?}{=} 5$.
$\frac{3}{5}$; $\frac{169}{25}$	$4(\_\_\_) + \sqrt{\_\_\_\_} \overset{?}{=} 5$.
$\frac{13}{5}$; yes	$\frac{12}{5} + \frac{\quad}{\_\_\_\_} \overset{?}{=} 5$. ___ (yes/no).

In Frames 279-286, solve the equation of Frame 279 and check.

Answer	Frame
	279. $(y^2 - 3y + 5)^{1/2} + (y + 2)^{1/2} = 0$.
$-(y + 2)^{1/2}$	280. $(y^2 - 3y + 5)^{1/2} = \_\_\_\_\_\_\_\_\_\_$.
$-(y + 2)^{1/2}$	281. $[(y^2 - 3y + 5)^{1/2}]^2 = [\_\_\_\_\_\_\_\_\_\_]^2$.
$y + 2$	282. $y^2 - 3y + 5 = \_\_\_\_\_\_$.
$y^2 - 4y + 3$	283. $\_\_\_\_\_\_\_\_\_\_ = 0$.
$y - 3$; $y - 1$	284. $(\_\_\_\_\_\_\_)(\_\_\_\_\_\_\_) = 0$.
3; 1	285. $y = \_\_\_$ or $\_\_\_$.
	286. Check: $(3^2 - 3 \cdot 3 + 5)^{1/2} + (3 + 2)^{1/2} \overset{?}{=} 0$.
5; 5; no	$(\_\_)^{1/2} + (\_\_)^{1/2} \overset{?}{=} 0$. ____ (yes/no).
1	$(1^2 - 3 \cdot 1 + 5)^{1/2} + (\_\_\_ + 2)^{1/2} \overset{?}{=} 0$.
3; no	$(\_\_)^{1/2} + (3)^{1/2} \overset{?}{=} 0$. ____ (yes/no).
$\emptyset$	Therefore, the solution set is ___.

In Frames 287-292, solve the equation of Frame 287 for x.

Answer	Frame
	287. $y = \frac{1}{\sqrt{1 - x}}$ $(x \neq 1)$.
$\frac{1}{1 - x}$	288. $y^2 = \frac{\quad}{\_\_\_\_\_\_}$.
$y^2 - xy^2$	289. $\_\_\_\_\_\_\_\_\_ = 1$.
$1 - y^2$	290. $-xy^2 = \_\_\_\_\_\_\_$.
$\frac{1 - y^2}{-y^2}$ or $\frac{y^2 - 1}{y^2}$	291. $x = \frac{\quad}{\_\_\_\_\_\_\_}$ $(y \neq 0)$.
$\frac{y^2 - 1}{y^2}$	292. Check: $y \overset{?}{=} \frac{1}{\sqrt{1 - \left(\frac{\quad}{\_\_\_\_\_\_}\right)}}$.
1; yes	$y \overset{?}{=} \frac{1}{\sqrt{\left(\frac{\_\_\_}{y^2}\right)}}$. ___ (yes/no).

6.7

EQUATIONS THAT ARE QUADRATIC IN FORM

In Frames 293-301, solve the equation of Frame 293 for y.

Answer	Frame
	293. $y + 3\sqrt{y} - 10 = 0$.
u^2	294. Let $\sqrt{y} = u$ and $y =$ ____.
	295. Substitute the results of Frame 294 into Frame 293 and obtain:
$u^2 + 3u - 10$	______________ $= 0$.
$u + 5$; $u - 2$	296. (________)(________) $= 0$.
-5; 2	297. $u =$ ____ or $u =$ ____.
	298. Substitute from Frame 294 into Frame 297 and obtain:
$\sqrt{y}$; $\sqrt{y}$	____ $= -5$ or ____ $= 2$.
25; 4	299. $y =$ ____ or $y =$ ____.
power	300. A check is needed, because to obtian Frame 299 both members of the equations of Frame 298 were raised to a _____.
no	$25 + 3\sqrt{25} - 10 \stackrel{?}{=} 0$. __ (yes/no).
yes	$4 + 3\sqrt{4} - 10 \stackrel{?}{=} 0$. ___ (yes/no).
{4}	301. The solution set is _____.

In Frames 302-312, solve the equation of Frame 302 for y.

Answer	Frame
	302. $y^2 - 5 - 5\sqrt{y^2 - 5} + 6 = 0$.
u^2	303. Let $\sqrt{y^2 - 5} = u$ and $y^2 - 5 =$ _____.
	304. Substitute from Frame 303 into Frame 302 and obtain:
$u^2 - 5u + 6$	______________ $= 0$.
$u - 3$; $u - 2$	305. (________)(________) $= 0$.
3; 2	306. $u =$ ___ or $u =$ ___.
	307. Substitute from Frame 303 into Frame 306 and obtain:
$\sqrt{y^2 - 5}$; $\sqrt{y^2 - 5}$	_________ $= 3$ or _________ $= 2$.
9; 4	308. $y^2 - 5 =$ ___ or $y^2 - 5 =$ ___.
14; 9	309. $y^2 =$ ____ or $y^2 =$ ___.
$\pm\sqrt{14}$; ± 3	310. $y =$ ______ or $y =$ ____.

Answers	Frames
	311. Check:
$\pm\sqrt{14}$	$(\pm\sqrt{14})^2 - 5 - 5\sqrt{(\_\_\_)^2 - 5} + 6 \stackrel{?}{=} 0.$
9; yes	$14 - 5 - 5\sqrt{\_\_} + 6 \stackrel{?}{=} 0.$ ____ (yes/no).
± 3; yes	$(\pm 3)^2 - 5 - 5\sqrt{(\_\_\_)^2 - 5} + 6 \stackrel{?}{=} 0.$ ____ (yes/no).
$\{\sqrt{14}, -\sqrt{14}, 3, -3\}$	312. The solution set is ______________.

In Frames 313-320, solve the equation of Frame 313 for x.

Answers	Frames
	313. $8x^{-2} + 7x^{-1} - 1 = 0.$
x^{-1}; x^{-2}	314. Let ____ $= u$ and ____ $= u^2$.
$8u^2 + 7u - 1$	315. Substitute from Frame 314 into Frame 313 and obtain: __________ $= 0.$
$8u - 1$; $u + 1$	316. (______)(______) = 0.
$\frac{1}{8}$; -1	317. $u =$ ___ or $u =$ ____.
x^{-1}; x^{-1}	318. Substitute from Frame 314 into Frame 317 and obtain: ____ $= \frac{1}{8}$ or ____ $= -1.$
$\frac{1}{x}$	319. $\frac{1}{x} = \frac{1}{8}$ or ___ $= -1.$
8; -1	320. Solve in Frame 319 for x and obtain: $x =$ ___ or $x =$ ____.
power	No checking is necessary here, since in no step above were both members raised to a ______.

In Frames 321-330, solve the equation of Frame 321 for x.

Answers	Frames
	321. $2x^{1/2} - x^{1/4} = 3.$
u^2	322. Let $x^{1/4} = u$ and $x^{1/2} =$ ____.
$2u^2 - u$	323. Substitute from Frame 322 into Frame 321 and obtain: ________ $= 3.$
$2u^2 - u - 3$	324. __________ $= 0.$
$2u - 3$; $u + 1$	325. (______)(______) = 0.
$\frac{3}{2}$; -1	326. $u =$ ___ or $u =$ ____.
-1	327. Substitute from Frame 322 into Frame 326 and obtain: $x^{1/4} = \frac{3}{2}$ or $x^{1/4} =$ ____.

328. Solve in Frame 327 for x and obtain:

1

$$x = \left(\frac{3}{2}\right)^4 = \frac{81}{16} \text{ or } x = \_\_\_.$$

329. Check: $2\left(\frac{81}{16}\right)^{1/2} - \left(\frac{81}{16}\right)^{1/4} \stackrel{?}{=} 3.$

yes

$2\left(\frac{9}{4}\right) - \frac{3}{2} \stackrel{?}{=} 3.$ _____ (yes/no).

no

$2(1)^{1/2} - (1)^{1/4} \stackrel{?}{=} 3.$ ____ (yes/no).

$\left\{\frac{81}{16}\right\}$

330. The solution set is ______.

6.8

QUADRATIC INEQUALITIES

0; undefined

331. **In an inequality of the form $Q(x) < 0$, $Q(x) \le 0$, $Q(x) > 0$, or $Q(x) \ge 0$, a is a critical number if $Q(x) =$ ___ or if $Q(x)$ is ___________ for $x = a$.**

In Frames 332-337, solve the inequality given in Frame 332 and represent the solution set on a line graph.

332. $(x - 2)(x - 5) > 0.$

2; 5

333. The critical numbers are ___ and ___.

are not

334. Graph the critical numbers with open dots to show that these points _________ (are/are not) in the graph.

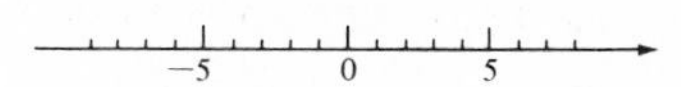

yes

no

yes

335. Substitute an arbitrary number of each of the three intervals thus obtained on the number line into the inequality, say 0, 3, and 6:

$(0 - 2)(0 - 5) \stackrel{?}{>} 0.$ _____ (yes/no).

$(3 - 2)(3 - 5) \stackrel{?}{>} 0.$ ____ (yes/no).

$(6 - 2)(6 - 5) \stackrel{?}{>} 0.$ _____ (yes/no).

$(5, +\infty)$

336. The intervals containing the numbers that satisfied $(x - 2)(x - 5) > 0$ in Frame 335 are the solutions that therefore are $(-\infty, 2) \cup$ _________.

337. The line graph of the solution set is

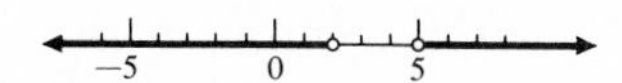

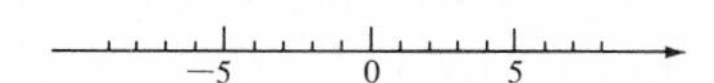

In Frames 338-344, solve and represent the solution set of the inequality given in Frame 338 on a line graph.

338. $x^2 - 5x + 6 \geq 0.$

$x - 3; \quad x - 2$

339. (_______)(_______) $\geq 0.$

3; 2

340. The critical numbers are ___ and ___.

are

341. Graph the critical numbers with closed dots to show that these points _____ (are/are not) in the graph.

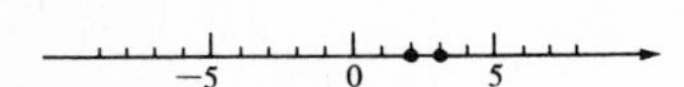

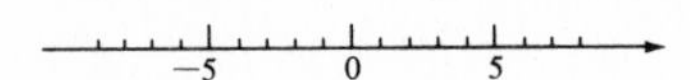

342. Substitute an arbitrary number from each of the three intervals thus obtained on the number line into the inequality, say 0, $\frac{5}{2}$, and 4:

0; yes

$0^2 - 5(\_\_\_) + 6 \overset{?}{\geq} 0.$ _____ (yes/no).

$\frac{5}{2}$; no

$\left(\frac{5}{2}\right)^2 - 5\left(\_\_\_\right) + 6 \overset{?}{\geq} 0.$ ____ (yes/no).

4; yes

$4^2 - 5(\_\_\_) + 6 \overset{?}{\geq} 0.$ _____ (yes/no).

$(-\infty, 2] \cup [3, +\infty)$

343. The intervals containing the numbers that satisfied $x^2 - 5x + 6 \geq 0$ in Frame 342 are the solutions that therefore are: _________________.

344. The line graph of the solution set is

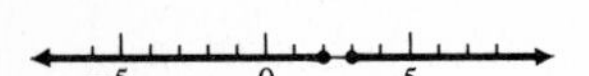

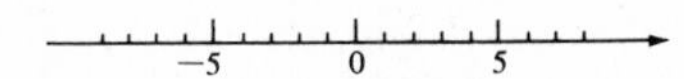

In Frames 345-349, solve and represent the solution set of the inequality given in Frame 345 on a line graph.

345. $x(x - 4) < 0.$

0; 4

346. The critical numbers are ___ and ___.

open

347. Graph the critical numbers with ______ (open/closed) dots.

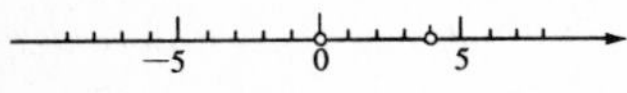

348. Substitute an arbitrary number from each of the three intervals thus obtained into the inequality, say -1, 1, and 5:

-1; no

$-1(\_\_\_\_ - 4) \overset{?}{<} 0.$ ____ (yes/no).

1; yes

$1(\_\_\_ - 4) \overset{?}{<} 0.$ _____ (yes/no).

5; no

$5(\_\_\_ - 4) \overset{?}{<} 0.$ ____ (yes/no).

(0, 4)

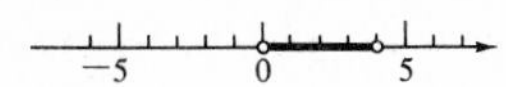

349. The solution set is ________ and the line graph of the solution set is

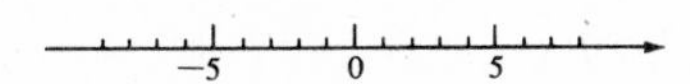

In Frames 350-355, solve and represent the solution set of the sentence given in Frame 339 on a line graph.

350. $\frac{x + 2}{x - 2} \geq 6.$

351. Express the inequality in the form $Q(x) \geq 0$, with $Q(x)$ as a single fraction:

0

$$\frac{x + 2}{x - 2} - 6 \geq \_\_\_$$

$x - 2$

$$\frac{x + 2 - 6(\_\_\_\_\_\_)}{x - 2} \geq 0$$

$-5x + 14$

$$\frac{\_\_\_\_\_\_}{x - 2} \geq 0\,.$$

undefined

0

352. 2 is a critical number because $\frac{-5x + 14}{x - 2}$ is ________ when $x = 2$. $\frac{14}{5}$ is a critical number because $\frac{-5x + 14}{x - 2}$ = ___ when $x = \frac{14}{5}$.

open; does

closed

353. Since 2 does not satisfy the inequality, it is graphed with a(n) ______ (open/closed) dot. Since $\frac{14}{5}$ ______ (does/does not) satisfy the inequality, it is graphed with a(n) ________ (open/closed) dot.

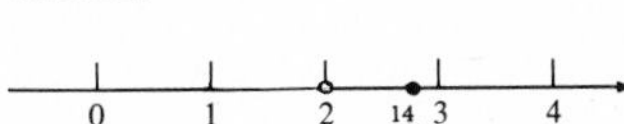

354. Substitute an arbitrary number from each of the three intervals thus obtained into the inequality, say, 1, $\frac{5}{2}$, and 3:

no

$\frac{5}{2}$; yes

3; no

$$\frac{1 + 2}{1 - 2} \overset{?}{\geq} 6.\_\_\_\_ \text{ (yes/no)}.$$

$$\frac{\_\_\_ + 2}{\frac{5}{2} - 2} \overset{?}{\geq} 6.\ \_\_\_\_ \text{ (yes/no)}.$$

$$\frac{\_\_\_ + 2}{3 - 2} \overset{?}{\geq} 6.\ \_\_\_\_ \text{ (yes/no)}.$$

$(2, \frac{14}{5}]$

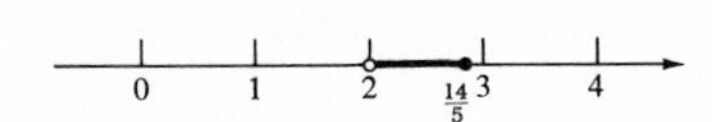

355. The solution set is ______________ and the line graph is

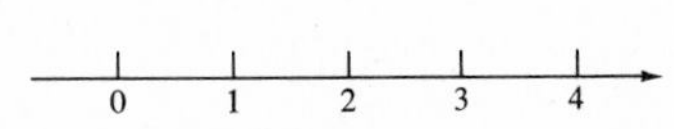

TEST PROBLEMS

In Problems 1-12, solve for x.

1. $5x^2 = 9x$.
2. $2x(x - 2) = x + 3$.
3. $\frac{2x^2}{7} = 8$.
4. $(2x + 1)^2 = 25$.
5. $3x^2 = 5x - 1$.
6. $\frac{x^2 - x}{2} + 1 = 0$.
7. $2x^2 - 3x + 2 = 0$.
8. $\sqrt{x - 3} + 5 = 0$.
9. $\sqrt{3x + 10} = x + 4$.
10. $4\sqrt{x} + \sqrt{16x + 1} = 5$.
11. $x^4 - 2x^2 - 24 = 0$.
12. $x^{2/3} - 2x^{1/3} = 35$.
13. Write an equation in standard form for the quadratic equation whose solutions are $-\frac{2}{3}$ and 2.
14. Solve $2x^2 + 3x - 2 = 0$ by completing the square.
15. Express each of the following in $a + bi$ or $a - bi$ form:
    a. $(2 - 3i) - (4 + 5i)$. b. $(2 - 3i)(4 + 5i)$.
    c. $\frac{4 + 5i}{2 - 3i}$.
16. Express each of the following in $a + bi$ or $a - bi$ form:
    a. $(3 + \sqrt{-12}) + (1 - \sqrt{-27})$.
    b. $(3 - \sqrt{-3})(1 - \sqrt{-3})$.
    c. $\frac{1}{2 - \sqrt{-3}}$.
17. Determine k so that $x^2 - kx + 9 = 0$ has one solution.
18. Determine k so that $x^2 - x + k - 2 = 0$ has imaginary solutions.

In Problems 19 and 20, solve and represent each solution set on a line graph.

19. $x^2 - 5x - 6 \geq 0$.
20. $\frac{3}{x - 6} > 8$.
21. Find two consecutive positive integers the sum of whose squares is 145.

22. A man sailed a boat across a lake and back in 2-1/2 hours. If his rate returning was 2 miles an hour less than his rate going, and if the distance each way was 6 miles, find his rate each way.

7

FUNCTIONS, RELATIONS, AND THEIR GRAPHS: PART I

7.1

SOLUTION OF AN EQUATION IN TWO VARIABLES

solution

1. If the substitution of the ordered pair (x,y) into an equation of the form $y = ax + b$ results in a true statement, then (x,y) is called a ________ of the equation.

In Frames 2-7, find each missing component so that each ordered pair is a solution of the equation.

0
5

2. $y = x + 5$, (0,?),
$y =$ ___ + 5,
$y =$ ___. The solution is (0,5).

0
-5; -5

3. $y = x + 5$, (?,0)
___ $= x + 5$,
___ $= x$. The solution is (___,0).

3
8; 3; 8

4. $y = x + 5$, (3,?)
$y =$ ___ + 5,
$y =$ ___. The solution is (___,___).

0
-2; (0,-2)

5. $2x - 3y = 6$, (0,?)
2(___) $- 3y = 6$,
$y =$ ____. The solution is ________.

0
3; (3,0)

6. $2x - 3y = 6$, (?,0)
$2x -$ 3(___) $= 6$,
$x =$ ___. The solution is ________.

2
2
$-\frac{2}{3}$; $\left(2,-\frac{2}{3}\right)$

7. $2x - 3y = 6$, (2,?)
2(___) $- 3y = 6$,
$-3y =$ ___,
$y =$ ___. The solution is ________.

In Frames 8-11, list the elements satisfying the given equation and having the given x-components.

2

8. $y = 2x + 4$; -1, 0, 1.

In $y = 2x + 4$, substitute -1 for x and obtain $y =$ ___.

(-1,2)
x; 4

9. Thus we have the ordered pair ______. Similarly, substitute 0 for ______ and obtain y = ______.

(0,4)
(1,6)

10. Hence we have the ordered pair ______. In a similar way, obtain the last ordered pair, ______.

(-1,2),(0,4),(1,6)

11. Therefore the desired ordered pairs are ______________.

In Frames 12-14, list the ordered pairs satisfying the given equation and having the given x-components.

12. $x + y = 4$; 0, 2, 4

x
4
(0, 4)

Substitute 0 for ___ in $x + y = 4$ and obtain y = ____. Thus we have the ordered pair ______.

(2, 2);
(4, 0)

13. Similarly, we obtain the ordered pairs ______ and ______.

(0, 4), (2, 2), (4, 0)

14. Therefore, the required ordered pairs are ______________.

In Frame 15, list the ordered pairs satisfying $2x + 3y = 6$ having -4, -2, 0 as x-components.

$(-4, \frac{14}{3})$, $(-2, \frac{10}{3})$, $(0, 2)$

(see below)

$$2(-4) + 3y = 6$$
$$3y = 14$$
$$y = \frac{14}{3}, \text{ etc.}$$

15. Use the method demonstrated in Frames 12-14 and obtain ______________.

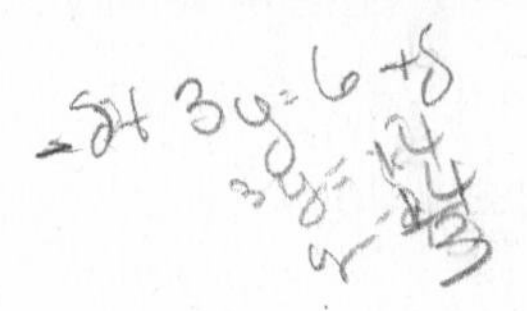

y

16. To transform an equation in two variables x and y to a form in which y is expressed explicitly in terms of x, write it as an equivalent equation in the form ___ = $f(x)$.

In Frames 17-18, transform the given equation into one in which y is expressed explicitly in terms of x.

$8 - 2x$
$\frac{8 - 2x}{x}$ $(x \neq 0)$

17. $2x + xy = 8$,
xy = ______,
y = ______________.

$9 + 9y^2$
$2x^3 - 9$
$\frac{2x^3 - 9}{9}$
$\pm \frac{\sqrt{2x^3 - 9}}{3}$

18. $9 = \frac{2x^3}{1 + y^2}$,
______ $= 2x^3$
$9y^2$ = ______,
y^2 = ______,
y = ______.

7.2

GRAPHS OF ORDERED PAIRS

Frames 19-23 refer to establishing a correspondence between an ordered pair (x, y) and a point in the plane.

vertical

19. The first component of the ordered pair denotes the directed distance from a ________ (horizontal/vertical) scaled axis to the point.

right
left

20. If the first component is positive, the point is to the ______ (right/left) of the vertical axis. If it is negative, the point is to the ______.

horizontal

21. The second component of the ordered pair denotes the directed distance from a __________ (horizontal/vertical) scaled axis to the point.

above

below

22. If the second component is positive, the point is _____ (above/below) the horizontal axis. If it is negative, the point is ______.

graph

23. The point obtained by the above correspondence is called the ______ of the ordered pair.

In Frame 24, graph the given set.

24. {(0, 2), (-3, 5), (2, -1), (-2, -2)}

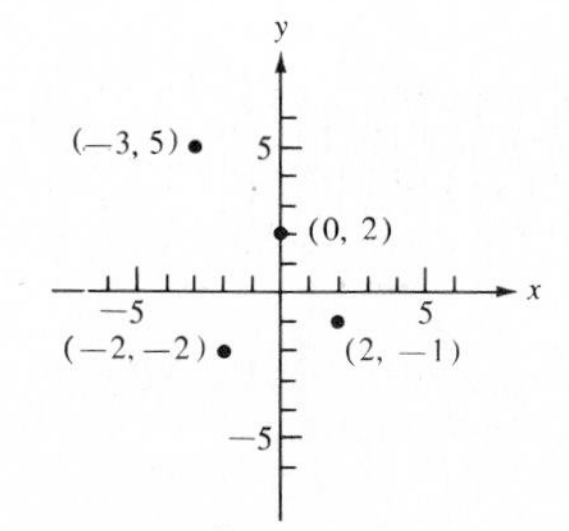

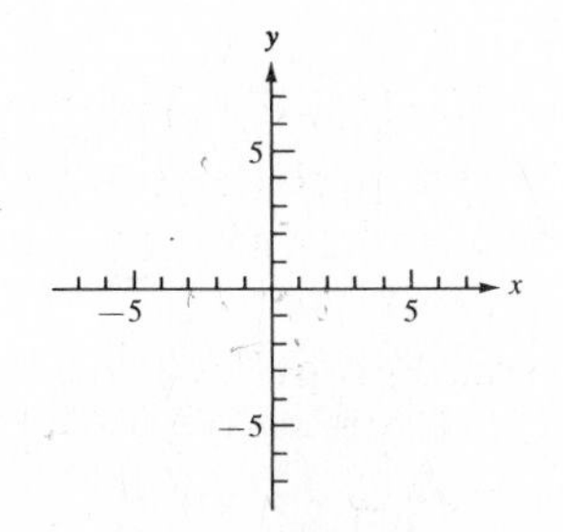

solution

25. The graph of an open sentence (equation or inequality) in two variables is the set of points which are the graphs of the members of its _________ set.

In Frames 26-27 graph the ordered pairs satisfying the given equation and having the given x-components.

26. $y = x - 2$; -3, -1, 1, 2, 3.

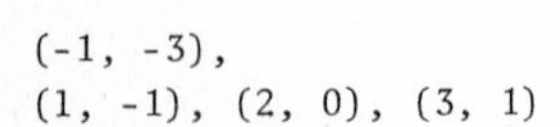

(-1, -3),
(1, -1), (2, 0), (3, 1)

The ordered pairs are (-3, -5), ___________ ______________.

The graph is:

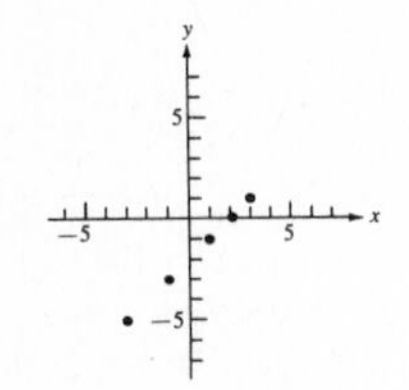

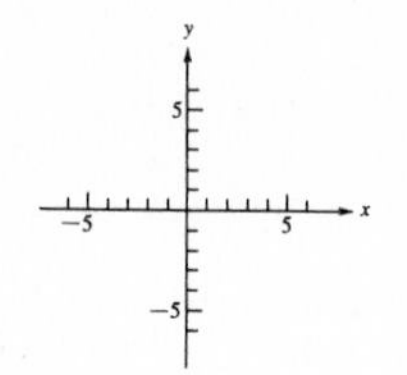

27. $y = 2x + 1$; -3, -1, 0, 1, 3.

(-3, -5), (-1, -1),
(0, 1), (1, 3) (3, 7)

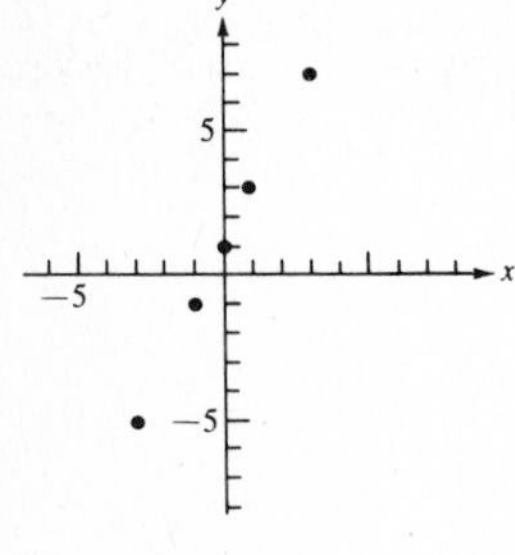

The ordered pairs are ________________ ______________.

The graph is:

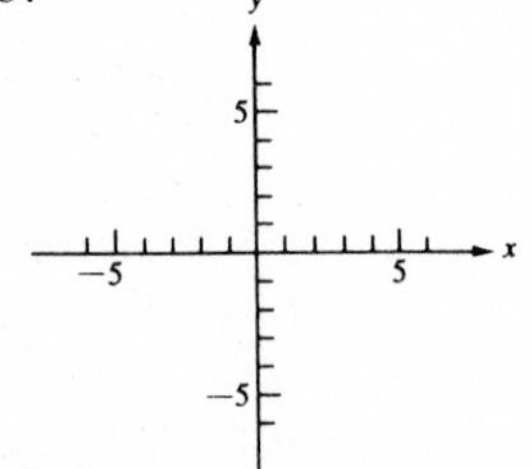

straight
linear

28. The graph of a first-degree equation in two variables is a _________ line. For this reason, such an equation is also called a ________ equation.

x

y

29. The first component, a, of the ordered pair $(a,0)$ in the solution set of an equation is called the ___-intercept of the graph of the equation. The second component, b, of the ordered pair $(0,b)$ in the solution set of an equation is called the ___-intercept of the graph of the equation.

vertical
k
horizontal; k

30. If k represents a constant (real number), then the graph of an equation of the form $x = k$ is a __________ line ___ units from the y-axis. An equation of the form $y = k$ is a ___________ line ___ units from the x-axis.

In Frames 31-36, graph each equation.

31. $y = x + 2$.

2; -2
2; -2

line

Determine the intercepts by observing that if $x = 0$, then $y =$ ___ and if $y = 0$, then $x =$ ____. So the y-intercept is ___ and the x-intercept is ____.

Graph (-2, 0) and (0, 2) and draw a ______ through these points.

The graph is

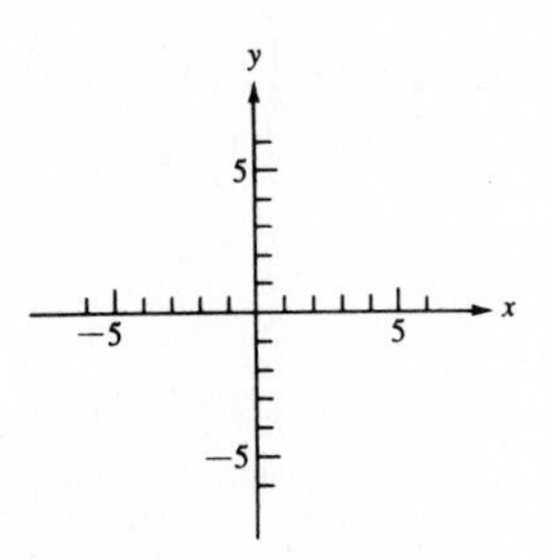

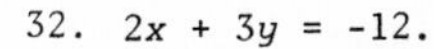

32. $2x + 3y = -12$.

-4; -6
-4; -6

If $x = 0$, then $y =$ ____ and if $y = 0$, then $x =$ ____. So the y-intercept is ____; the x-intercept is ____. Therefore, the graph is

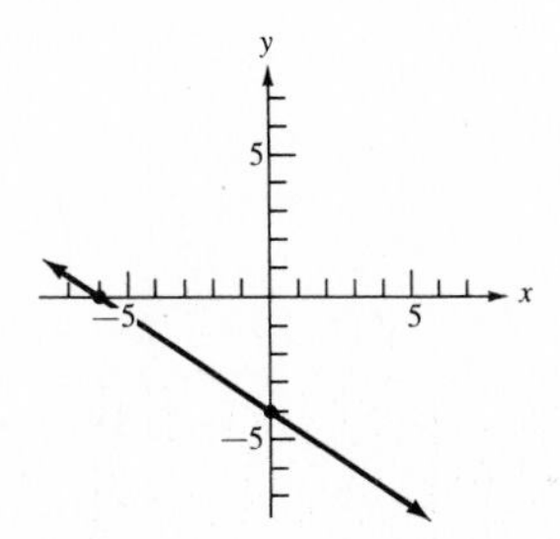

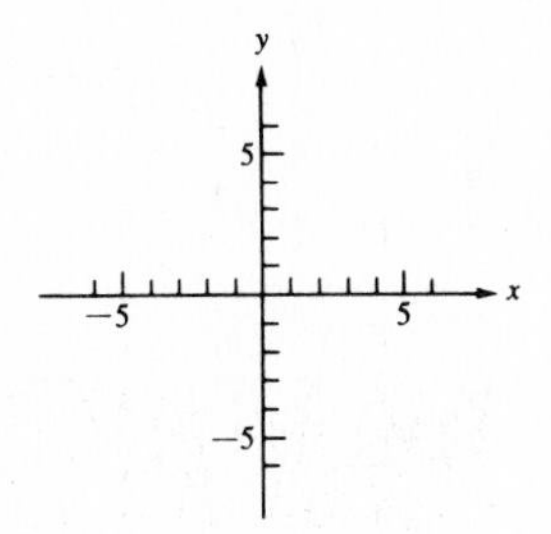

33. $x = -3$.

vertical; 3; left

The equation is of the form $x = k$, so its graph is a _________ line ___ units to the ______ (left/right) of the y-axis. The required graph is

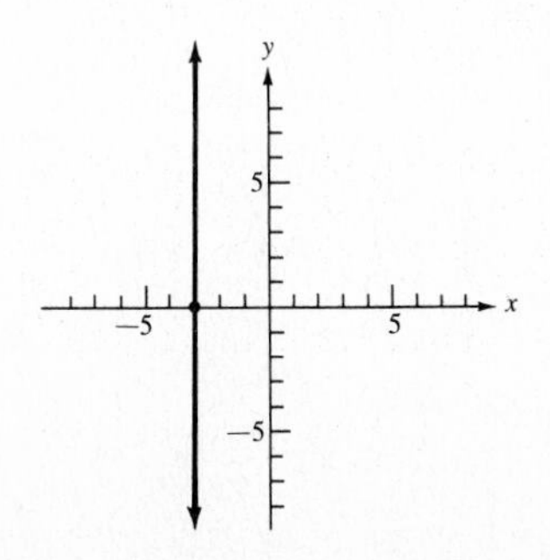

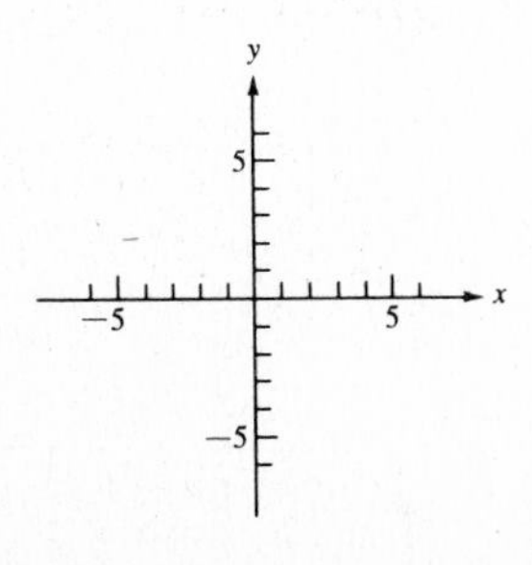

1; -2

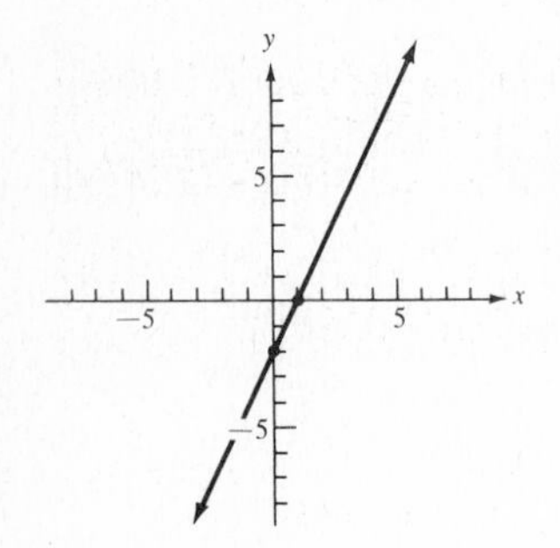

34. $y = 2x - 2$.

The x- and y-intercepts are ___ and ____, respectively.
The required graph is

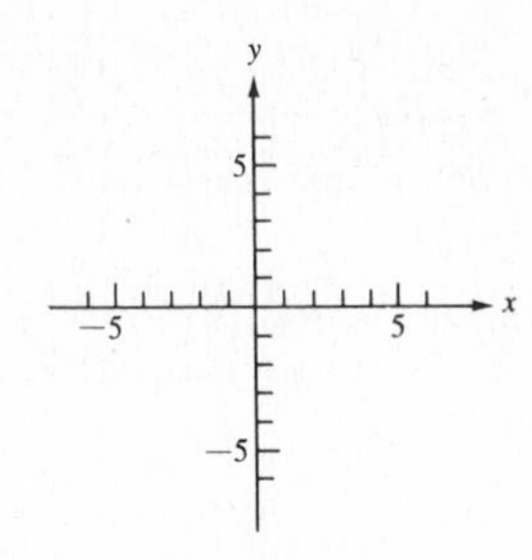

7

5; $-\frac{15}{4}$

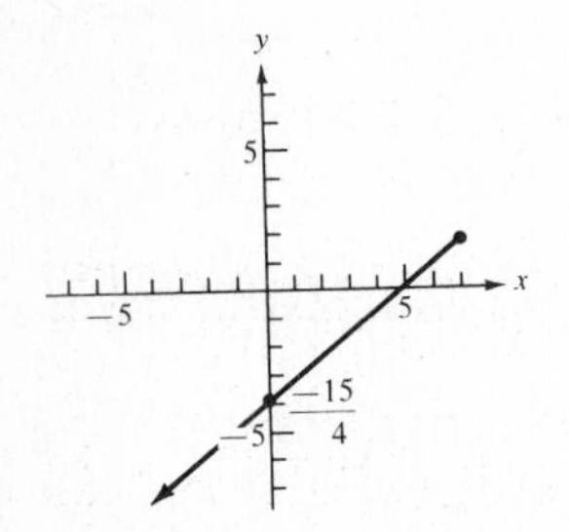

35. $3x - 4y = 15$ for $x < 7$.

Graph the equation but delete all those points whose first coordinate is greater than or equal to ___. The x- and y-intercepts are ___ and ____, respectively.

The required graph is

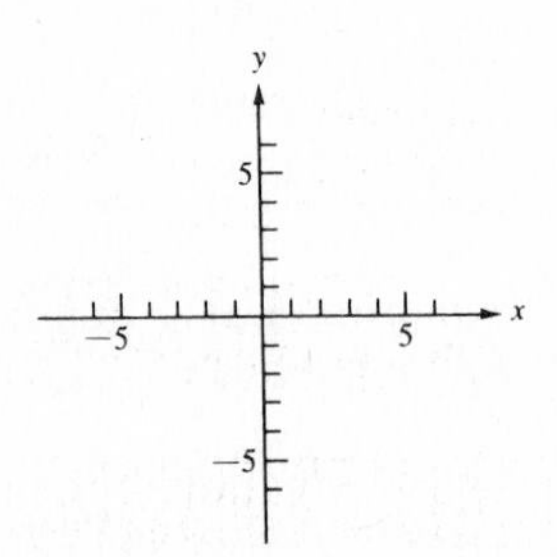

$x - 1$

$-(x - 1)$

$x < 1$

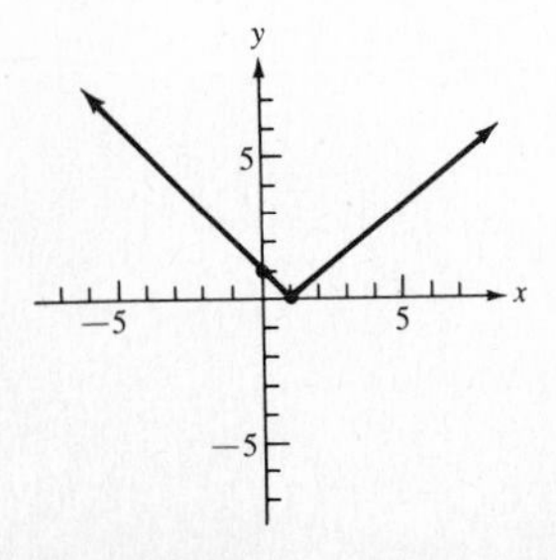

36. $y = |x - 1|$.

By the definition of absolute value,

$y = |x - 1|$ is equivalent to

$y = x - 1$ for _______ ≥ 0 $(x \geq 1)$ or

$y =$ _________ for $x - 1 < 0$ $(x < 1)$.

Graph $y = x - 1$ for $x \geq 1$.
Graph $y = -(x - 1)$ for _______.

The required graph is the union of the two graphs.

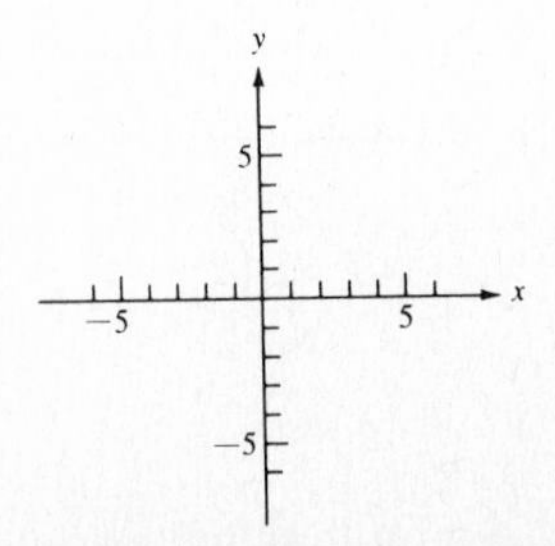

7.3

RELATIONS AND FUNCTIONS

relation

37. **Any set of ordered pairs is a ________.**

domain
range

38. **The set of all first components of the ordered pairs in a relation is the ________ of the relation, and the set of all second components is the ______ of the relation.**

function

39. **If, in a relation, no value of x is paired with more than one value of y, the relation is a __________.**

In Frames 40-41, specify the range and domain of each relation and state whether or not each relation is a function.

40. {(1,2),(-1,3),(2,2),(3,3)}.

first
1,-1,2,3; second
2,3

function

The domain is the set of _______ (which?) components and is {__________}. The range is the set of _______ components and is {_____}. Since no two ordered pairs in the relation have the same first component, the relation is a __________.

41. {(0,0),(1,2),(2,1),(1,3)}.

{0,1,2}; {0,1,2,3}

first

The domain is _________. The range is __________. The relation is not a function, because (1,2) and (1,3) have the same _______ component.

solution

42. **The graph of a relation which is defined by an equation is the graph of the _________ set of the equation.**

function

43. **If no vertical line would intersect the graph of a relation at more than one point, the graph is that of a _________.**

In Frames 44-45, graph the relation defined by the given equation.

44. $x - 3y = 6$.

6; -2
first
straight

The x-intercept is ___, the y-intercept is ____. Since the equation is of the _______ degree, the graph will be a __________ line.

45. The graph is

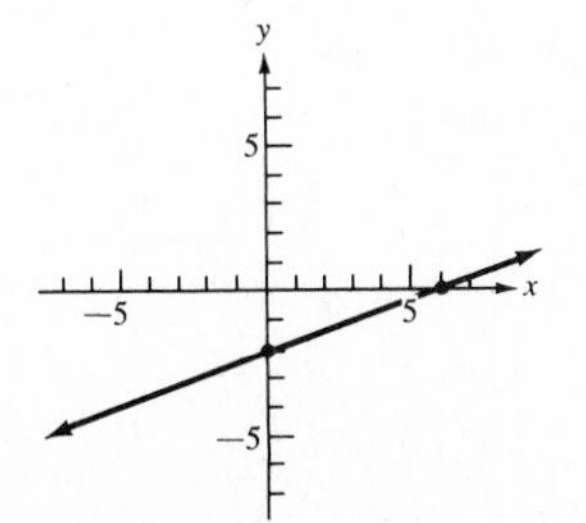

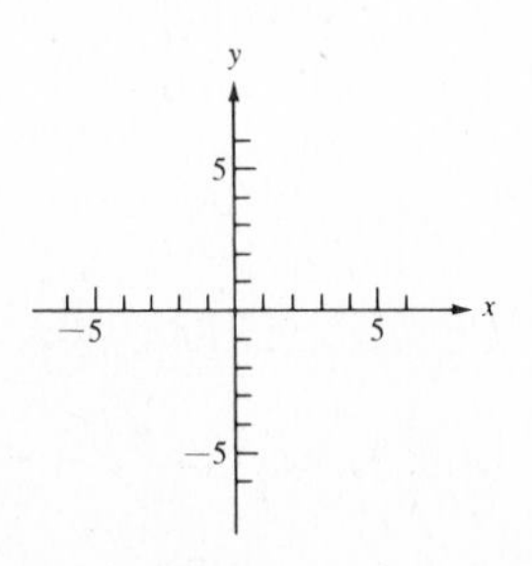

In Frames 46-47, graph the relation defined by the given equation.

46. $x = 4$.

The equation is of the form $x = k$, so the graph is a ________ line ___ units to the right of the y-axis.

vertical; 4

47. The graph is

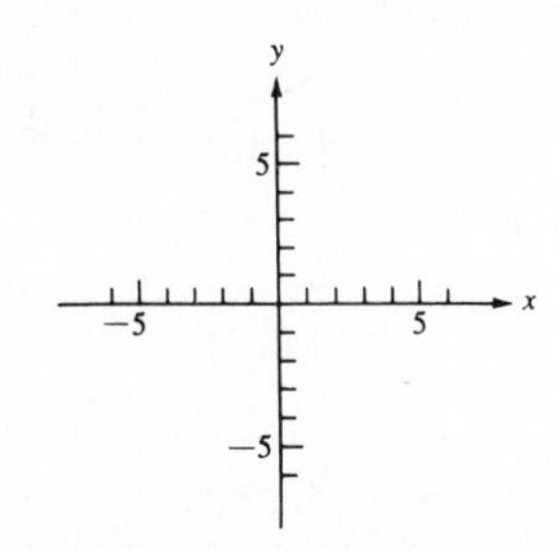

48. **If a function is designated by f, then the element in the range of f associated with the element x in the domain is designated by ______.**

$f(x)$

49. **$f(x)$ is read "f of x" or "the ______ of the function f at x."**

value

50. **The directed distance from the point $(x,0)$ to the graph of a function f can be denoted by ______.**

$f(x)$

In Frames 51-53, find the value indicated.

51. $f(2)$ if $f(x) = x^2 + 2x - 1$.

$f(2) = (\_\_\_)^2 + 2(\_\_\_) - 1$.

$f(2) = \_\_\_$.

2; 2
7

52. $g(4) - g(2)$ if $g(x) = 3x + 2$.

$g(4) = 3(\_\_\_) + 2 = \_\_\_\_,$

$g(2) = 3(\_\_\_) + 2 = \_\_\_,$

$g(4) - g(2) = \_\_\_\_ - \_\_\_ = \_\_\_.$

4; 14
2; 8
14; 8; 6

53. $\dfrac{f(x+h) - f(x)}{h}$ if $f(x) = x^2 - 2x$.

$f(x + h) = (x + h)^2 - 2(x + h)$

$= \_\_\_\_\_\_\_\_\_\_\_\_\_\_\_\_.$

$f(x + h) - f(x) = x^2 + 2hx + h^2 - 2x - 2h - (\_\_\_\_\_\_\_\_)$

$= \_\_\_\_\_\_\_\_\_\_\_\_$

$\dfrac{f(x+h) - f(x)}{h} = \dfrac{h(\_\_\_\_\_\_\_\_)}{h}$

$= \_\_\_\_\_\_\_\_\_\_.$

$x^2 + 2hx + h^2 - 2x - 2h$
$x^2 - 2x$
$2hx + h^2 - 2h$
$2x + h - 2$
$2x + h - 2$

In Frames 54-55, graph $f(x) = 2x - 1$ and represent on this graph $f(2)$ and $f(-3)$ by drawing line segments from (2, 0) to $[2, f(2)]$ and from (-3, 0) to $[-3, f(-3)]$.

54. The graph of $f(x) = 2x - 1$ is the same as the graph of $y = 2x - 1$ and is

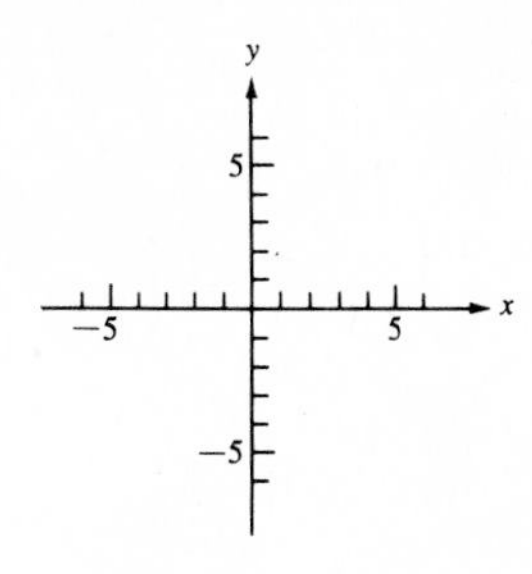

Answer:

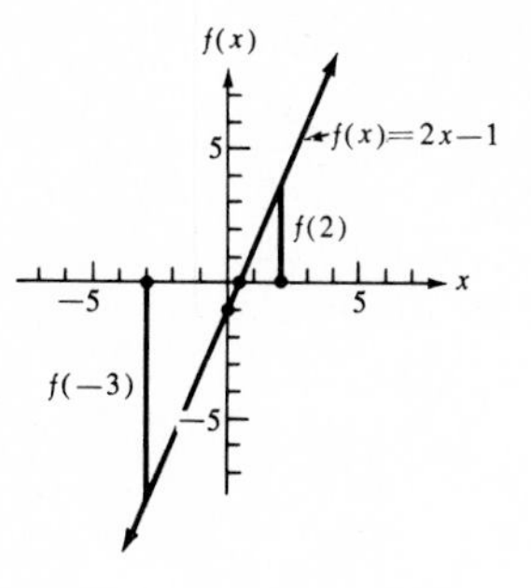

55. At (2,0) and (-3,0) draw vertical line segments terminating on the ______.

Answer: graph (see above)

7.4

DISTANCE AND SLOPE FORMULAS

56. The distance between two points whose coordinates are (x_1,y_1) and (x_2,y_2) is given by the formula $d =$ ________________.

Answer: $\sqrt{(x_2 - x_1)^2 + (y_2 - y_1)^2}$

57. The slope of the line containing the points (x_1,y_1) and (x_2,y_2) is given by the formula $m =$ ______ $(x_1 \neq$ ____ $)$.

Answer: $\dfrac{y_2 - y_1}{x_2 - x_1}$; x_2

In Frames 58-61, find the distance between each of the given pairs of points, and find the slope of the line joining them.

58. (3,10),(-3,2).

Take $x_1 = 3$, $y_1 =$ ____ and $x_2 = -3$, $y_2 =$ ____.

$d = \sqrt{(-3 - 3)^2 + (\_\_\_\_)^2} = \sqrt{36 + \_\_\_\_} = \_\_\_\_.$

$m = \dfrac{2 - 10}{\_\_\_\_} = \dfrac{-8}{\_\_\_\_} = \_\_\_\_.$

Answers: 10; 2
2 - 10; 64; 10
-3 - 3; -6; $\frac{4}{3}$

59. (0,0),(3,-2).

Take $x_1 = 3$, $y_1 =$ ____ and $x_2 =$ ____, $y_2 =$ ____.

$d \sqrt{(\_\_\_\_)^2 + (\_\_\_\_)^2} = \sqrt{\_\_\_\_}.$

$m = \dfrac{\_\_\_\_}{-3} = \dfrac{-2}{3}.$

Answers: -2; 0; 0
0 - 3; 0 - (-2); 13
2

60. (1,-4),(6,6).

Take $x_1 = 1$, $y_1 =$ ____ and $x_2 =$ ____, $y_2 =$ ____.

$d = \sqrt{(\_\_)^2 + (\_\_)^2} = \sqrt{\_\_\_\_} = 5\sqrt{\_\_}.$

$m = \dfrac{\_\_\_\_}{5} = \_\_\_\_.$

Answers: -4; 6; 6
5; 10; 125; 5
10; 2

61. (5,-3),(5,6).

$d = \sqrt{(\_\_\_)^2 + (\_\_\_)^2} = \sqrt{\_\_\_} = \_\_\_$.

Since $x_2 - x_1 =$ ___, the slope is __________.

0; 9; 81; 9

0; undefined

62. If a line is horizontal, its slope is ___.

0

63. If a line is vertical, its slope is __________.

undefined

In Frame 64, sketch the line through P(-2, 2) whose slope is -3.

64. To obtain a second point on the line, draw a horizontal segment ___ unit(s) long to the right from P followed by a vertical segment ___ unit(s) down, arriving at point A.

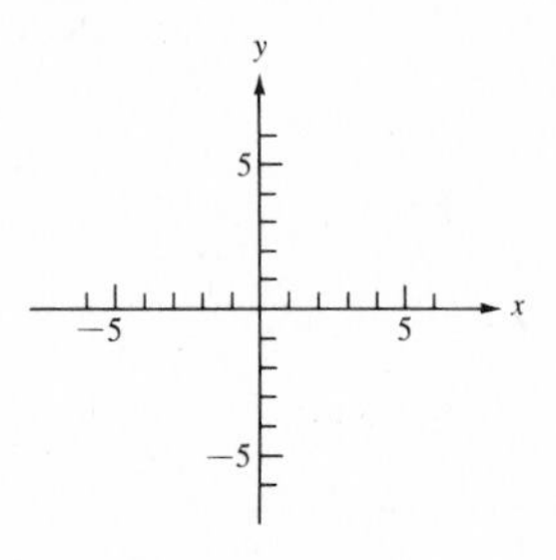

1

3

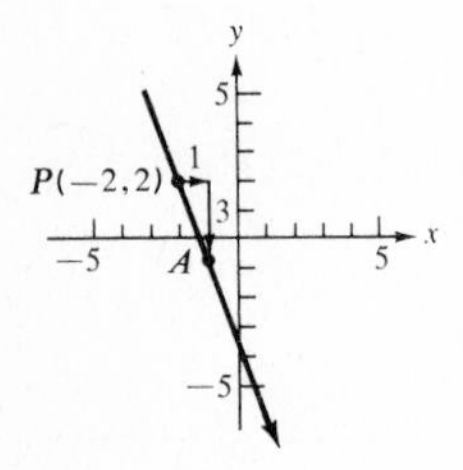

In Frame 65, sketch the line through (0, -1) whose slope is 1/2 and write its equation in standard form.

65. The sketch is

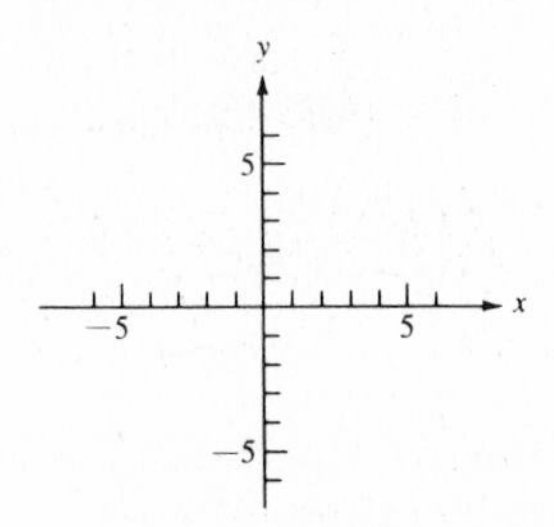

66. **Two lines with slopes m_1 and m_2 are parallel if and only if $m_1 =$ ____ and they are perpendicular if and only if $m_1 m_2 =$ ____ or, equivalently, $m_1 = -\frac{1}{m_2}$.**

m_2

-1

In Frames 67-71, show that the points (-5,4),(7,-11),(12,25), and (0,40) are the vertices of a parallelogram.

67. The slope of the line containing (-5,4) and (7,-11) is

$\frac{15}{-12}$ or $-\frac{5}{4}$

$\frac{-15}{12}$ or $-\frac{5}{4}$

68. The slope of the line containing (12,25) and (0,40) is ____.

$\frac{36}{5}$

69. The slope of the line containing (-5,4) and (0,40) is ____.

$\frac{36}{5}$

70. The slope of the line containing (7,-11) and (12,25) is ____.

equal

71. The given points are vertices of a parallelogram because two pairs of sides have _______ slopes and so are parallel.

In Frames 72-74, show that the two line segments whose end points are (4,-6),(2,0) and (7,2),(-2,-1) are perpendicular.

-3

72. The slope of the line containing (4,-6) and (2,0) is ____.

$\frac{3}{9}$ or $\frac{1}{3}$

73. The slope of the line containing (7,2) and (-2,-1) is ___.

-1

74. The lines are perpendicular because $(-3)\frac{1}{3}$ = ____.

7.5 FORMS OF LINEAR EQUATIONS

$ax + by + c = 0$ (a and b not both 0)

75. **Standard form for a linear equation is ____________________.**

$y - y_1 = m(x - x_1)$

76. **The point-slope form for the equation of the line containing the point (x_1,y_1) and whose slope is m is ____________.**

In Frames 77-78, write an equation in standard form of the line that goes through the point (-2, 2) and whose slope is -3.

2, -3, -2

77. In point-slope form, the equation is
y - (___) = ___(x - ___).

6; $3x + y + 4$

78. To obtain standard form, first write
$y - 2 = -3x$ - ___ and then ________ =0.

In Frames 79-80, write the equation in standard form of the line that goes through (0, -1) and whose slope is $\frac{1}{2}$.

-1; $\frac{1}{2}$; 0

79. Point-slope: y - ___ = ___(x - ___).

$x - 2y - 2$

80. Standard form: __________ = 0.

$y = mx + b$
b

81. The slope-intercept form for a linear equation is __________. In this form, m is the slope and the y-intercept is ___.

In Frames 82-83, write each of the given equations in slope-intercept form and specify the slope and the y-intercept.

82. $2x + 3y = 1$.

$y = -\frac{2}{3}x + \frac{1}{3}$

Solve explicitly for y and obtain __________.

$-\frac{2}{3}$

The coefficient of x, which is ___, is the slope and the

$\frac{1}{3}$

constant term ___ is the y-intercept.

83. $3x - 8y = 0$.

$y = \frac{3}{8}x$

Solve explicitly for y and obtain ______.

$\frac{3}{8}$; 0

The slope is ___ and the y-intercept is ___.

In Frames 84-88, write the equation of the line which has the same slope as the graph of $4x - 3y = 6$, and passes through the origin.

$\frac{4}{3}x - 2$

84. To find the slope of the given line, write its equation in slope-intercept form, which is y = ______.

$\frac{4}{3}$

85. The slope is ___.

$(0,0)$; $\frac{4}{3}$

86. The graph of the required equation will contain the origin, whose coordinates are ______, and have slope ___.

$y - 0 = \frac{4}{3}(x - 0)$ or $y = \frac{4}{3}x$

87. In point-slope form, its equation is

______________________________.

$4x - 3y = 0$

88. In standard form, it is __________.

7.6 GRAPHS OF FIRST-DEGREE INEQUALITIES

$ax + by + c = 0$

broken

89. The graph of the relation defined by an open sentence such as $ax + by + c \leq 0$ is obtained by first graphing the equality ____________ with a solid line. If the open sentence being graphed involves $<$ rather than $\leq$, the graph is a ______ line. Similar statements hold for graphing open sentences involving $\geq$ or $>$.

90. To complete the graph referred to in Frame 89, determine which of the two half-planes must be shaded by choosing any _______ in either half-plane (the origin is usually convenient). Shade the half-plane containing this point if its coordinates _________ the inequality. If they do not, shade the other ____________.

point
satisfy
half-plane

In Frames 91-94, graph the solution set of $y \leq 2x + 3$.

91. Graph the equality $y = 2x + 3$ with a _______ (broken/solid) line.

solid

92. The test point (0,0) has coordinates which ____ (do/do not) satisfy $y \leq 2x + 3$.

do

93. Therefore, shade the half-plane which contains the point corresponding to (___,___).

0; 0

94. The graph is

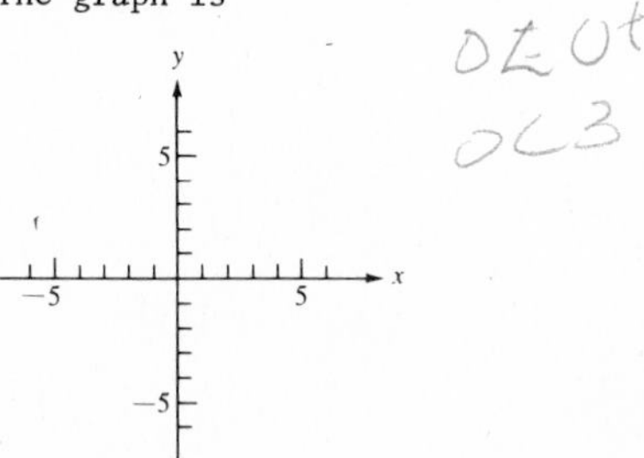

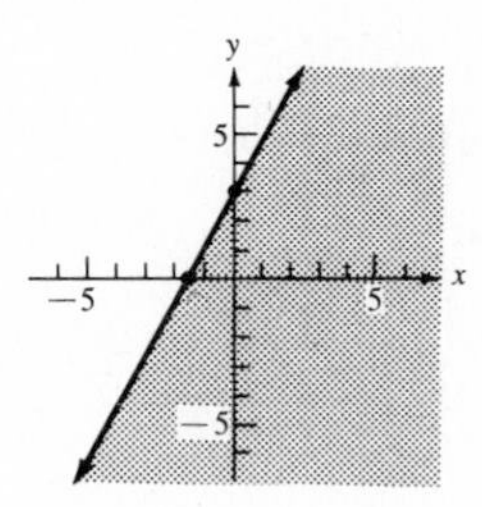

In Frames 95-98, graph the solution set of $2x > y$.

95. Graph the equality $2x = y$ with a ________ (solid/broken) line.

broken

96. The test point (0,1), chosen arbitrarily, has coordinates which ________ (do/do not) satisfy $2x > y$.

do not

97. Therefore, shade the half-plane which __________ (does/does not) contain this point.

does not

98. The graph is

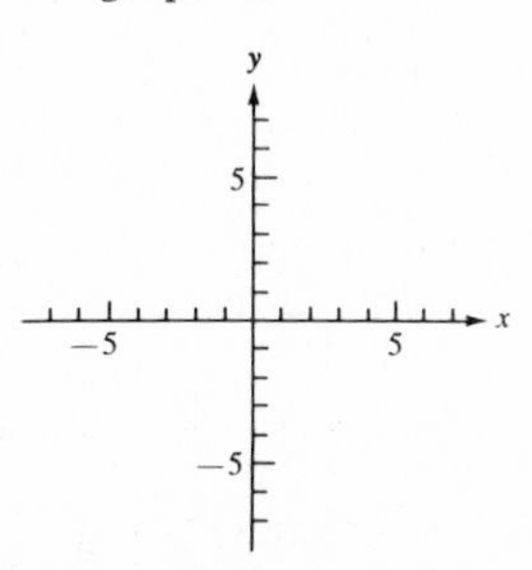

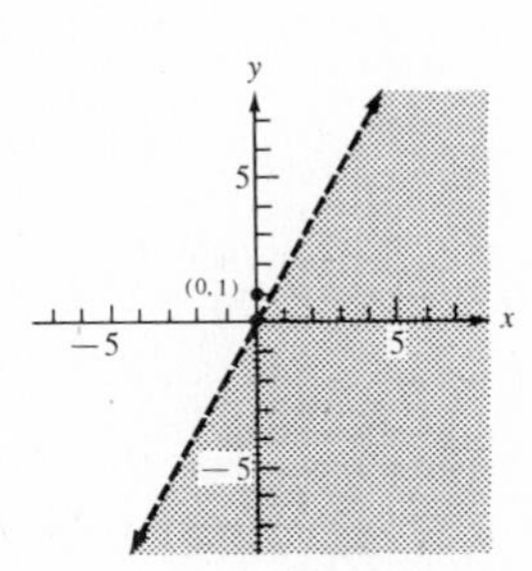

In Frames 99-101, graph $\{(x, y) \mid |y| > 2\}$.

99. Rewrite the given inequality equivalently as $y > 2$ or ________. (See Section 4.6.)

$y < -2$

100. Thus, it is seen that we seek the set of points whose y-coordinates are greater than 2 or ______ (less/greater) than -2.

less

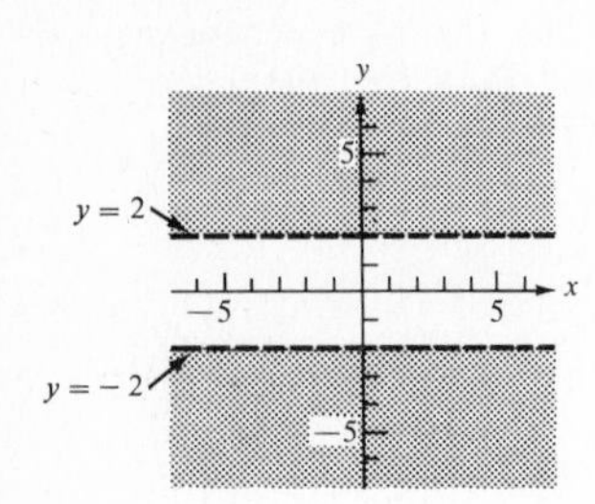

101. The graph is

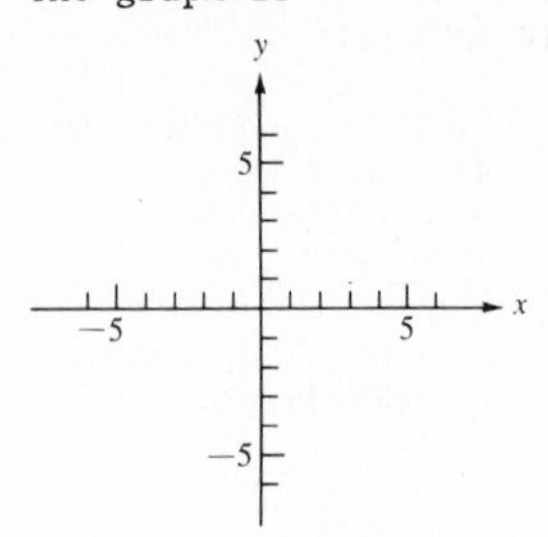

In Frames 102-105, graph $\{(x, y) \mid |x - 1| \leq 2\}$.

-2

102. Write $|x - 1| \leq 2$ equivalently as ____ $\leq x - 1 \leq 2$. (See Section 4.6.)

-1

103. Solve this inequality and obtain ____ $\leq x \leq 3$.

-1; 3

104. Thus, it is seen that we seek the set of points in the plane whose x-coordinates are between ____ and ___ inclusive.

105. The graph is

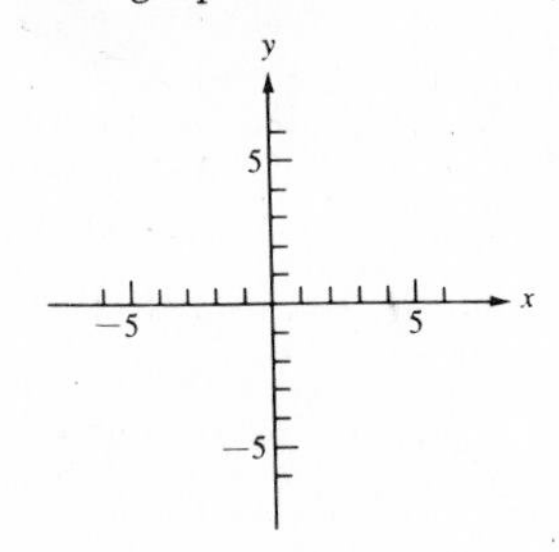

In Frames 106-110, graph $\{(x, y) \mid |x - 1| \leq 2\} \cap \{(x, y) \mid |y| > 2\}$ *using double shading.*

106. Graph $\{(x,y) \mid |x - 1| \leq 2\}$ as in Frames 102-105.

107. Graph $\{(x,y) \mid |y| > 2\}$ as in Frames 99-101.

108. Shade each graph differently.

common

109. The required graph is the region shaded both ways, since, by definition of $\cap$, the solution set must be ________ to both solution sets.

110. The graph is

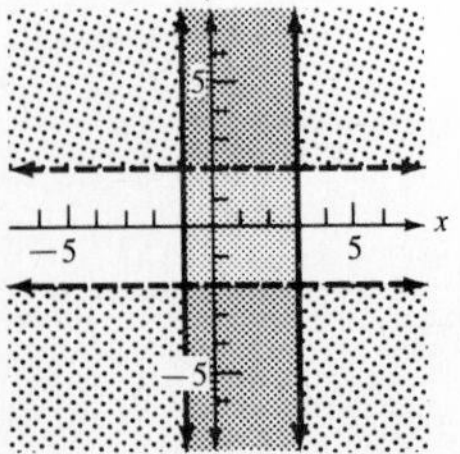
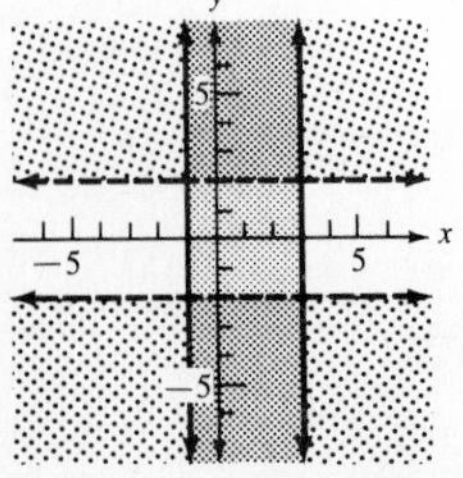

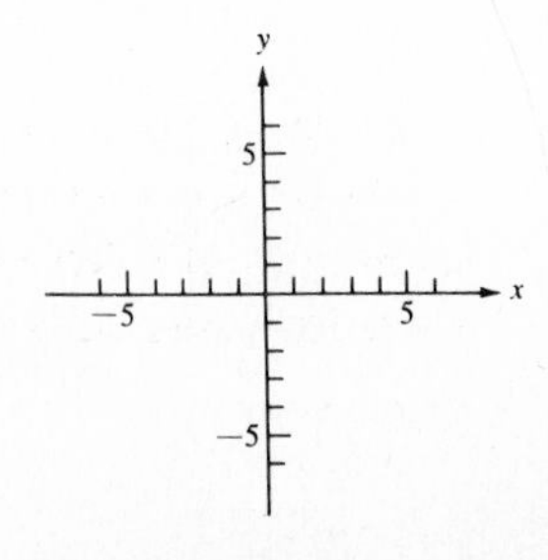

TEST PROBLEMS

1. Find the missing component in each solution of $2x - 9y = 18$.

   a. $(0,?)$ b. $(?,0)$ c. $(2,?)$

2. In the equation $\frac{2x^4}{3 + 2y^2} = 5$, express y explicitly in terms of x.

3. Find k if $kx - 3y = 9$ has $(1,-3)$ as a solution.

4. Graph $y = 2x - 1$, if x is -2, -1, 0, 1, and 2.

5. Graph $y = -4x + 2$.

6. Graph the equation $3x - 2y = 6$, $x < 4$.

7. Graph $y = |2x - 6|$.

8. Specify the domain and range of each relation and state whether or not the relation is a function:

   a. $\{(3,1),(4,1),(5,1),(6,1)\}$.

   b. $\{(1,3),(1,4),(1,5),(1,6)\}$.

9. Given that $h(x) = 2x^2 + 3x - 2$, find:

   a. $h(3)$ b. $h(1) - h(0)$

10. Given that $f(x) = x^2 + 2x$, find:

    a. $f(x + h)$ b. $\frac{f(x + h) - f(x)}{h}$

11. Given $(2,-3)$ and $(-2,-1)$, find:

    a. the distance between the two points.

    b. the slope of the line segment joining the two points.

12. Find the equation in standard form of the line through $(2,-1)$ with slope equal to $\frac{3}{4}$.

13. Given the equation $3x - 2y = 5$:

    a. Write the equation in slope-intercept form.

    b. Specify the slope of its graph.

    c. Specify the y-intercept of its graph.

14. Write in standard form the equation of the line through $(0, 5)$ that is perpendicular to the graph of $3x - 2y + 5 = 0$.

15. Graph the relation $y < 2x - 4$.

16. Graph the relation $\{(x,y) | x < 2\} \cap \{(x, y) | y < 3\}$

8

FUNCTIONS, RELATIONS, AND THEIR GRAPHS: PART II

8.1

GRAPHS OF QUADRATIC RELATIONS

1. The graph of a quadratic equation of the form $y = ax^2 + bx + c$ is called a ____________.

parabola

2. Since with each x, an equation of the form $y = ax^2 + bx + c$ associates only one y, such an equation defines a ____________.

function

In Frames 3-4, find solutions of $y = x^2 + 3$, using integral values for x where $-4 \le x \le 4$, and use the solutions to graph the equation.

3. The following pairs are in the solution set: (-4,19),(-3,12),(-2,7),(-1,4),(0,3),(1,4),(2,___), (3,____),(4,____).

7
12; 19

4. The graph is

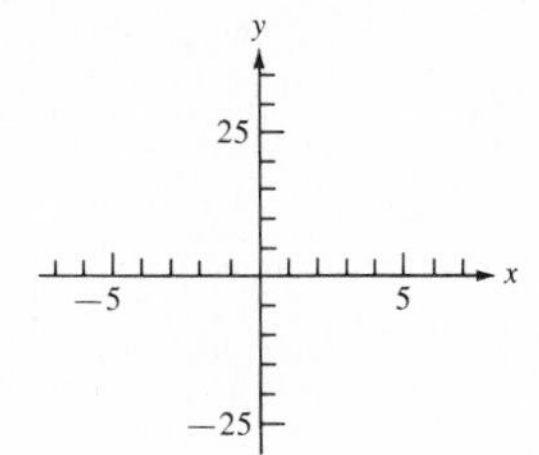

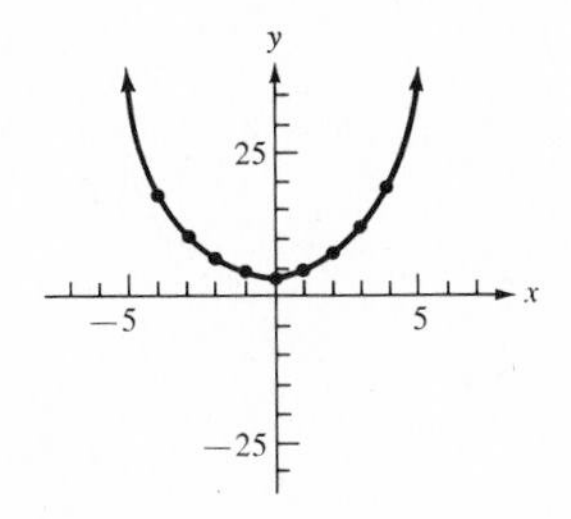

In Frames 5-6, apply the instructions for Frames 3-4 to $y = -x^2 + 3$.

5. The following pairs are in the solution set: (-4,-13),(-3,-6),(-2,-1),(-1,2),(0,3),(1,___),(2,____), (3,____),(4,_____).

2; -1
-6; -13

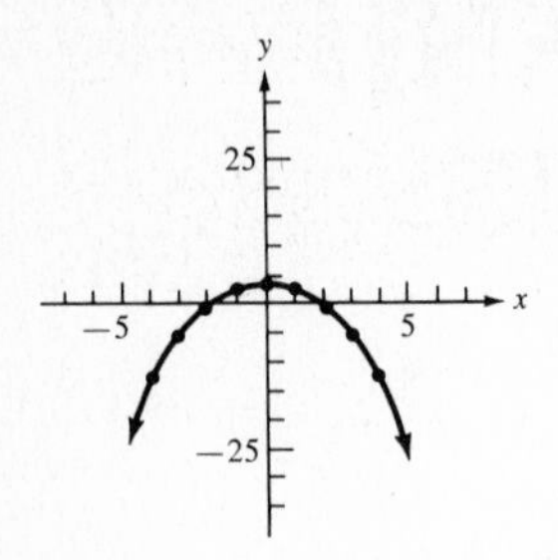

6. The graph is

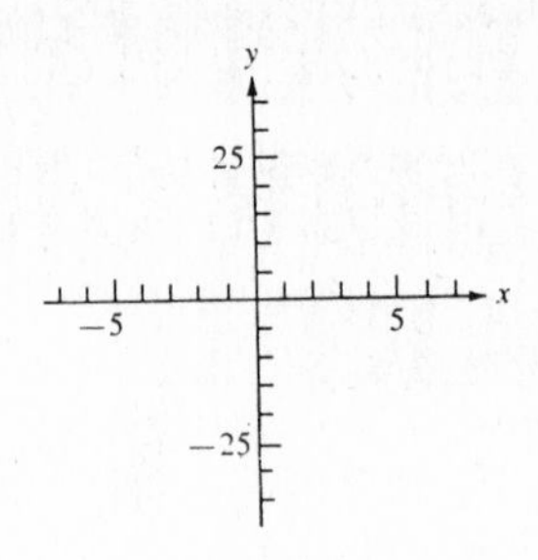

In Frames 7-8, apply the instructions for Frames 3-4 to $y = 4x^2 - x$.

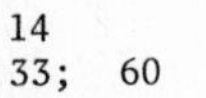
14
33; 60

7. The following pairs are in the solution set: (-4,68),(-3,39),(-2,18),(-1,5),(0,0),(1,3),(2,____),

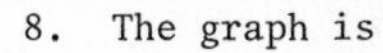
(3,____),(4,____).

8. The graph is

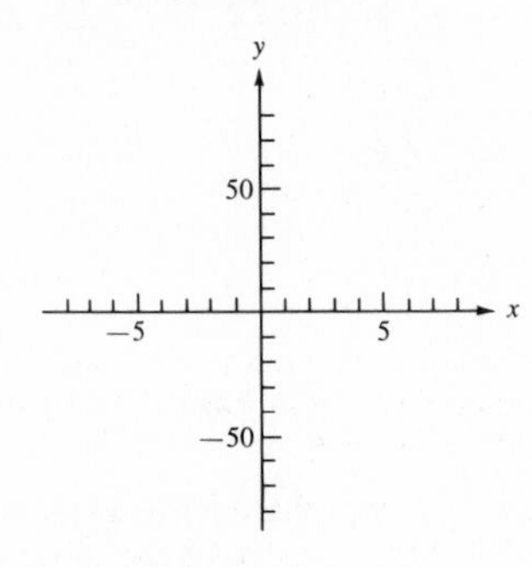

In Frames 9-10, apply the instructions of Frames 3-4 to $f(x) = -2x^2 + x - 2$.

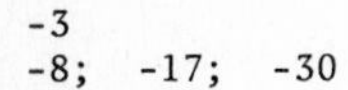
-3
-8; -17; -30

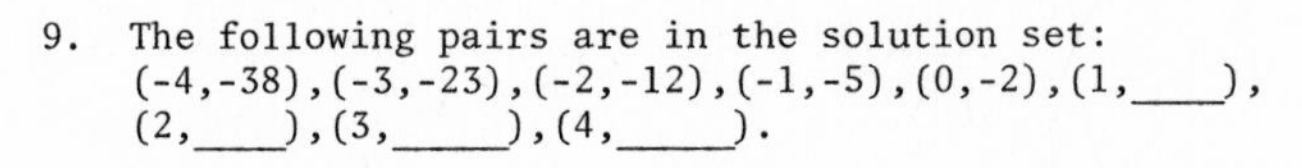
9. The following pairs are in the solution set: (-4,-38),(-3,-23),(-2,-12),(-1,-5),(0,-2),(1,____), (2,____),(3,_____),(4,_____).

10. The graph is

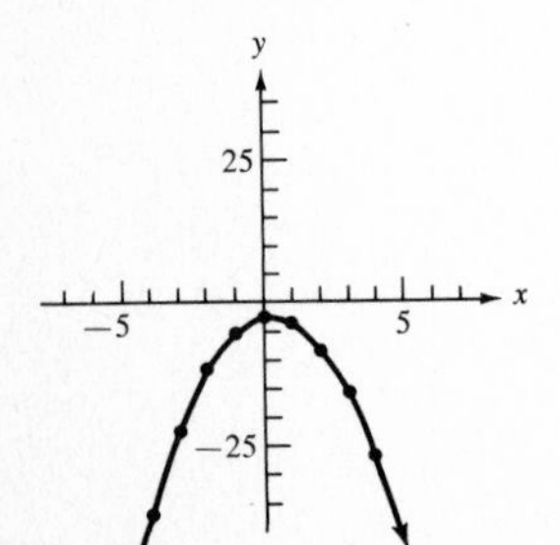

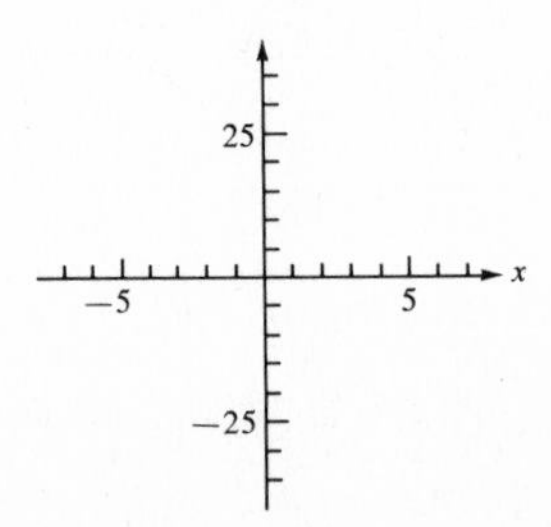

parabola

11. The graph of an equation of the form $x = ay^2 + by + c$ is called a ________.

two
function

12. Since with each x in the domain there are associated ______ y's, $x = ay^2 + by + c$ does not define a ________.

In Frames 13-14, find solutions of $x = y^2 - 3$ by using $-4 \leq y \leq 4$ and graph the equation.

-2; 1
6; 13

13. The following pairs are in the solution set: (13,-4), (6,-3), (1,-2), (-2,-1), (-3,0), (____,1), (___,2), (___,3) (____,4).

14. The graph is

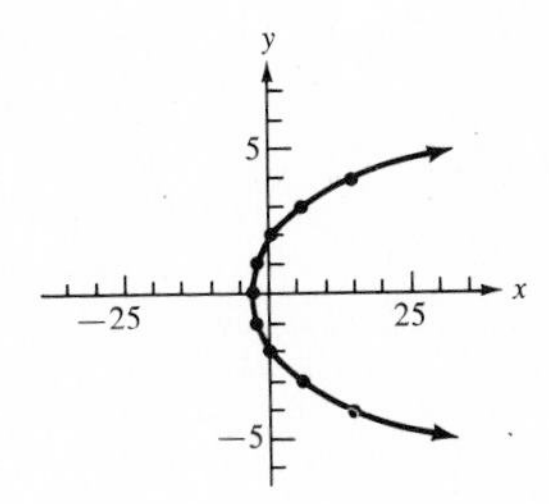

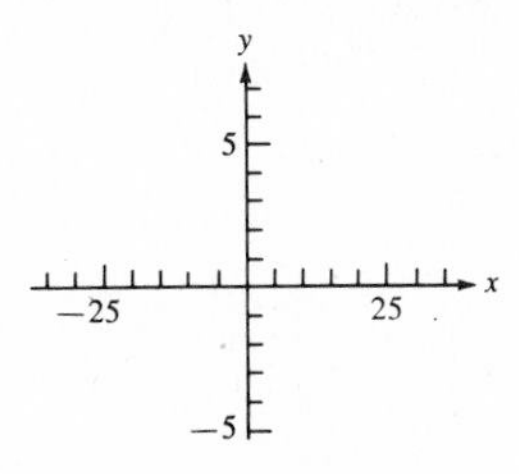

In Frames 15-16, apply the instructions of Frames 13-14 to $x = -y^2 - y + 2$.

0; -4;
-10; -18

15. The following pairs are in the solution set: (-10,-4), (-4,-3), (0,-2), (2,-1), (2,0), (___,1) (____,2), (_____,3), (_____,4).

16. The graph is

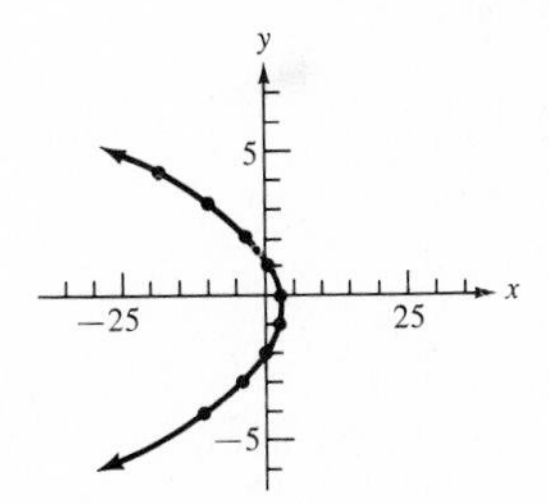

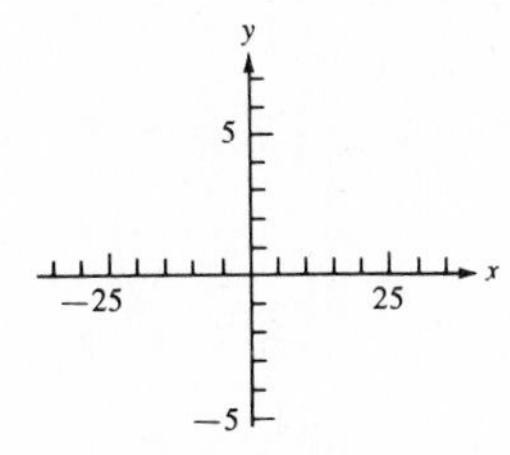

8.2

SKETCHING PARABOLAS

17. The x-intercepts of the graph of the function defined by $f(x) = ax^2 + bx + c$ are found, if they exist, by identifying those values of x for which $ax^2 + bx + c =$ ___.

0

18. The y-intercept of the graph of the function defined by $f(x) = ax^2 + bx + c$ is found by computing $f($___$)$.

0

In Frames 19-21, find the x- and y-intercepts of the graph of $f(x) = x^2 - x - 6$.

19. The x-intercepts are the solutions to $x^2 - x - 6 =$ _______.

0

20. The x-intercepts are ___ and ____.

3; -2 (detail below)

$x^2 - x - 6$
$= (x - 3)(x + 2)$, etc.

21. The y-intercept is $f(0) = 0^2 - 0 - 6 =$ ____.

-6

In Frames 22-24, find the x- and y-intercepts of the graph of $y = -3x^2 - 5x + 2$.

22. The x-intercepts are the solutions to ___________________.

$-3x^2 - 5x + 2 = 0$

23. The x-intercepts are ___ and ____.

$\frac{1}{3}$; -2 (detail below)

$3x^2 + 5x - 2$
$= (3x - 1)(x + 2) = 0$, etc.

24. The y-intercept is $-3(0)^2 - 5(0) + 2 =$ ___.

2

25. The elements of the solution set of the equation $ax^2 + bx + c = 0$ in addition to being called x-intercepts are also called _______ of the function defined by $y = ax^2 + bx + c$.

zeros

26. The highest or lowest points on the graph of $y = ax^2 + bx + c$ are called the maximum or _______ points, respectively. The maximum or minimum point is also called the ___________.

minimum

vertex

27. The x-coordinate of the vertex of the graph of $y = ax^2 + bx + c$ is given by the expression _____. The y-coordinate is found by substituting the value of $-\frac{b}{2a}$ for x in the equation ____________ and solving for y.

$-\frac{b}{2a}$

$y = ax^2 + bx + c$

28. **The graph of $y = ax^2 + bx + c$ opens downward if $a < 0$ and it opens upward if ________.**

$a > 0$

In Frames 29-31, find the coordinates of the vertex of the graph of $f(x) = x^2 - x - 6$ and determine if the graph opens upward or downward.

29. $a = 1$, $b =$ ____. Hence the x-coordinate of the vertex is $-\frac{\_\_\_}{2(1)} =$ ________.

-1
-1; $\frac{1}{2}$

30. Substitute $\frac{1}{2}$ for x in $y = x^2 - x - 6$ and obtain $y = (\_\_\_)^2 - (\_\_\_) - 6 =$ ______.

$\frac{1}{2}$; $\frac{1}{2}$; $\frac{-25}{4}$

31. Since $a = 1 > 0$, the graph opens _______ and hence has a minimum point at $(\frac{1}{2}, \frac{-25}{4})$.

upward

In Frames 32-34, find the coordinates of the vertex of the graph of $y = -3x^2 - 5x + 2$ and determine if the graph opens upward or downward.

32. $a =$ ____, $b =$ ____. Hence the x-coordinate of the vertex is $-\frac{\_\_\_}{2(-3)} =$ ____.

-3; -5
-5; $\frac{-5}{6}$

33. Substitute ___ for x in $y = -3x^2 - 5x + 2$ and obtain $y = -3(\frac{-5}{6})^2 - 5(\_\_\_) + 2 =$ ____.

$\frac{-5}{6}$

$\frac{-5}{6}$; $\frac{49}{12}$

(detail below)

$$y = -3(\frac{25}{36}) + \frac{25}{6} + 2$$

$$= \frac{-25}{12} + \frac{50}{12} + \frac{24}{12} = \frac{49}{12}$$

34. Since $a = -3 < 0$, the graph opens _________ and hence has a maximum point at $(\frac{-5}{6}, \frac{49}{12})$.

downward

In Frames 35-36, use the information about intercepts and maximum or minimum point to sketch the graph of $f(x) = x^2 - x - 6$.

35. From Frames 20, 21, and 31, the x-intercepts are ___ and ____, the y-intercept is ____, and there is a minimum point at ________.

3
-2; -6
$\left(\frac{1}{2}, \frac{-25}{4}\right)$

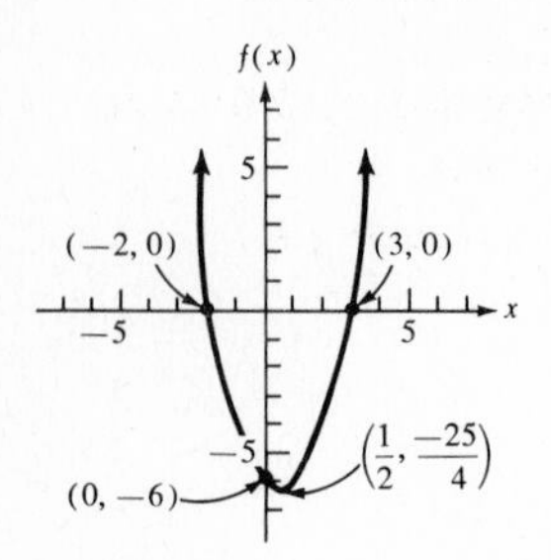

36. The graph is

In Frames 37-38, use the information about intercepts and maximum or minimum points to sketch the graph of $y = -3x^2 - 5x + 2$.

$\frac{1}{3}$

-2; 2

$\left(\frac{-5}{6}, \frac{49}{12}\right)$

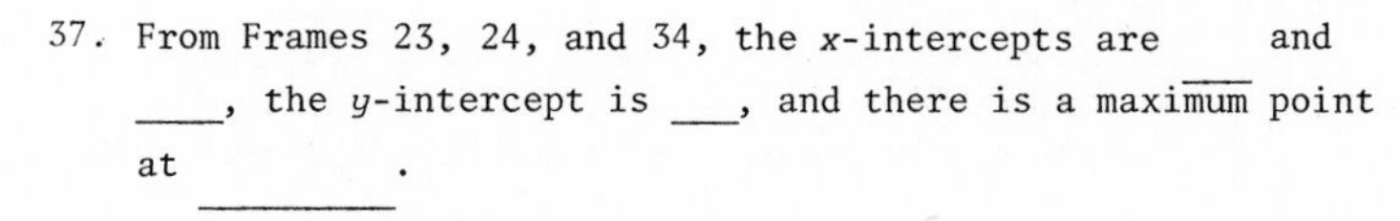

37. From Frames 23, 24, and 34, the x-intercepts are ____ and ____, the y-intercept is ___, and there is a maximum point at ________.

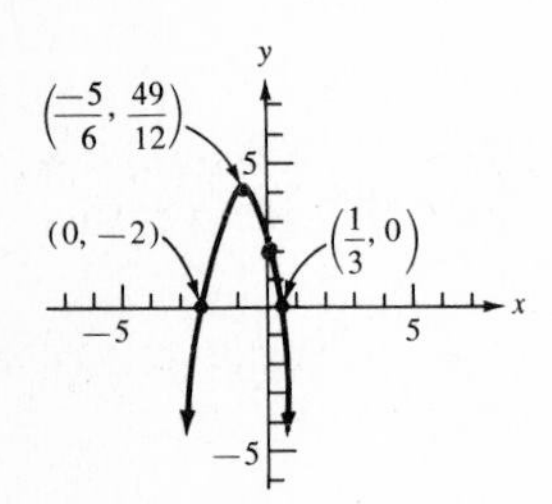

38. The graph is

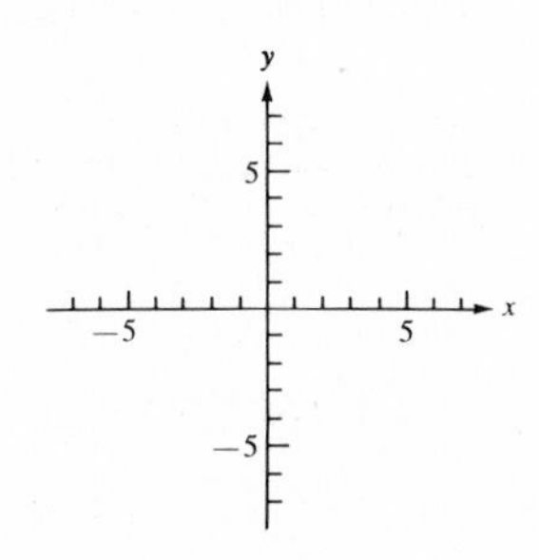

no

39. If the equation $ax^2 + bx + c = 0$ has no real solutions, the graph of $y = ax^2 + bx + c$ will have ____ (how many?) crossings of the x-axis.

tangent

40. If the equation $ax^2 + bx + c = 0$ has one real solution, the graph of $y = ax^2 + bx + c$ will be ________ to the x-axis.

2

41. If the equation $ax^2 + bx + c = 0$ has two real solutions, the graph of $y = ax^2 + bx + c$ will have ___ crossings of the x-axis.

In Frames 42-46, graph the inequality $y < 2x^2$.

are not

42. First sketch the graph of the equality $y = 2x^2$; but, since the relationship "less than" (<) is involved in the required graph, make the sketch with a broken line to show that points on this curve ________ (are/are not) included.

parabola; upward

43. Using previously discussed techniques, observe that $y = 2x^2$ will graph as a _________ opening ________ (upward/downward).

$0 < 2$

44. Substitute the coordinates of the test point (1,0) into $y < 2x^2$ and obtain the statement _______.

true

45. Shade in the region containing this point because the statement in Frame 44 is ______ (true/false).

46. The required graph is

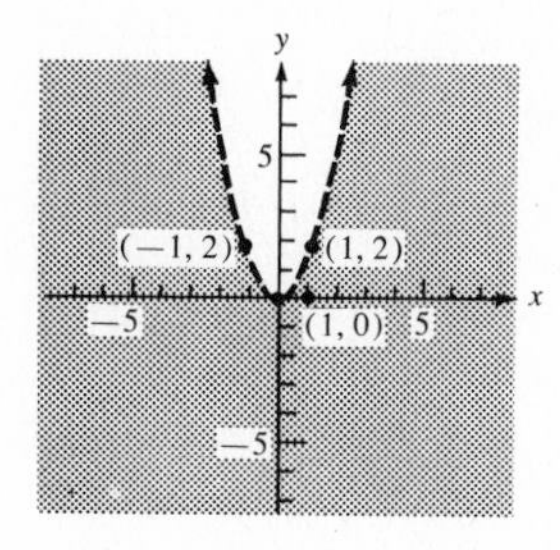

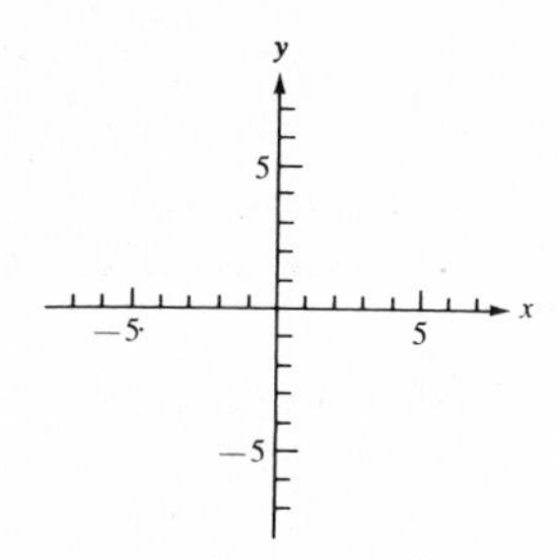

In Frames 47-51, graph the inequality $-2x^2 + 3x - 1 \geq y$.

$y = -2x^2 + 3x - 1$

are

47. Sketch the graph of the equality ________________ and, since the relationship $\geq$ is involved in the required graph, make the sketch with a solid line to show that points on this curve _____ (are/are not) included.

parabola
downward; 1/2; 1
-1

48. Observe that $y = -2x^2 + 3x - 1$ will graph as a ___________ opening __________ with x-intercepts of _____ and ___ and a y-intercept of ____.

$-1 \geq 0$

49. Use (0,0) as a test point and substitute its coordinates into $-2x^2 + 3x - 1 \geq y$ and obtain the statement __________.

false

50. Shade the region not containing (0,0), because the statement of Frame 49 is _______ (true/false).

51. The required graph is

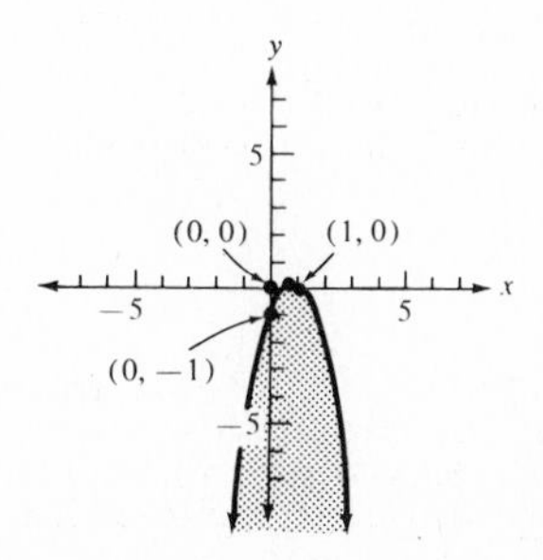

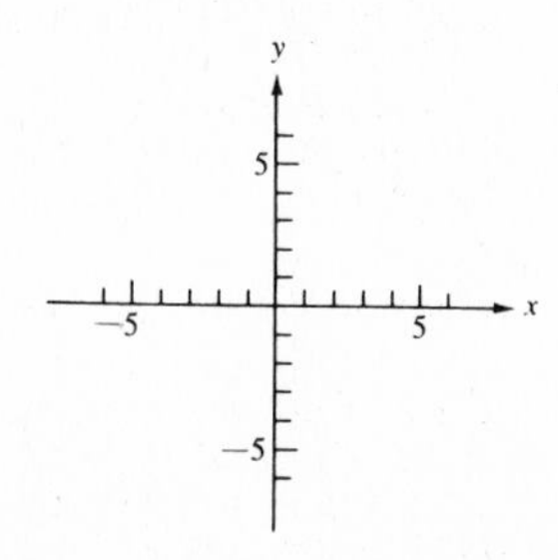

In Frames 52-54, graph $\{(x, y) | y = x^2 - 4\}$ $\{(x, y) | y = -x^2 + 1\}$.

parabola
upward; 2
-2; -4
-4

52. The graph of $\{(x,y) | y = x^2 - 4\}$ is a ________ opening ________ (upward/downward) with x-intercepts of ____ and ____, and y-intercept of ____. The maximum point is given by (0, ____). (See Frame 27.)

parabola
downward; 1
-1; 1
1

53. The graph of $\{(x,y) | y = -x^2 + 1\}$ is a ________ opening ________ (upward/downward) with x-intercepts of ____ and ____, and y-intercept of ___. The minimum point is given by (0, __).

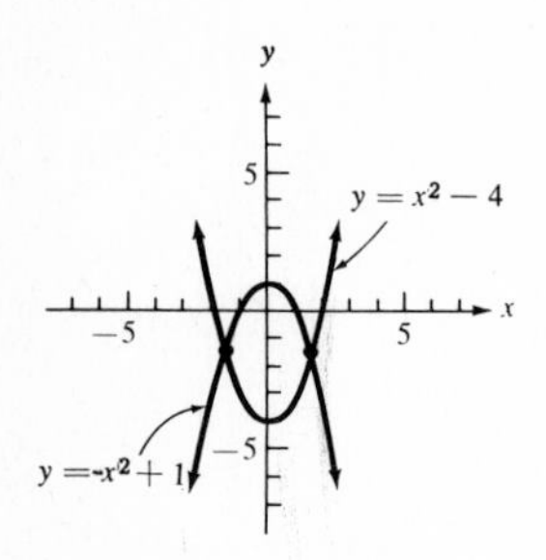

54. The graphs are

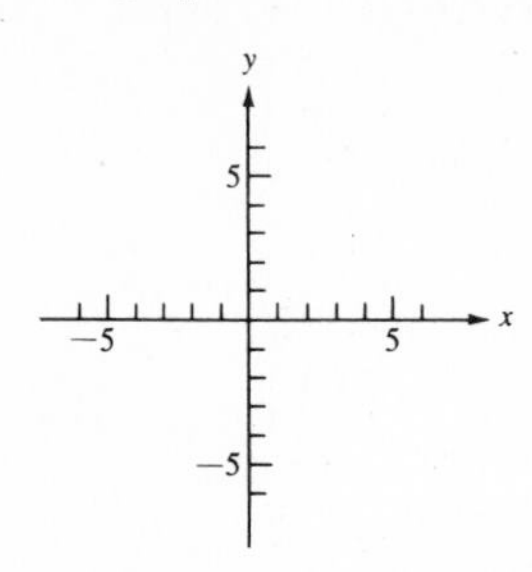

2

The graph of the intersection of the given sets consists of the ___ (how many?) points where the graphs intersect.

In Frames 55- 59, graph $\{(x,y) | y \geq x^2 - 4\} \cup \{(x,y) | y \leq -x^2 + 1\}$.

55. As in Frames 52 and 53 graph $\{(x,y) | y = x^2 - 4\}$ and $\{(x,y) y = -x^2 + 1\}$.

56. Since neither graph contains the origin, use (0,0) as a test point for testing each inequality.

true
above

57. Substituting (0,0) into $y \geq x^2 - 4$, we obtain $0 \geq 0^2 - 4$, which is ______ (true/false). Shade the region containing (0,0), that is the region _______ (above/below) the graph of $y = x^2 - 4$.

below

58. Similarly, we test $y \leq -x^2 + 1$ with (0,0) and find that we shade the region _______ (above/below) the graph of $y = -x^2 + 1$ with a different kind of shading from that referred to in Frame 57.

59. The graphs are

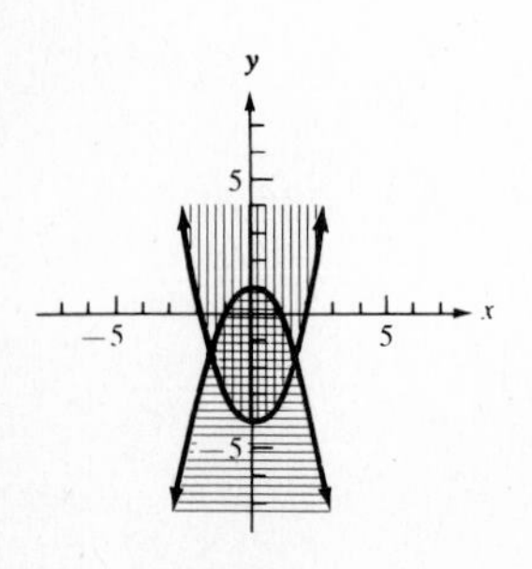

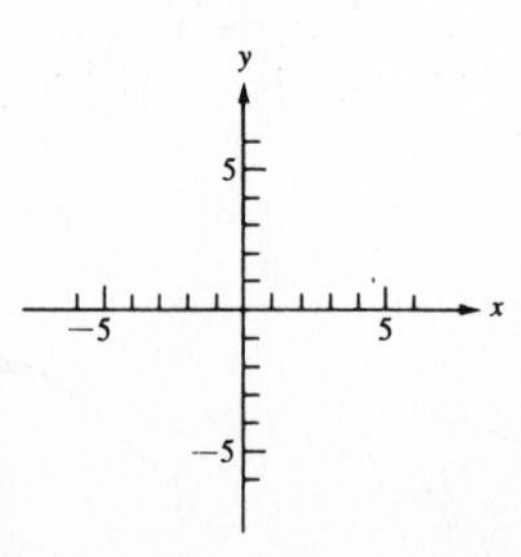

The graph of the union of the given sets is the region shaded both ways.

8.3

CIRCLES, ELLIPSES, AND HYPERBOLAS

In Frames 60-64, graph $x^2 + y^2 = 16$.

$\pm\sqrt{16 - x^2}$

60. Rewriting $x^2 + y^2 = 16$ with y as the left-hand member, obtain y = ___________.

0; $\{x|-4 \le x \le 4\}$

61. The domain of this relation is the solution set of $16 - x^2 \ge$ ___ which is _______________.

$\pm 2\sqrt{3}$; $\pm 2\sqrt{3}$; $\pm\sqrt{15}$
$\pm\sqrt{15}$; ± 4

62. Obtaining solutions to $y = \pm\sqrt{16 - x^2}$, restricting substitutions for x to the above domain, we have (-4,0), (4,0), $(-3,\pm\sqrt{7})$, $(3,\pm\sqrt{7})$, (-2,______), (2,______), (-1,______), (1,______), (0,____).

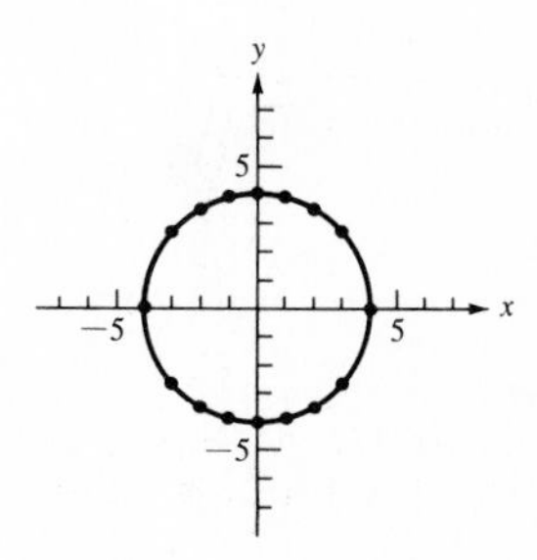

63. The graph is

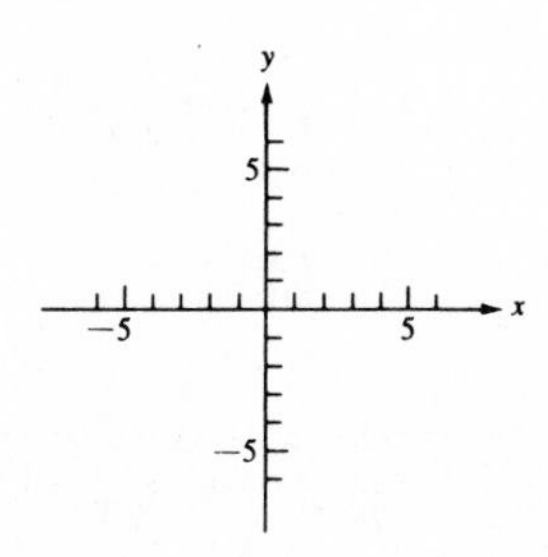

circle

64. The curve we have obtained is a ________.

In Frames 65-69, graph $x^2 + 4y^2 = 36$.

$y = \pm\frac{1}{2}\sqrt{36 - x^2}$

65. Rewriting $x^2 + 4y^2 = 36$ with y as the left-hand member, obtain _______________.

$36 - x^2 \ge 0$
$\{x|-6 \le x \le 6\}$

66. The domain of this relation is the solution set of _____________ which is _______________.

0; 0

$\pm 2\sqrt{2}$

$\pm\frac{1}{2}\sqrt{35}$; $\pm\frac{1}{2}\sqrt{35}$; ± 3

67. Obtaining solutions to $y = \pm\frac{1}{2}\sqrt{36 - x^2}$, restricting substitutions for x to the above domain, we have (-6,___), (6,___), $\left(-5,\pm\frac{1}{2}\sqrt{11}\right)$, $\left(5,\pm\frac{1}{2}\sqrt{11}\right)$, $(-4,\pm\sqrt{5})$, $(4,\pm\sqrt{5})$, $\left(-3,\pm\frac{3}{2}\sqrt{3}\right)$, $\left(3,\pm\frac{3}{2}\sqrt{3}\right)$, $(-2,\pm 2\sqrt{2})$, (2,______), (-1,________), (1,________), (0,____).

68. The graph is

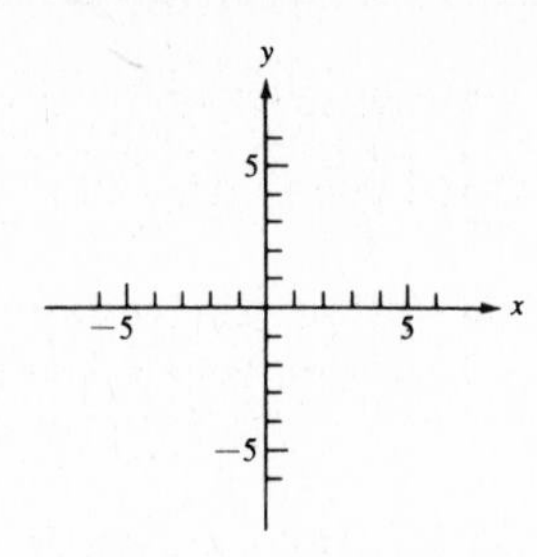

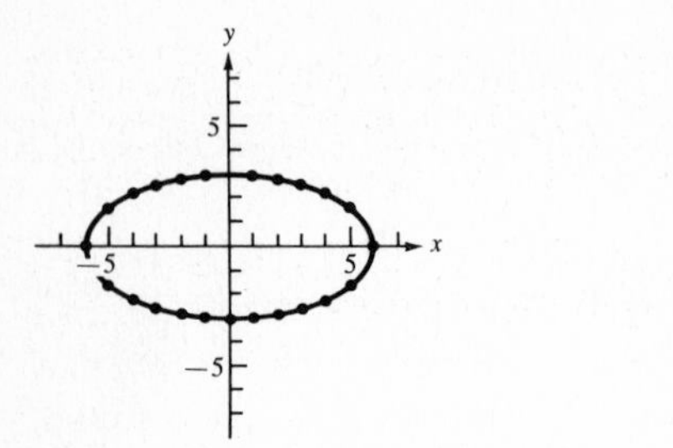

69. The curve we have obtained is called an ________.

ellipse

In Frames 70-74, graph $x^2 - y^2 = 16$.

70. Rewriting $x^2 - y^2 = 16$ with y as the left-hand member, obtain ______________.

$y = \pm\sqrt{x^2 - 16}$

71. The domain of this relation is the solution set of ____________ and is ________________________.

$x^2 - 16 \geq 0$
$\{x|x > 4\}$ or $\{x|x \leq -4\}$

72. Obtaining solutions to $y = \pm\sqrt{x^2 - 16}$, restricting solutions to the above domain, we have $(-4,0)$, $(-5,\pm 3)$, $(-6,\pm 2\sqrt{5})$, $(4,\_\_\_)$, $(5,\_\_\_\_)$, $(6,\_\_\_\_\_\_)$.

0; ± 3; $\pm 2\sqrt{5}$

73. The graph is

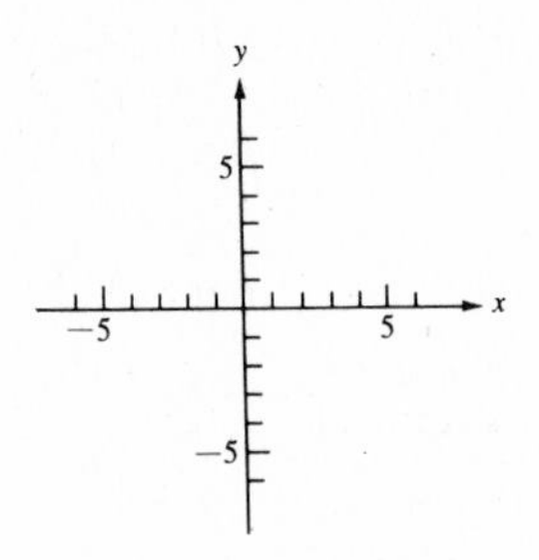

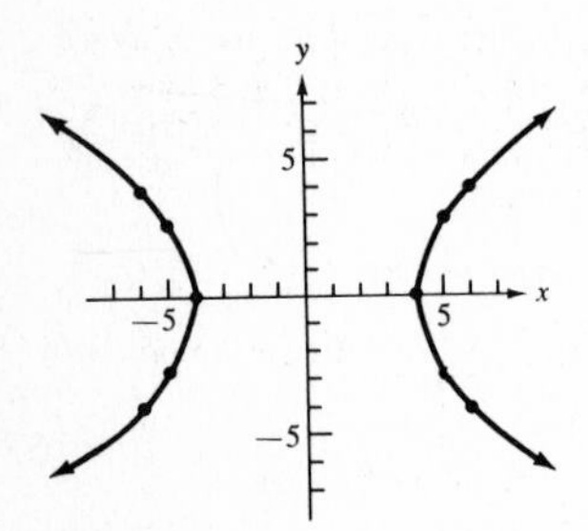

74. The curve we have obtained is called a __________.

hyperbola

75. The graphs of the equations dealt with in this section together with the parabola are called conics, or conic sections, because they are intersections of a plane and a ______.

cone

8.4

SKETCHING GRAPHS OF CONIC SECTIONS OTHER THAN PARABOLAS

76. If an equation of the form $ax^2 + by^2 = c$ $(a^2 + b^2 \neq 0)$ is graphed and $a = b$ and a, b, and c have like signs, the graph will be a ________.

circle

77. If an equation of the form $ax^2 + by^2 = c$ $(a^2 + b^2 \neq 0)$ is graphed and $a \neq b$ and a, b, and c have like signs, the graph is an ________.

ellipse

78. If an equation of the form $ax^2 + by^2 = c$ $(a^2 + b^2 \neq 0)$ is graphed and a and b have opposite signs and $c \neq 0$, the graph is a ___________.

hyperbola

79. If an equation of the form $ax^2 + by^2 = c$ $(a^2 + b^2 \neq 0)$ is graphed and a and b have opposite signs and $c = 0$, the graph is two distinct _______ through the ________.

lines; origin

80. If an equation of the form $ax^2 + by^2 = c$ $(a^2 + b^2 \neq 0)$ is graphed and a and b are both positive or both negative and $c = 0$, the graph is a _______.

point

81. If an equation of the form $ax^2 + by^2 = c$ $(a^2 + b^2 \neq 0)$ is to be graphed and a and b are both positive and c is negative or if both a and b are ≤ 0 and $c > 0$, the graph __________ (does/does not) exist.

does not

82. Once the general form of the curve is recognized, the graph can be sketched with knowledge of a few points which should include the ___________, if they exist.

intercepts

In Frames 83-84, name and sketch the graph of $x^2 + y^2 = 1$.

83. $a = b =$ ___ and a, b, and c have the ______ (same/opposite) signs. (See Frame 76.)

1; same

84. Therefore, the graph is a ________ with x- and y-intercepts of ____.

circle
±1

85. The sketch is

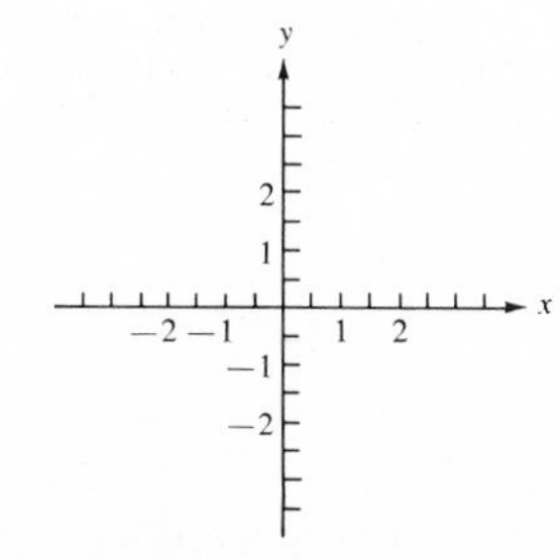

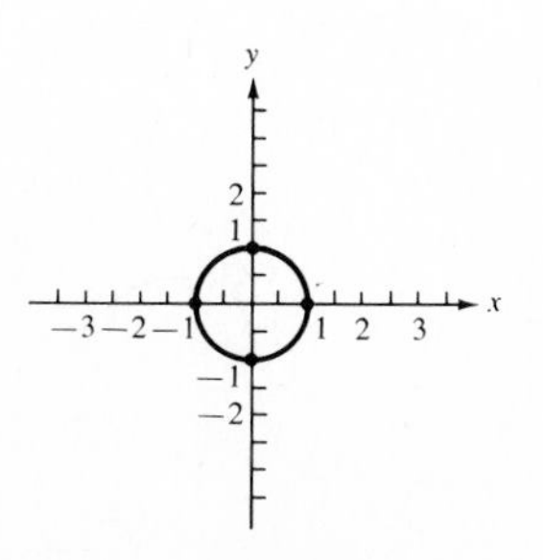

In Frames 86– 89, name and sketch the graph of $2y^2 = 8 - x^2$.

$x^2 + 2y^2 = 8$

86. Express the equation in the form $ax^2 + by^2 = c$ and obtain ______________.

1; 2; 8
same
ellipse

87. From Frame 86, observe that $a =$ ___, $b =$ ___, and $c =$ ___; that is, $a \neq b$ and a, b, and c have ______ (same/opposite) signs, so the graph is an _________. (See Frame 77.)

2; -2
$2\sqrt{2}$; $-2\sqrt{2}$

88. The y-intercepts are ___ and ____, and the x-intercepts are _____ and ______.

89. The sketch is

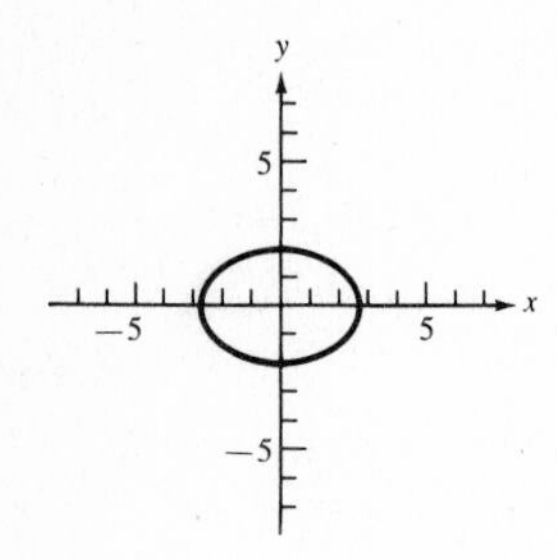

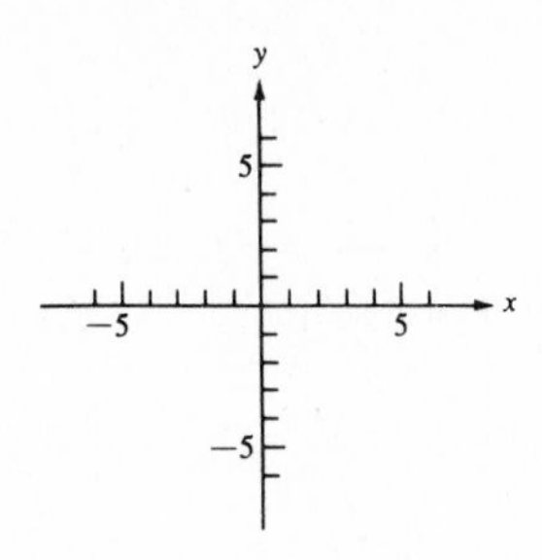

In Frames 90– 94 name and sketch the graph of $3x^2 = 12 + 4y^2$.

$3x^2 - 4y^2 = 12$

90. Express the equation in the form $ax^2 + by^2 = c$ and obtain ______________.

3; -4
12; opposite
0
hyperbola

91. From Frame 90, observe that $a =$ ____, $b =$ _____, and $c =$ ____; that is, a and b are __________ (same/opposite) in sign and $c \neq$ ___. So, the graph is a ____________.

2; -2

92. The only intercepts are x-intercepts of ___ and ____.

3; -3
3; -3

93. Other convenient points in the graph are (4,___),(4,____), (-4,___),(-4,____).

94. The sketch is

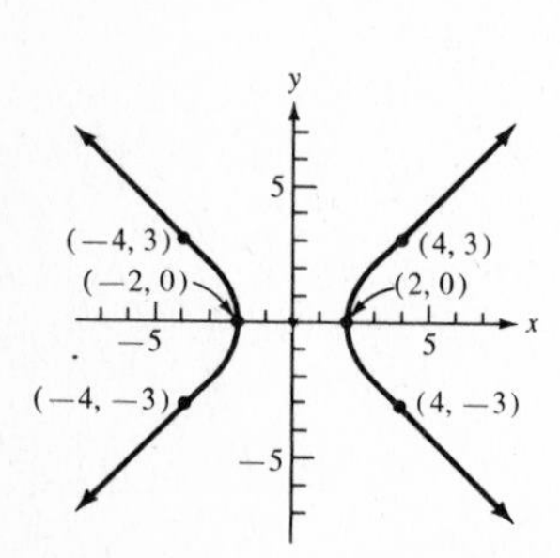

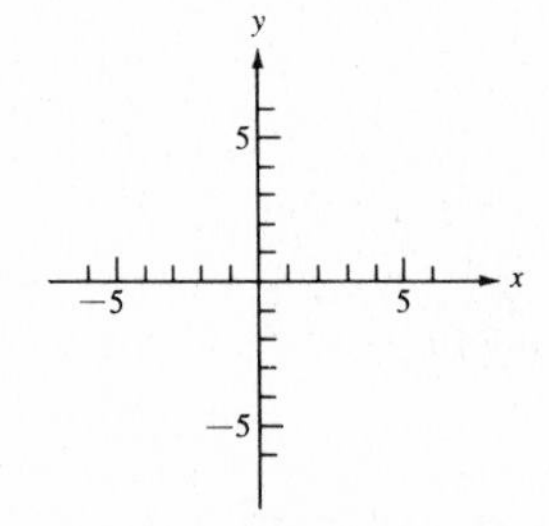

In Frames 95-97, name and sketch the graph of $9x^2 - 4y^2 = 0$.

9; -4; 0
opposite
lines
origin

95. Compare the equation with the form $ax^2 + by^2 = c$ and observe that a = ___, b = ___, and c = ___, that is, a and b are __________(same/opposite) in sign, and that $c = 0$; so the graph is two distinct _______ through the ________. (See Frame 79.)

$3x + 2y$; $3x - 2y$
$3x - 2y = 0$

96. $9x^2 - 4y^2$ = (_________)(_________) and graph each of the equations $3x + 2y = 0$ and ____________.

97. The graph is

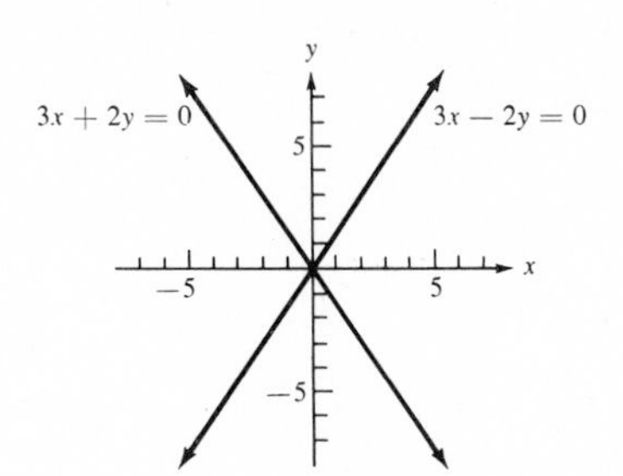

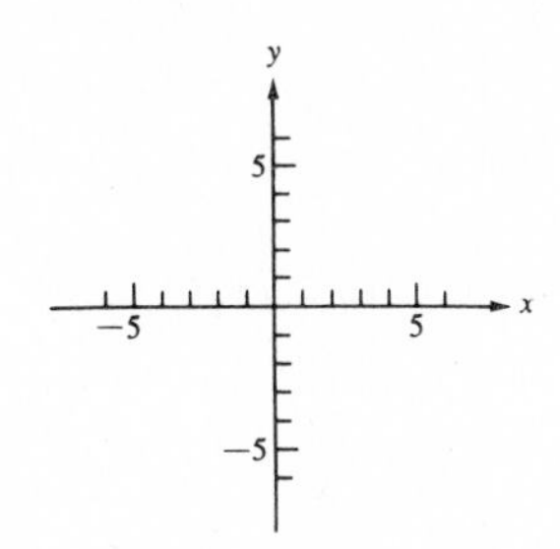

In Frames 98-100, name and sketch the graph of $2x^2 + 3y^2 + 6 = 0$.

$2x^2 + 3y^2 = -6$

98. Express the equation in the form $ax^2 + by^2 = c$.

2; 3
-6; < 0

99. From Frame 98, observe that a = ___, b = ___, and c = ____; that is, $a > 0$, $b > 0$, and c _____. (See Frame 81.)

does not

100. Therefore, the graph _________ (does/does not) exist.

ellipse

hyperbola

x
y

101. The form $\frac{x^2}{a^2} + \frac{y^2}{b^2} = 1$ is standard form for the equation of the ___________. The forms $\frac{x^2}{a^2} - \frac{y^2}{b^2} = 1$ or $\frac{y^2}{b^2} - \frac{x^2}{a^2} = 1$ are standard forms for the equation of the __________. When the equations are in this form, a and $-a$ are the ___-intercepts when such intercepts exist and b and $-b$ are the ___-intercepts when such intercepts exist.

In Frames 102-105, express $9x^2 + 4y^2 = 25$ in standard form and sketch the graph.

25; 1

102. Divide each term by the right-hand member ___ and obtain $\frac{9x^2}{25} + \frac{4y^2}{25} = $ ___.

9; 4

$\frac{5}{2}$

103. Divide the numerator and denominator of each of the fractions by the coefficient of the numerator to obtain

$$\frac{\frac{9x^2}{9}}{\frac{25}{(\quad)}} + \frac{\frac{4y^2}{4}}{\frac{25}{(\quad)}} = 1$$

$$\frac{x^2}{(\frac{5}{3})^2} \quad \frac{y^2}{(\quad)^2} = 1$$

$\frac{5}{3}$; $\frac{-5}{3}$

$\frac{5}{2}$; $\frac{-5}{2}$

104. Hence the x-intercepts are ___ and ___ and the y-intercepts are ____ and ____.

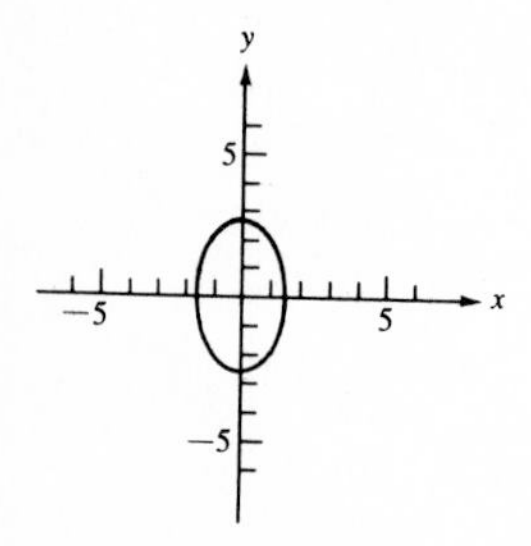

105. The graph is

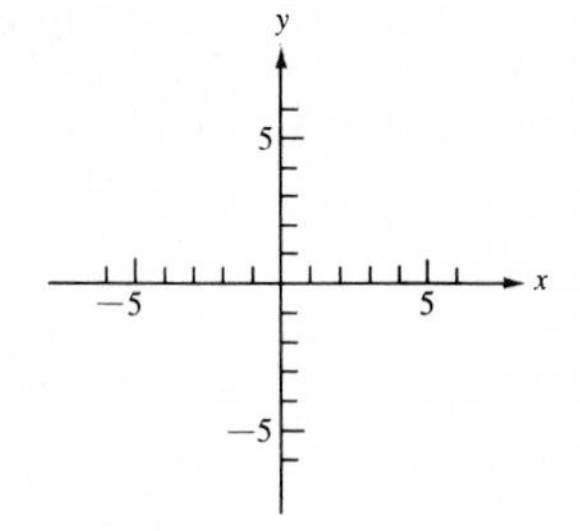

0

0

asymptotes

106. The graph of the hyperbola $ax^2 - by^2 = c$ will "approach" the two straight lines that are the graph of $ax^2 - by^2 = $ ___, and the graph of the hyperbola $ay^2 - bx^2 = c$ will "approach" the two straight lines that are the graph of $ay^2 - bx^2 = $ ____ $(a, b, c > 0)$. The straight lines "approached" are called ____________.

In Frames 107-112, express $4x^2 - y^2 = 9$ in standard form and sketch the graph.

9; 1

107. From Frame 78, the graph is a hyperbola. Divide each term by the right-hand member ___ and obtain $\frac{4x^2}{9} - \frac{y^2}{9} = $ ___.

108. Divide the numerator and denominator of $\frac{4x^2}{9}$ by the coefficient of the numerator to obtain

$$\frac{\frac{4x^2}{4}}{\frac{9}{(\_\_\_)}} - \frac{y^2}{9} = 1 \quad \text{or} \quad \frac{x^2}{(\_\_\_)^2} - \frac{y^2}{3^2} = 1.$$

$4; \frac{3}{2}$

109. It is seen that when $x = 0$, $y^2 = -9$. Hence y is an __________ number and there is no y-intercept.

imaginary

110. The x-intercepts are ___ and ___.

$\frac{3}{2}$; $\frac{-3}{2}$

111. The asymptotes are the graph of $4x^2 - y^2 =$ ___, which can be written as $y^2 =$ ___ or $y =$ ____.

0
$4x^2$; $\pm 2x$

112. The asymptotes are graphed first. Then, the x-intercepts are graphed. The sketch is completed using the asymptotes as guides. The graph is

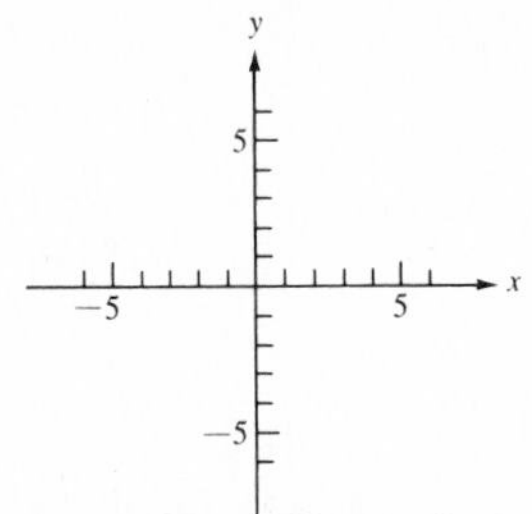

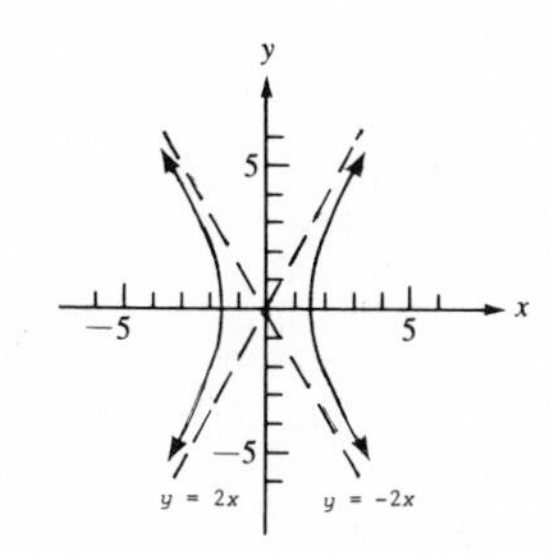

8.5

VARIATION AS A FUNCTIONAL RELATIONSHIP

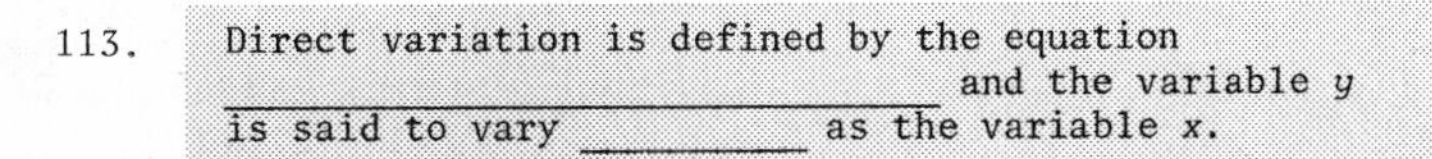

113. Direct variation is defined by the equation ________________________ and the variable y is said to vary __________ as the variable x.

$y = kx$ (k a positive constant)
directly

114. In general, $y = kx^n$ (k a positive constant and $n > 0$) asserts that y varies __________ as the _____ power of ___.

directly; nth
x

In Frames 115-117, write an equation expressing the relationship between the variables if the interest (I) varies directly as the time (t) and I = 50 when t = 2.

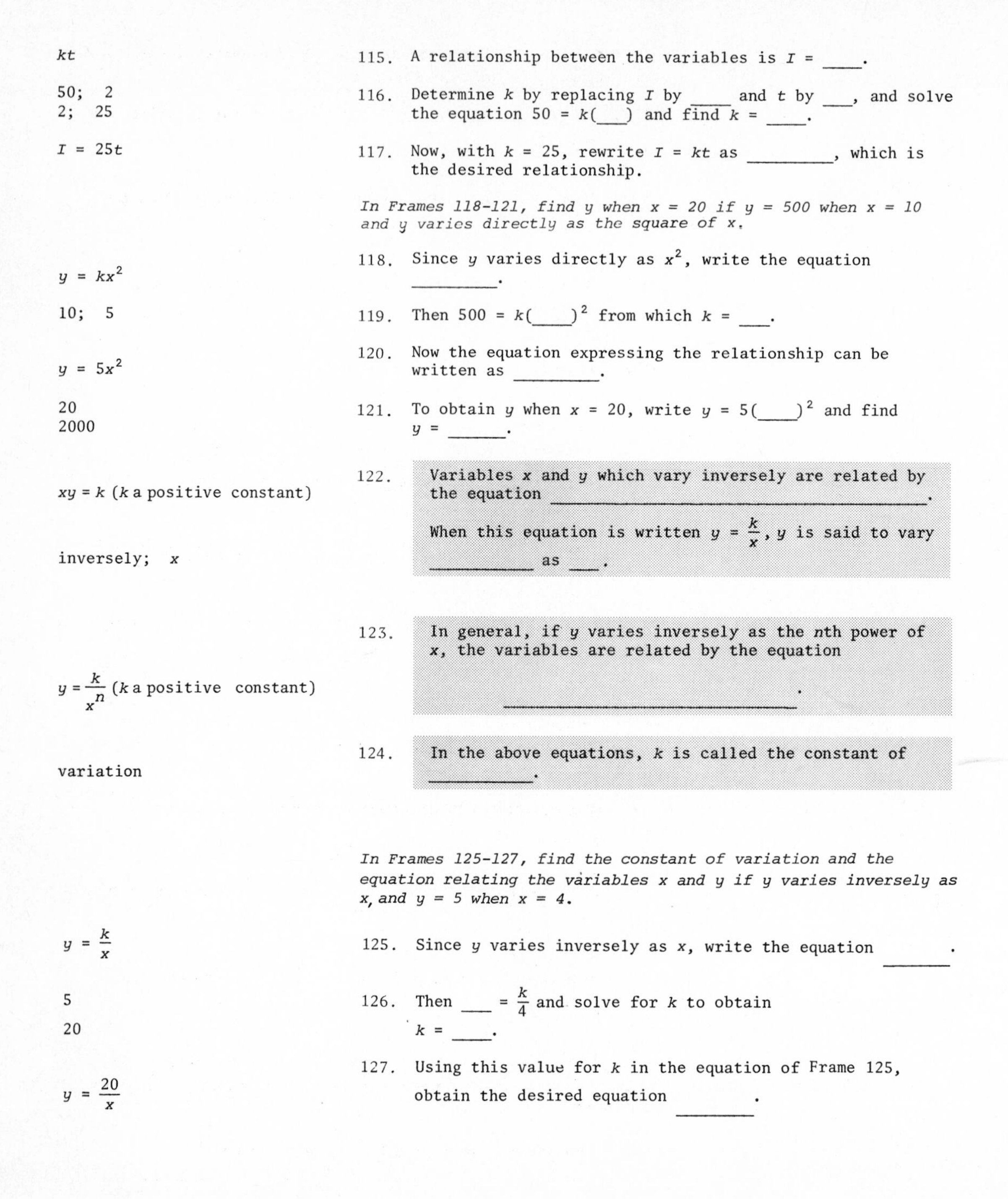

Answer	Frame
kt	115. A relationship between the variables is I = ____.
50; 2 2; 25	116. Determine k by replacing I by ____ and t by ___, and solve the equation $50 = k($___$)$ and find k = ____.
$I = 25t$	117. Now, with $k = 25$, rewrite $I = kt$ as _________, which is the desired relationship.

In Frames 118-121, find y when $x = 20$ if $y = 500$ when $x = 10$ and y varies directly as the square of x.

Answer	Frame
$y = kx^2$	118. Since y varies directly as x^2, write the equation _________.
10; 5	119. Then $500 = k($____$)^2$ from which k = ___.
$y = 5x^2$	120. Now the equation expressing the relationship can be written as _________.
20 2000	121. To obtain y when $x = 20$, write $y = 5($____$)^2$ and find y = ______.
$xy = k$ (k a positive constant) inversely; x	122. **Variables x and y which vary inversely are related by the equation ________________. When this equation is written $y = \frac{k}{x}$, y is said to vary __________ as ___.**
$y = \frac{k}{x^n}$ (k a positive constant)	123. **In general, if y varies inversely as the nth power of x, the variables are related by the equation ________________.**
variation	124. **In the above equations, k is called the constant of __________.**

In Frames 125-127, find the constant of variation and the equation relating the variables x and y if y varies inversely as x, and $y = 5$ when $x = 4$.

Answer	Frame
$y = \frac{k}{x}$	125. Since y varies inversely as x, write the equation _______.
5 20	126. Then ___ $= \frac{k}{4}$ and solve for k to obtain k = ____.
$y = \frac{20}{x}$	127. Using this value for k in the equation of Frame 125, obtain the desired equation ________.

In Frames 128- 131 find y when $x = 20$ if y varies inversely as the square of x and $y = 4$ when $x = 10$.

$y = \frac{k}{x^2}$

128. Since y varies inversely as the square of x, write the equation ________.

4; 10

129. Then determine the constant of variation, k, from the equation ___ $= \frac{k}{(\_\_\_)^2}$.

400

$y = \frac{400}{x^2}$

130. Solve for k and obtain $k =$ _____; so the equation of Frame 128 can be written as ________.

20; 1

131. When $x = 20$, the equation of Frame 130 says $y = \frac{400}{(\_\_\_)^2}$. So $y =$ ___.

joint

132. In the event that one variable varies as the product of two or more variables, the relationship is referred to as _______ variation.

$kuvw$

133. If y varies jointly as u, v, and w, then $y =$ ________.

In Frames 134-137, find y when $u = 2$, $v = 3$, and $w = 4$ if y varies jointly as u, v, and w and $y = 24$ when $u = \frac{1}{2}$, $v = 4$, and $w = 8$.

$y = kuvw$

134. Since y varies jointly as u, v, and w, write the equation __________.

$\frac{1}{2}$; 4; 8

135. Then determine the constant of variation, k, from the equation $24 = k(\_\_\_)(\_\_\_)(\_\_\_)$.

$\frac{24}{16}$ or $\frac{3}{2}$

$y = \frac{3}{2}uvw$

136. From Frame 135, $k =$ ______. So the equation of Frame 134 can be written as __________.

2; 3; 4; 36

137. When $u = 2$, $v = 3$, and $w = 4$, $y = \frac{3}{2}uvw$ becomes $y = \frac{3}{2}(\_\_\_)(\_\_\_)(\_\_\_)$. So $y =$ ____.

In Frames 138-141, find y when $x = 10$ and $z = 3$ if y varies directly as x and inversely as the cube of z, and $y = 4$ when $x = 2$ and $z = 4$.

$y = \frac{kx}{z^3}$

138. The given statement defines the equation ________.

Answer	Frame
4	139. The constant of variation, k, is determined by $4 = \dfrac{k(2)}{(\_\_\_)^3}$.
128; $y = \dfrac{128x}{z^3}$	140. Solving this equation for k, obtain $k =$ _____ and rewrite the equation of Frame 138 as __________.
3; $\dfrac{1280}{27}$	141. Then, $y = \dfrac{128(10)}{(\_\_\_)^3}$. So $y =$ ______.
proportional	142. **Another way of describing direct or inverse variation is to say that y is directly or inversely __________ to x.**

In Frames 143-148, represent the following relationship as a proportion and then solve for the required variable. If y varies directly as x and inversely as the cube of z, and $y = 4$ when $x = 2$ and $z = 4$, find y when $x = 10$ and $z = 3$. (Note: this is the same problem as that worked in Frames 138-141.)

Answer	Frame
$\dfrac{kx}{z^3}$	143. Express the variation relationship by means of the equation $y =$ ____.
$\dfrac{yz^3}{x}$	144. Solve the above equation explicitly for k and obtain $k =$ _____.
$\dfrac{y_2 z_2^3}{x_2}$	145. Let (x_1, y_1, z_1) and (x_2, y_2, z_2) be two solutions of $k = \dfrac{yz^3}{x}$ and obtain the two equations $k = \dfrac{y_1 z_1^3}{x_1}$ and $k =$ ______.
$\dfrac{y_1 z_1^3}{x_1} = \dfrac{y_2 z_2^3}{x_2}$	146. Equating these two expressions for k, obtain the required proportion ____________.
4; 3; $\dfrac{y(3)^3}{10}$	147. Now, in the above proportion, from the given statement take (x_1, y_1, z_1) as $(2, 4, \_\_\_)$ and (x_2, y_2, z_2) as $(10, y, \_\_\_)$ and obtain $\dfrac{4 \cdot 4^3}{2} =$ ______.
$\dfrac{1280}{27}$	148. Solve for y and obtain $y =$ ______.

8.6

THE INVERSE OF A FUNCTION

149. If the components of each ordered pair in a given relation are interchanged, the resulting relation and the given relation are said to be ________ of each other.

inverses

150. If q names a relation, then the inverse of q is denoted by ______.

q^{-1}

151. If $q = \{(1, 2), (-1, -1), (2, -3), (4, 5)\}$, then $q^{-1} =$ ______________________.

$\{(2, 1)(-1, -1), (-3, 2)(5, 4)\}$

152. The domain and range of q are the ________ and ________, respectively, of q^{-1} if q and q^{-1} are functions.

range
domain

153. In Frame 151, the domain of q is {________} and the range of q is {________}, while the domain of q^{-1} is {________} and the range of q^{-1} is {__________}.

1,-1,2,4
2,-1,-3,5
2,-1,-3,5; 1,-1,2,4

154. If $y = q(x)$ defines a relation q, then q^{-1} is defined by the equation $x =$ ________ which, when solved explicitly for y, it is written $y =$ __________.

$q(y)$

$q^{-1}(x)$

155. If q is defined by $y = 3x-1$, then q^{-1} is defined by $x =$ ________ or, $y =$ __________.

$3y - 1$ $\frac{1}{3}(x + 1)$

156. If q is defined by $y = 4x^2$, then q^{-1} is defined by $x =$ __________ or, $y =$ ________.

$4y^2$ $\pm\frac{1}{2}\sqrt{x}$

157. The graphs of $y = q(x)$ and $y = q^{-1}(x)$ are always located symmetrically with respect to the graph of $y =$ ___.

x

158. In order for a function f to have an inverse which is a function, each element in its range must be associated with but one element in its ______. Such a function is called a one-to-____ function.

domain
one

In Frames 159-161, refer to $2x - 3y = 6$, which defines a relation q.

$2y - 3x = 6$

159. q^{-1} is defined by the equation ________.

160. The graphs q and q^{-1} are

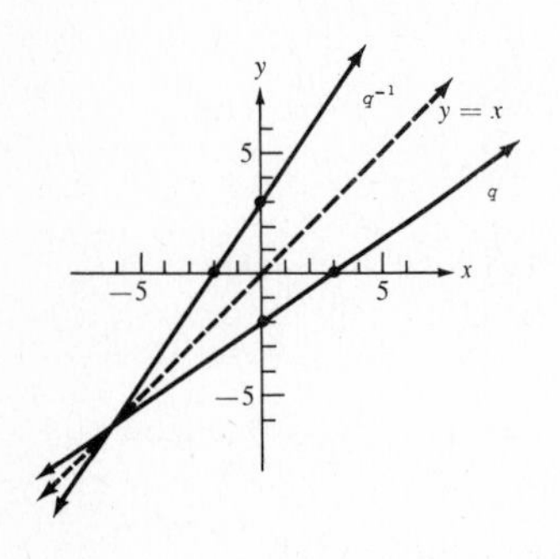

function

161. Because the graph of q^{-1} (or the equation) shows that each element in the domain of q^{-1} is associated with only one element in its range, q^{-1} is a ________.

Frames 162-164 refer to $y = x^2 - 1$, which defines a relation p.

$x = y^2 - 1$

162. p^{-1} is defined by the equation ________.

163. The graphs of p and p^{-1} (obtained by the methods of Section 8.2) are

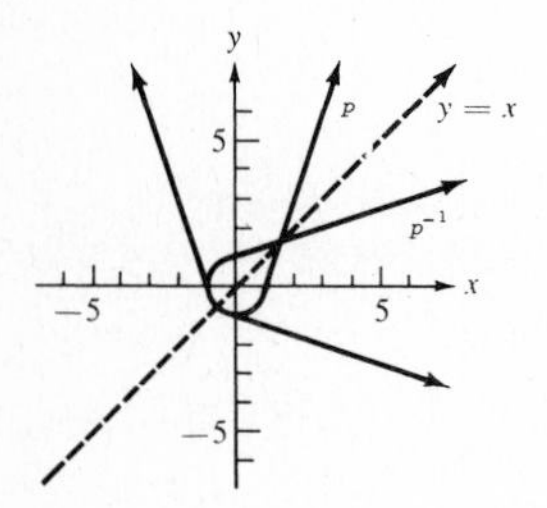

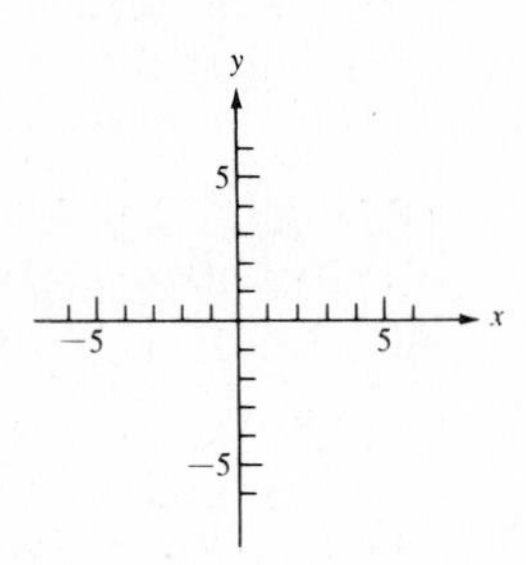

two

164. p^{-1} is not a function, because the graph (or the equation) of p^{-1} shows that each element in the domain of p^{-1} is paired with _____(how many?) elements in the range.

In Frames 165-167, refer to $x^2 + 9y^2 = 36$, which defines a relation t.

$y^2 + 9x^2 = 36$

165. t^{-1} is defined by the equation ________.

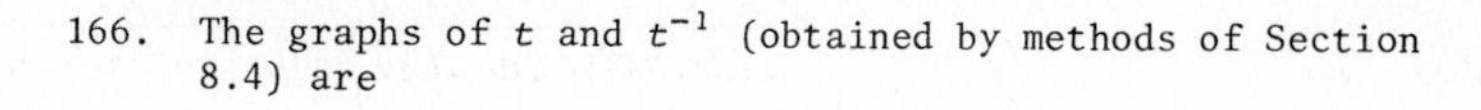

166. The graphs of t and t^{-1} (obtained by methods of Section 8.4) are

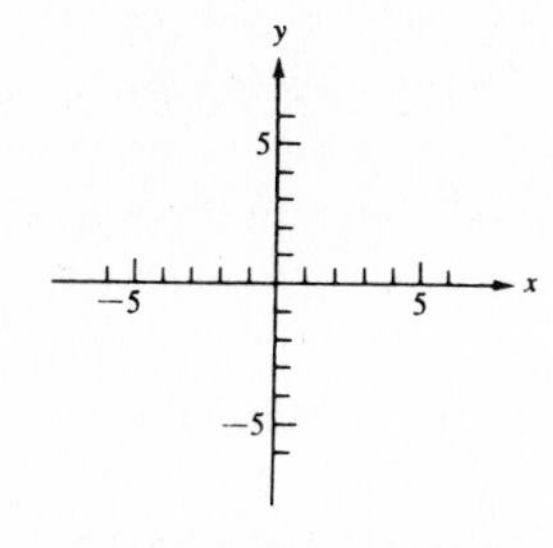

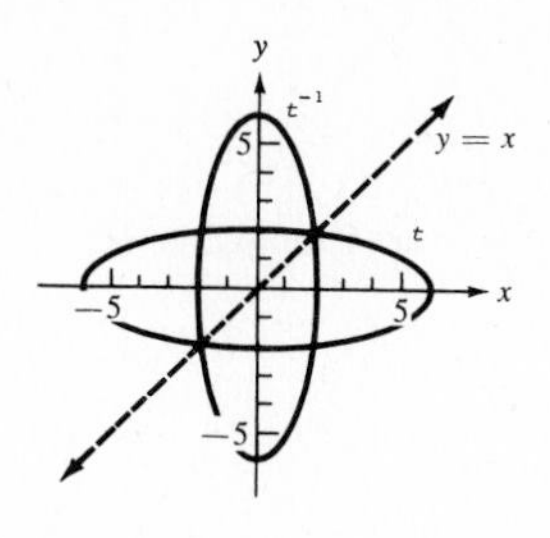

167. t^{-1} is not a function, because the graph (or the equation) shows that each element in the domain (except $x = 2, -2$) of t^{-1} is paired with ____ (how many?) in the range.

two

168. If f is a one-to-one function, then $f^{-1}[f(x)] = f[f^{-1}(x)] =$ ____.

x

In Frames 169-176, use $4x + 3y = 12$, which defines a one-to-one function f, and show that $f^{-1}[f(x)] = f[f^{-1}(x)] = x$.

169. Solve $4x + 3y = 12$ explicitly for y and obtain $y =$ ________, and since this equation defines f, write $f(x) =$ ________.

$\frac{-4}{3}x + 4$

$\frac{-4}{3}x + 4$

170. Now find the equation defining f^{-1} by interchanging the roles of x and y in $4x + 3y = 12$ to obtain ________ and then solve explicitly for y and let $y = f^{-1}(x)$ to obtain $f^{-1}(x) =$ ________.

$4y + 3x = 12$

$-\frac{3}{4}x + 3$

171. $f^{-1}[f(x)]$ is obtained by replacing every x in the equation defining f^{-1} by ________. Therefore,

$f^{-1}[f(x)] = -\frac{3}{4}[\quad\quad] + 3$ or, since $f(x) = -\frac{4}{3}x + 4$,

$f^{-1}[f(x)] = -\frac{3}{4}[\quad\quad] + 3$.

$f(x)$

$f(x)$

$-\frac{4}{3}x + 4$

172. Simplify the last equation and obtain

$f^{-1}[f(x)] = (x - 3) + 3 =$ ___.

x

173. Frame 172 establishes that $f^{-1}[f(x)] = x$. Now, to show that $f[f^{-1}(x)] = x$ in the equation defining f substitute $f^{-1}(x)$ for every x to obtain $f[f^{-1}(x)] = -\frac{4}{3}[\quad\quad] + 4$.

$f^{-1}(x)$

174. Since $f^{-1}(x) = -\frac{3}{4}x + 3$, $f[f^{-1}(x)]$ can be written as

$f[f^{-1}(x)] = -\frac{4}{3}[\quad\quad] + 4$.

$-\frac{3}{4}x + 3$

175. Simplifying the last equation, obtain $f[f^{-1}(x)] = (\quad\quad) + 4 =$ ___.

$x - 4$; x

176. Comparing the results of Frames 172 and 175, it can be seen that $f^{-1}[f(x)] = f[f^{-1}(x)] =$ ___.

x

TEST PROBLEMS

1. Find the x- and y-intercepts of the graph of $f(x) = x^2 + x - 6$.

2. Find the coordinates of the maximum point on the graph of $y = -3x^2 + 2x$.

3. Graph the function of Problem 2 above.

4. State the domain of the function defined by $x^2 + y^2 = 49$.

5. Name the graph of each of the following equations.
   a. $x^2 = 3 + 5y^2$.
   b. $3y - 2x^2 + 2 = 0$.
   c. $x^2 = 9 - 5y^2$.

In Problems 6-9, graph each of the given relations.

6. $9x^2 + y^2 = 36$.

7. $4x^2 - 9y^2 = 36$.

8. $xy = -6$.

9. $9y^2 - 4x^2 = 0$.

10. If y varies inversely as t^2 and $y = 2$ when $t = 2$, find y when $t = 3$.

11. The resistance (R) of a wire varies directly as the length (l) and inversely as the square of its diameter (d). Twenty-five feet of wire of diameter 0.006 inches has a resistance of 100 ohms. What is the resistance of 100 feet of the same type of wire whose diameter is 0.012 inches?

In Problems 12-15, graph each of the given relations.

12. $\{(x,y)\,|\,y = x + 4\} \cap \{(x,y)\,|\,x^2 + y^2 = 49\}$.

13. $\{(x,y)\,|\,y = x + 4\} \cup \{(x,y)\,|\,9x^2 + y^2 = 36\}$.

14. $\{(x,y)\,|\,4x^2 + y^2 \geq 36\}$.

15. $2x - 3y = 6$ defines a relation h in $r \times r$.
    a. Write an equation defining h^{-1}.
    b. Graph h and h^{-1} on the same axis.
    c. State whether or not h^{-1} is a function.

9

EXPONENTIAL AND LOGARITHMIC FUNCTIONS

9.1

THE EXPONENTIAL FUNCTION

x

1. The exponential function is defined by the equation $f(x) = b^{(\ \ )}$ $(b > 0,\ b \neq 1)$.

solution

2. The graph of the exponential function is obtained by graphing ordered pairs in the equation's ________ set.

In Frames 3-5, find the second component of each ordered pair (0,?),(1,?),(2,?), that makes the pair a solution of $y = 6^x$.

1

3. Replace x by 0 in $y = 6^x$ and obtain $y = 6^0 =$ ___.

1; 6

4. Replace x by 1 in $y = 6^x$ and obtain $y = 6^{(\ \ )} =$ ___.

2; 36

5. In $y = 6^x$, replace x by ___ and obtain $y =$ ____.

In Frames 6-8, find the second component of each ordered pair (-2,?),(0,?),(2,?) that makes the pair a solution of $y = 3^x$.

-2; $\frac{1}{9}$

6. In $y = 3^x$, replace x by ____ and obtain $y = 3^{-2} =$ ___.

0; 1

7. In $y = 3^x$, replace x by ___ and obtain $y =$ ___.

2; 9

8. In $y = 3^x$, replace x by ___ and obtain $y =$ ___.

In Frames 9-13, find the second component of each ordered pair (-2,?),(-1,?),(0,?),(1,?),(2,?) that makes the pair a solution of $y = \left(\frac{1}{4}\right)^x$.

9. In $y = \left(\frac{1}{4}\right)^x$, replace x by -2 and obtain

$$y = \left(\frac{1}{4}\right)^{(\_\_)} = \frac{1}{\left(\frac{1}{4}\right)^2} = \_\_\_\_.$$

-2 ; 16

10. In $y = \left(\frac{1}{4}\right)^x$, replace x by -1 and obtain

$$y = \left(\frac{1}{4}\right)^{(\_\_)} = \_\_\_\_.$$

-1 ; 4

11. In $y = \left(\frac{1}{4}\right)^x$, replace x by 0 and obtain $y =$ ____.

1

12. In $y = \left(\frac{1}{4}\right)^x$, replace x by 1 and obtain $y =$ ____.

$\frac{1}{4}$

13. In $y = \left(\frac{1}{4}\right)^x$, replace x by 2 and obtain $y =$ ____.

$\frac{1}{16}$

In Frames 14-15, graph $y = 6^x$.

14. For selected integral values of x, say -2, -1, 0, 1, 2, determine ordered pairs in the solution set of $y = 6^x$ and obtain ______, ______, (0,1),(1,6)(2,36).

$\left(-2,\frac{1}{36}\right)$; $\left(-1,\frac{1}{6}\right)$

15. Draw a smooth curve through the graphs of these ordered pairs and obtain

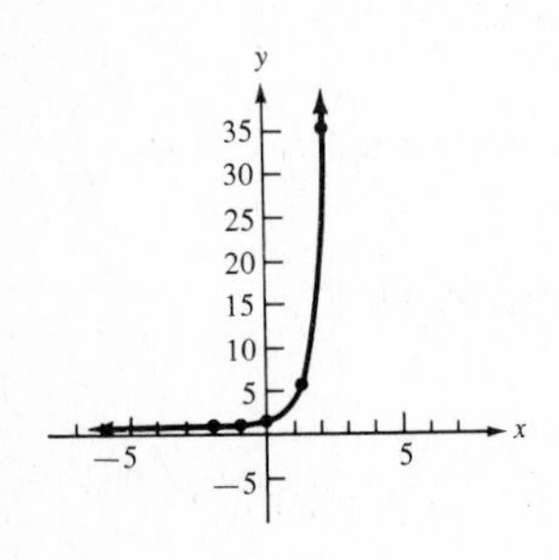

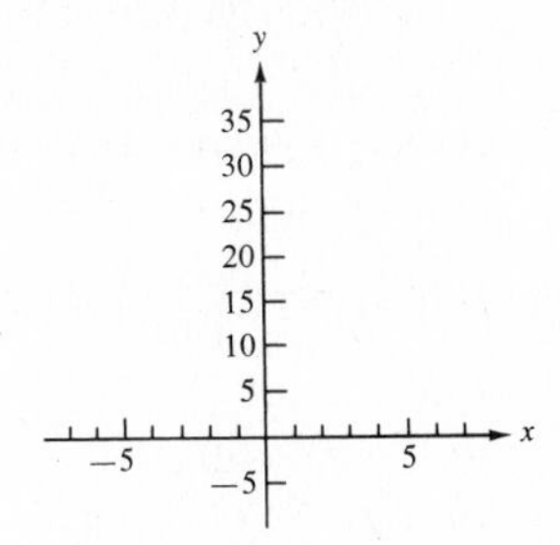

In Frames 16-17, graph $y = 6^{-x}$.

16. For selected integral values of x, say -2,-1,0,1,2, determine ordered pairs in the solution set of $y = 6^{-x}$ and obtain (-2,36), ______, (0,1), $\left(1, \frac{1}{6}\right)$, ______.

(-1,6); $\left(2,\frac{1}{36}\right)$

17. The graph is

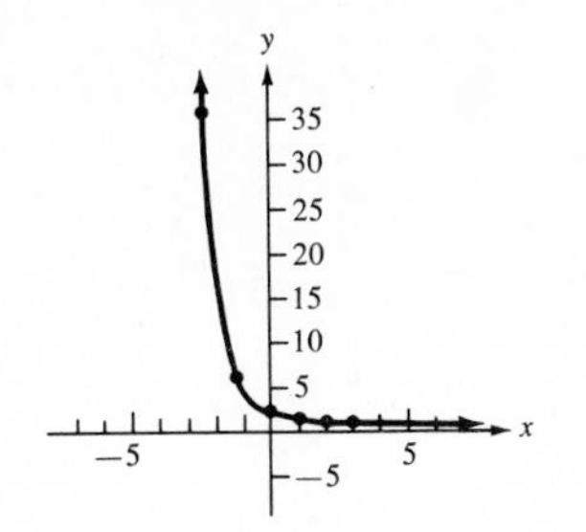

9.2

THE LOGARITHMIC FUNCTION

b^y
$\log_b x$

18. The logarithmic function is defined by the equation $x =$ ____, $(b > 0, b \neq 0)$, or, when solved explicitly for y, by the equation $y =$ ________ $(x > 0, b > 0, b \neq 1)$.

b^y

19. It is important to recognize that the same function is defined by $y = \log_b x$ and $x =$ ____.

3

20. In logarithmic notation, $2^3 = 8$ is equivalent to $\log_2 8 =$ ___.

1000

21. In logarithmic notation, $10^3 = 1000$ is equivalent to $\log_{10}$ (______) $= 3$.

$\log_{10} (.001) = -3$

22. In logarithmic notation, $10^{-3} = .001$ is equivalent to ____________________.

$1/2$; 3

23. In logarithmic notation, $\left(\frac{1}{2}\right)^3 = \frac{1}{8}$ is equivalent to $\log_{(\_\_\_\_)} \frac{1}{8} =$ ___.

$\log_{27} \left(\frac{1}{3}\right) = \frac{-1}{3}$

24. In logarithmic notation, $27^{-1/3} = \frac{1}{3}$ is equivalent to __________________.

2; 36

25. In exponential notation, $\log_6 36 = 2$ is equivalent to $6^{(\_\_\_)} =$ ____.

4; 64

26. In exponential notation, $\log_4 64 = 3$ is equivalent to $(\_\_\_)^3 =$ ____.

-2; 25

27. In exponential notation, $\log_{1/5} 25 = -2$ is equivalent to $\left(\frac{1}{5}\right)^{(\_\_\_\_)} =$ ____.

$10^{-4} = .0001$

28. In exponential notation, $\log_{10} .0001 = -4$ is equivalent to ________________.

Answer	Frame
1; 1	29. The value of $\log_6 6 =$ ___, because $6^{(\ )} = 6$.
0; 0	30. The value of $\log_{10} 1 =$ ___, because $10^{(\ )} = 1$.
-1; -1	31. The value of $\log_6 \frac{1}{6} =$ ___, because $6^{(\ )} = \frac{1}{6}$.
$\frac{1}{2}$; 1/2	32. The value of $\log_7 \sqrt{7} =$ ___, because $7^{(\ )} = \sqrt{7}$.
3; 3	33. The value of $\log_5 125 =$ ___, because $5^{(\ )} = 125$.

In Frames 34-38, solve for the unknown value.

34. $\log_7 49 = y$
In exponential form, this is equivalent to ________; by inspection, obtain $y =$ ___.

$7^y = 49$; 2

35. $\log_b 81 = 4$
In exponential form, this is equivalent to ________. Then write $(b^4)^{1/4} = (81)^{1/4}$ to obtain $b =$ ___.

$b^4 = 81$
3

36. $\log_4 \frac{1}{64} = y$
In exponential form, this is equivalent to ________; and since $\frac{1}{64} = \frac{1}{4^3} = 4^{(\ )}$, the exponential equation can be rewritten as $4^y = 4^{-3}$, from which it is seen that $y =$ ____.

$4^y = \frac{1}{64}$
-3
-3

37. $\log_3 x = -2$
In exponential form, this is equivalent to ________ and in simple form $x =$ ___.

$3^{-2} = x$
$\frac{1}{9}$

38. $\log_b 4 = \frac{1}{3}$
In exponential form, this is equivalent to ________. Then write $(b^{1/3})^3 = 4^3$ and obtain $b =$ ____.

$b^{1/3} = 4$
64

Answer	Frame
x	39. $b^{\log_b x} =$ ___.
5	40. $8^{\log_8 5} =$ ___.
16	41. $10^{\log_{10} 16} =$ ____.
5	42. $10^{\log_{10} 5} =$ ___.
3.2	43. $7^{\log_7 3.2} =$ _____.

In Frames 44-47, simplify $\log_{10} [\log_4(\log_2 16)]$.

44. Find the value within the innermost parentheses first and obtain $\log_2 16 =$ ___.

4

4

45. Then $\log_{10}[\log_4(\log_2 16)] = \log_{10}[\log_4 (\_\_\_)]$.

1

46. Again, find the value within the innermost parentheses and obtain $\log_{10}[\log_4 4] = \log_{10}[\_\_\_]$.

0

0

47. Since $\log_{10} 1 = \_\_\_$, we have
$\log_{10}[\log_4(\log_2 16)] = \log_{10}[\log_4 4] = \log_{10}[1] = \_\_\_$.

9.3

PROPERTIES OF LOGARITHMS

The three laws of logarithms stated in Frames 48, 50, and 52 are valid for positive real numbers b, $(b \neq 1)$, x_1, x_2, and all real numbers m.

$\log_b x_1 + \log_b x_2$

48. $\log_b(x_1 x_2) = \_\_\_\_\_\_\_\_\_\_$.

$\log_5 7$

49. $\log_5(3 \cdot 7) = \log_5 3 + \_\_\_\_\_\_$.

$\log_b x_2 - \log_b x_1$

50. $\log_b\left(\frac{x_2}{x_1}\right) = \_\_\_\_\_\_\_\_\_\_$.

$\log_5 3$

51. $\log_5\left(\frac{7}{3}\right) = \log_5 7 - \_\_\_\_\_\_$.

$m \log_b x_1$

52. $\log_b(x_1)^m = \_\_\_\_\_\_$.

3

53. $\log_5 7^3 = (\_\_\_) \log_5 7$.

In Frames 54-57, assume all variables denote positive real numbers and express each given expression as a sum or difference of simpler logarithmic quantities.

$\log_b y^2$; 2

54. $\log_b x^2y^2 = \log_b x^2 + \_\_\_\_\_\_ = 2 \log_b x + \_\_\_ \log_b y$.

$1/2$; $\frac{1}{2}$

$\log_b y$

55. $\log_b\sqrt{xy} = \log_b(xy)^{(\_\_\_)} = \_\_\_ \log_b(xy)$
$= \frac{1}{2}\left[\log_b x + \_\_\_\_\_\_\right]$.

$1/4$

$\frac{1}{4}$

$\log_{10} y^3$

2; 3

$\frac{1}{2}$; $\frac{3}{4}$

56. $\log_{10} \sqrt[4]{x^2y^3} = \log_{10}(x^2y^3)^{(\_\_\_)}$
$= \left(\_\_\_\right) \log_{10}(x^2y^3)$
$= \frac{1}{4}\left[\log_{10} x^2 + \_\_\_\_\_\_\right]$
$= \frac{1}{4}\left[(\_\_\_)\log_{10} x + (\_\_\_)\log_{10} y\right]$
$= \left(\_\_\_\right) \log_{10} x + \left(\_\_\_\right) \log_{10} y$.

$y^{1/2}$; $\log_{10} z^2$

$\frac{1}{2}$; $2 \log_{10} z$

57. $\log_{10} \frac{\sqrt[3]{x^2}\sqrt{y}}{z^2} = \log_{10}(x^{2/3}\_\_\_\_) - \_\_\_\_\_\_\_\_$

$= \frac{2}{3} \log_{10} x + (\_\_\_)\log_{10} y - \_\_\_\_\_\_\_\_.$

In Frames 58-60, write each given expression as a single logarithm with a coefficient of 1. Assume all variables denote positive real numbers.

2

$\frac{x}{y}$; 2

58. $2 \log_b x - 2 \log_b y = \_\_\_(\log_b x - \log_b y)$

$= 2 \log_b(\_\_\_) = \log_b\left(\frac{x}{y}\right)^{(\_\_\_)}.$

4; 2

x^3y^4

$\frac{x^3y^4}{z^2}$

59. $3 \log_b x + 4 \log_b y - 2 \log_b z$

$= \log_b x^3 + \log_b y^{(\_\_\_)} - \log_b z^{(\_\_\_)}$

$= \log_b(\_\_\_\_\_) - \log_b z^2$

$= \log_b \_\_\_\_\_.$

5; 3

$\log_{10} y^5 + \log_{10} z^3$

y^5z^3

$\frac{x}{y^5z^3}$; $\log_{10}\sqrt{\frac{x}{y^5z^3}}$

60. $\frac{1}{2}(\log_{10} x - 5 \log_{10} y - 3 \log_{10} z)$

$= \frac{1}{2}\left[\log_{10} x - \log_{10} y^{(\_\_\_)} - \log_{10} z^{(\_\_\_)}\right]$

$= \frac{1}{2}\left[\log_{10} x - (\_\_\_\_\_\_\_\_\_\_\_\_)\right]$

$= \frac{1}{2}\left[\log_{10} x - \log_{10}(\_\_\_\_\_)\right]$

$= \frac{1}{2} \log_{10}\left(\_\_\_\_\_\right) = \_\_\_\_\_\_\_\_$

In Frames 61-63, verify that $\log_2(\log_{10} 100) = (\log_2 4) - 1$.

2

2; 1

61. Since $\log_{10} 100 = \_\_\_$, it follows that $\log_2(\log_{10} 100) = \log_2(\_\_\_) = \_\_\_$.

2

2; 1

62. Since $\log_2 4 = \_\_\_$, it follows that $(\log_2 4) - 1 = (\_\_\_) - 1 = \_\_\_$.

1

63. The given equation is true because it has been demonstrated that both its members equal ___.

In Frames 64-66, solve the equation $\log_{10}(x - 1) - \log_{10} 4 = 2$.

$\log_{10} \frac{x - 1}{4} = 2$

64. Write the left member as a single logarithm with coefficient 1 and obtain the equation __________.

$10^2 = \frac{x - 1}{4}$

401

65. In exponential form, this equation is __________ and solving it for x, find $x = \_\_\_\_$.

2

66. Check:

$\log_{10} \frac{401 - 1}{4} = \log_{10} \frac{400}{4} = \log_{10} 100 = \_\_\_.$

In Frames 67-69, solve the equation
$\log_{10}(x + 3) - \log_{10}(x - 1) = 1$.

$\log_{10} \frac{x + 3}{x - 1}$

67. Write the left member as a single logarithm with coefficient 1 and obtain the equation ________.

$10^1 = \frac{x + 3}{x - 1}$

$\frac{13}{9}$

68. In exponential form, this equation is ________ and solving it for x, find x = __.

1

69. Check (using the results of Frame 67):

$$\log_{10} \frac{\frac{13}{9} + 3}{\frac{13}{9} - 1} = \log_{10} \frac{\frac{40}{9}}{\frac{4}{9}} = \log_{10} 10 = \_\_.$$

9.4

LOGARITHMS TO THE BASE 10

characteristic
mantissa

70. The logarithm of a number consists of an integral part called the ________ and a nonnegative decimal fraction part called the ________.

characteristic

table*

71. When writing the logarithm to the base 10 of a number, express the number in scientific notation. Then, the exponent on the 10 is the ________, while the mantissa is the logarithm to the base 10 of the other factor, which is found by referring to the ______ of values of $\log_{10} x$ for $1 < x < 10$.

8 - 10 or 7 - 9, etc.

72. A characteristic such as -2 is frequently written in the form ________.

1

1
.8525

1.8525

73. $\log_{10} 71.2 = \log_{10}(7.12 \times 10^{(\_\_)})$.
So the characteristic is ___ and the mantissa is $\log_{10} 7.12$, which, from the table*, is ______.
$\log_{10} 71.2$ = ______.

5.84; $^{-3}$

-3 or (7 - 10)
.7664

7.7664 - 10

74. $\log_{10} 0.00584 = \log_{10}(\_\_\_\_ \times 10^{(\_\_)})$.
So the characteristic is ________ and the mantissa is $\log_{10} 5.84$, which, from the table, is ______.
$\log_{10} 0.00584$ = ________.

4.98; ^{2}

2
.6972

2.6972

75. $\log_{10} 498 = \log_{10}(\_\_\_\_ \times 10^{(\_\_)})$.
So the characteristic is ___ and the mantissa is $\log_{10} 4.98$, which, from the table, is ______.
$\log_{10} 498$ = ______.

*The table is in the Wooton-Drooyan text, *Intermediate Algebra*.

In Frame 76, find $\log_{10}$ 0.4982 using the nearest table entry.

three

0.498; 9.6972 - 10

76. The table enables us to find the logarithm of a number that has _____ (how many?) significant figures. Hence, before referring to the table, round off 0.4982 to _______. Then, using the table, $\log_{10}$ 0.498 = ______.

77. **If $\log_{10} x$ is given, then x is called the antilog$_{10}$ of the given number.**

498

78. If $\log_{10}$ 498 = 2.6972, then antilog$_{10}$ 2.6972 = _____.

0.00584

79. If $\log_{10}$ (0.00584) = 7.7664 - 10, then antilog$_{10}$ (7.7664 - 10) = _________.

In Frames 80-82, find antilog$_{10}$ 3.8663.

7.35

80. From the table, the number whose logarithm to base 10 has a mantissa of 0.8663 is ______.

3
3

81. Since the given logarithm, 3.8663, has a characteristic of ___, the required antilog can be written in scientific notation as $7.35 \times 10^{(\_\_\_)}$.

7350

82. antilog$_{10}$ 3.8663 = ______.

In Frames 83-85, find antilog$_{10}$ 8.5092 - 10.

3.23

83. Locate 0.5092 in the table among the mantissa entries and observe that it is the logarithm to the base 10 of _______.

(8 - 10) or -2
3.23×10^{-2}

84. Since the given logarithm, 8.5092 - 10, has a characteristic of _______________, the required antilog can be written in scientific notation as _____________.

.0323

85. Antilog$_{10}$ 8.5092 - 10 = _______.

In Frames 86-87, find the antilog$_{10}$ 0.9812 using the nearest table entry.

0.9814

86. From the table, the mantissa entry closest to 0.9812 is __________.

9.58

87. Hence, to the nearest table entry, $\log_{10}$ 0.9812 = _______.

9.5

COMPUTATIONS WITH LOGARITHMS

In Frames 88-89, two assumptions useful in making computations with logarithms are stated for M, N > 0.

$\log_{10} N$

88. **If $M = N$, then $\log_{10} M$ = ________.**

N

89. **If $\log_{10} M = \log_{10} N$, then M = ___.**

In Frames 90-94, compute by means of logarithms $\frac{34.5 \times 162}{.0158}$ *to the nearest table entry.*

$\frac{34.5 \times 162}{.0158}$

90. Let $p = \frac{34.5 \times 162}{.0158}$ and write $\log_{10} p = \log_{10}\left(\_\_\_\_\_\_\_\_\right)$.

$\log_{10} 162 - \log_{10} .0158$

91. Rewrite the right member as a sum or difference of simpler logarithmic quantities and obtain the equation $\log_{10} p = \log_{10} 34.5 +$ ________________.

2.2095; 8.1987 - 10

92. Use the table of logarithms and replace each logarithm in the right member by its value and obtain $\log_{10} p = 1.5378 +$ ________ - ____________.

5.5486 (detail below)

$$\begin{array}{r} 3.7473 \\ -\ 8.1987 - 10 \\ \hline \end{array} = \begin{array}{r} 13.7473 - 10 \\ -\ 8.1987 - 10 \\ \hline 5.5486 \end{array}$$

93. Combine the terms of the right member and obtain the equation $\log_{10} p$ = ________.

354,000

94. Then using the nearest table entry, $p = \text{antilog}_{10}$ 5.5486 = ____________.

In Frames 95-98, compute $\sqrt[6]{85}$ *by means of logarithms.*

1/6; $\frac{1}{6}$

95. Let $p = \sqrt[6]{85}$ and write $\log_{10} p = \log_{10}(85)^{(\_\_\_\_)} = (\_\_\_)\log_{10} 85$.

1.9294

96. Using the table, find $\log_{10} 85$ = ________.

0.3216

97. Then write $\log_{10} p = \frac{1}{6}(1.9294)$ = ________.

2.10

98. Using the nearest table entry, $p = \text{antilog}_{10}$ 0.3216 = _________.

In Frames 99-103, compute $\sqrt{\frac{(0.0678)^3(5.86)}{.0825}}$ *by means of logarithms.*

99. Let $p = \sqrt{\dfrac{(0.0678)^3(5.86)}{0.0825}}$ and write

$\log_{10} p = \log_{10}\left(\dfrac{(0.0678)^3(5.86)}{0.0825}\right)^{(\_\_\_\_)}$.

1/2

100. Expand the right member using the laws of logarithms and obtain the equation

$\log_{10} p$

$= \dfrac{1}{2}\left[3 \log_{10}(0.0678) + \_\_\_\_\_\_\_\_\_\_\_\_\_\_\_\_\_\_\_\_\right]$.

$\log_{10} 5.86 - \log_{10} 0.0825$

101. Refer to the table to replace each logarithm above with its value and obtain

$\log_{10} p = \dfrac{1}{2}\left[3(8.8312 - 10) + \_\_\_\_\_\_\_\_\_\_\_\_\_\_\_\_\_\_\_\_\right]$.

0.7679 - (8.9165 - 10)

102. Perform all indicated operations in the right member and obtain

$\log_{10} p = \dfrac{1}{2}\left[\_\_\_\_\_\_\_\_\_\_\_\_\right] = \_\_\_\_\_\_\_\_\_\_$.

18.3450 - 20; 9.1725 - 10

103. Using the nearest table entry, $p = \text{antilog}_{10}$ $(9.1725 - 10) = \_\_\_\_\_\_\_\_$.

0.149

In Frames 104-108, compute $\dfrac{\sqrt{51.9}\,\sqrt[3]{657}}{\sqrt[4]{4.35}}$ *by means of logarithms.*

104. Let $p = \dfrac{\sqrt{51.9}\sqrt[3]{657}}{\sqrt[4]{4.35}}$ and write

$\log_{10} p = \log_{10}\left(\_\_\_\_\_\_\_\_\right)$.

$\dfrac{\sqrt{51.9}\sqrt[3]{657}}{\sqrt[4]{4.35}}$

105. Expand the right member using the laws of logarithms and obtain

$\log_{10} p = \dfrac{1}{2} \log_{10} 51.9 + \_\_\_\_\_\_\_\_\_\_\_\_\_\_\_\_\_\_\_\_$.

$\dfrac{1}{3} \log_{10} 657 - \dfrac{1}{4} \log_{10} 4.35$

106. Refer to the table to replace each logarithm above with its value and obtain

$\log_{10} p = \dfrac{1}{2}(1.7152) + \dfrac{1}{3}(\_\_\_\_\_\_) - \dfrac{1}{4}(\_\_\_\_\_\_)$.

2.8176; 0.6385

107. $\log_{10} p = \_\_\_\_\_\_\_\_$.

1.6372

108. Using the nearest table entry, $p = \text{antilog}_{10}$ $1.6372 = \_\_\_\_\_\_\_\_$.

43.37

9.6

LOGARITHMS TO THE BASE e; EXPONENTIAL EQUATIONS

$\log_e x$

109. **ln x is another symbol for ______.**

110. **ln x = 2.303 $\log_{10} x$.**

9; 0.9542

2.20

111. ln 9 = 2.303 $\log_{10}$ _____ = 2.303 (_______)

= _______, to the nearest hundredth.

2.303; 108

2.0334; 4.68

112. ln 108 = (_____) $\log_{10}$ ______

= 2.303 (_____) = ______ to the nearest hundredth.

exponent

113. An exponential equation is an equation in which the variable occurs in an _____________.

In Frames 114-116, solve $4^x = 3$ using logarithms to the base 10. Express the solution in logarithmic and decimal form to the nearest hundredth.

$\log_{10} 3$

114. Use the assumption of Frame 88 and write

$\log_{10} 4^x$ _________.

$x \log_{10} 4$

115. Use the third law of logarithms on the left member to obtain the equation

___________ = $\log_{10} 3$.

$\frac{\log_{10} 3}{\log_{10} 4}$

0.4771

116. Multiply both members by $\frac{1}{\log_{10} 4}$ to obtain

x = ________. To obtain decimal form, we have

$x = \frac{\rule{2cm}{0.4pt}}{0.6021} \approx 0.79$

In Frames 117-120, solve $4^{x-2} = 9$ using logarithms to the base 10. Express the solution in logarithmic and decimal form to the nearest hundredth.

$\log_{10} 9$

117. Write $\log_{10} 4^{x-2}$ = _________.

$(x - 2)\log_{10} 4$

118. Use the third law of logarithms on the left member to obtain the equation

________________ = $\log_{10} 9$.

$\frac{\log_{10} 9}{\log_{10} 4}$

119. Multiply both members by $\frac{1}{\log_{10} 4}$ to obtain

$x - 2$ = _________.

$\frac{\log_{10} 9}{\log_{10} 4} + 2$

0.9542

120. Then, x = _____________ or in decimal form

$x = \frac{\rule{2cm}{0.4pt}}{0.6021} + 2 = 3.58.$

In Frames 121-124, solve $7 = e^x$ using logarithms to the base e. Express the solution in logarithmic and decimal form to the nearest hundredth.

Answer	Frame
$\ln e^x$	121. Write $\ln 7 =$ ________.
x	122. Use the third law of logarithms on the right-hand member to obtain the equation $\ln 7 =$ ___ $\ln e$.
1; x	123. Because $\ln e =$ ___, this last equation can be written as $\ln 7 =$ ___. Hence the solution is $x = \ln 7$.
0.8451; 1.95	124. For decimal approximation to the solution we have $\ln 7 = 2.303 \log_{10} 7 = 2.303$ (______) = ______, to the nearest hundredth.

In Frames 125-129, solve $32.7 = e^{x+2}$ using logarithms to the base e. Express the solution in logarithmic form and decimal form to the nearest hundredth.

Answer	Frame
$\ln e^{x+2}$	125. Write $\ln 32.7 =$ ________
$x + 2$	126. $\ln 32.7 =$ (____) $\ln e$.
1; $x + 2$	127. Because $\ln e =$ ___, $\ln 32.7 =$ _____.
-2	128. Hence, $x =$ ___ $+ \ln 32.7$.
	129. For a decimal approximation we have
32.7	$x = -2 + \ln 32.7 = -2 + 2.303 \log_{10}$ _______
1.5145	$= -2 + 2.303$ (______) $= -2 + 3.4879$
1.49	= _____, to the nearest hundredth.

In Frames 130-133, solve $y = Cx^{-2n}$ for n using logarithms to the base 10. Leave the results in the form of an equation equivalent to the given equation.

Answer	Frame
Cx^{-2n}	130. Write $\log_{10} y = \log_{10}($_______$)$.
$\log_{10} C - 2n \log_{10} x$	131. Using the first and third laws of logarithms, expand the right side and obtain the equation $\log_{10} y =$ __________________.
$\log_{10} C - \log_{10} y$	132. Then $2n \log_{10} x =$ __________________.
$\dfrac{\log_{10} C - \log_{10} y}{2 \log_{10} x}$	133. Multiply both members of the equation by $\dfrac{1}{2 \log_{10} x}$ and obtain $n =$ ______________.

In Frames 134-136, solve $y = Ce^{-kt}$ for t using logarithms to the base e. Leave the results in the form of an equation equivalent to the given equation.

Ce^{-kt}

e^{-kt}; $-kt$

134. $\ln y = \ln$ ________

$= \ln C + \ln$ ____ $= \ln C + ($ ____ $)\ln e$.

1

135. Since $\ln e =$ ____, the last equation in Frame 134 can be written as

kt

$\ln y = \ln C -$ ____.

$\ln y$

136. Hence, $kt = \ln C -$ ____, from which

$\dfrac{\ln C - \ln y}{k}$

$t =$ ____________.

9.7 APPLICATIONS

The problems for Frames 137-147 are applications of the compound interest formula

$$A = p\left(1 + \frac{r}{t}\right)^{tn},$$

which says that p dollars invested at interest rate r compounded t times a year for n years yields A dollars.

In Frames 137-142, find the rate of interest (to the nearest $\frac{1}{2}$ percent) *if \$40 compounded 4 times a year for 3 years yields \$50.90.*

12

137. Substitute the given data into $A = p\left(1 + \frac{r}{t}\right)^{tn}$ and obtain

$50.90 = 40\left(1 + \frac{r}{4}\right)^{\_\_\_}$.

12 $\log_{10}$

$\frac{1}{12}\left(\log_{10} 50.90 - \log_{10} 40\right)$

138. Take $\log_{10}$ of both members and obtain

$\log_{10} 50.90 = \log_{10} 40 +$ ________ $\left(1 + \frac{r}{4}\right)$, which can then be written as

$\log_{10}\left(1 + \frac{r}{4}\right) =$ ____________________.

1.7067
1.6021

139. Using the table, $\log_{10} 50.90 =$ ________ and $\log_{10} 40 =$ ________.

.0087

140. Substitute the information from Frame 139 into the equation of Frame 138 and perform the indicated computations in the right member to obtain

$\log_{10}\left(1 + \frac{r}{4}\right) =$ ______.

1.020

141. Therefore, $1 + \frac{r}{4} = \text{antilog}_{10}(.0087) =$ ______.

.080; 8

142. Solve for r and obtain $r =$ ______ $=$ ___ %.

In Frames 143-147, find (to the nearest year) how long it would take for \$60 compounded semiannually at 4 percent to yield \$116.

$2n$

143. Substitute the given data into $A = p\left(1 + \frac{r}{t}\right)^{tn}$ and obtain $116 = 60(1 + \frac{0.04}{2})^{\_\_\_\_}$.

$2n \log_{10}(1.02)$

144. Take $\log_{10}$ of both members and obtain $\log_{10} 116 = \log_{10} 60 +$ ____________.

$\frac{\log_{10} 116 - \log_{10} 60}{2 \log_{10} 1.02}$

145. Add $(-\log_{10} 60)$ to both members and then multiply both members by $\frac{1}{2 \log_{10} 1.02}$ to obtain $n =$ ____________.

2.0645; 1.7782
0.0086

146. Evaluate the logarithms and obtain $\log_{10} 116 =$ ________, $\log_{10} 60 =$ ________, and $\log_{10} 1.02 =$ ________.

17

147. Substitute these values into the equation and perform the indicated operations to obtain $n =$ ____, rounded off to the nearest year.

The problems of Frames 148-154 are applications of the formula

$$pH = \log_{10} \frac{1}{[H^+]},$$

by which the chemist defines the hydrogen potential (pH) of a solution where $[H^+]$ is a numerical value for the concentration of hydrogen ions in an aqueous solution in moles per liter.

In Frames 148- 150, calculate the pH of a solution with hydrogen ion concentration 5.7×10^{-7}.

7

148. Substitute 5.7×10^{-7} for $[H^+]$ in the formula and obtain $pH = \log_{10} \frac{1}{5.7 \times 10^{-7}} = \log_{10} \frac{1}{5.7} \times 10^{(\_\_)}$.

.175
10^6

149. $\frac{1}{5.7} \approx .175$, so $pH \approx \log_{10}(\_\_\_\_\_ \times 10^7)$ or, writing the last number in scientific notation, $pH \approx \log_{10}(1.75 \times \_\_\_\_)$.

6.2

150. Use the table to determine $\log_{10}(1.75 \times 10^6)$ and find pH = _____ (rounded off to tenths).

In Frames 151-154, find $[H^+]$ if $pH = 6.9$.

6.9

151. Substitute 6.9 for pH in the formula to obtain $\log_{10} \frac{1}{[H^+]} =$ _____.

6.9; 7.9×10^6

152. Equate antilog_{10} of each member, to obtain $\frac{1}{[H^+]} = \text{antilog}_{10}$ _____ = __________, to the nearest tenth.

153. Solve $\frac{1}{[H^+]} = 7.9 \times 10^6$ for $[H^+]$ and obtain

10^{-6}

$[H^+] = \frac{1}{7.9 \times 10^6} = \frac{1}{7.9} \times$ _____.

154. Since $\frac{1}{7.9} \approx 0.13$, in scientific notation

1.3×10^{-7}

$[H^+] =$ __________.

The problem of Frames 155-165 is an application of the formula

$$V = 100(1 - e^{-0.5t})$$

where V is the voltage across a capacitor in a certain circuit and t is the time in seconds.

In Frames 155-165, determine how much time must elapse for the voltage to reach 60 volts.

155. In $V = 100(1 - e^{-0.5t})$, replace V by 60 and obtain the equation

60

_____ $= 100(1 - e^{-0.5t})$

t

156. We must now solve the above equation for _____.

157. Divide both sides of the equation of Frame 155 by 100 and obtain

0.6

_____ $= 1 - e^{-0.5t}$.

158. Subtract 1 from both sides of $0.6 = 1 - e^{-0.5t}$ and obtain ________ or equivalently, $0.4 = e^{-0.5t}$

$-0.4 = -e^{-0.5t}$

159. Because the base in the exponential term is e, we shall use __________ (base 10/base e) logarithms to solve $0.4 = e^{-0.5t}$ for t.

base e

$\ln e^{-0.5t}$

160. $\ln 0.4 =$ ________ from which,

$-0.5t$

$\ln 0.4 =$ (_____)$\ln e$.

1

$\ln 0.4 = -0.5t$

161. Since $\ln e =$ ___, the last equation in Frame 160 can be written as ______________.

$\log_{10} 0.4$

$9.6021 - 10$

162. Then, since $\ln 0.4 = 2.303$ (________), we have, using the table of logarithms, $\ln 0.4 = 2.303$(________).

163. Since $9.6021 - 10 = -0.3979$,

-0.3979; -0.9164

$\ln 0.4 = 2.303$(________) = ________.

164. Replacing $\ln 0.4$ by -0.9164 in the last equation of Frame 161, we obtain

-0.9164

____________ $= -0.5t$

165. Dividing both sides of $-0.9164 = -0.5t$ by -0.5 we obtain = _____, to the nearest tenth. Hence it would take 1.8 seconds.

1.8

TEST PROBLEMS

In Problems 1 and 2, find the second component of each of the ordered pairs that makes the ordered pair a solution of the given equation.

1. $y = 4^x$: a. $(-1,?)$, b. $(0,?)$, c. $(2,?)$.

2. $y = \left(\frac{1}{4}\right)^x$: a. $(1,?)$, b. $(0.?)$, c. $(2,?)$.

In Problems 3 and 4, graph the function defined by each of the given equations. Use selected integral values for x between -4 and 4.

3. $y = 3^x$. 4. $y = 3^{-x}$.

5. Express each of the following in logarithmic notation:
   a. $2^4 = 16$. b. $10^{-2} = 0.01$.

6. Express each of the following in exponential notation:
   a. $\log_{10}(0.00001) = -5$. b. $\log_{1/4} 16 = -2$.

7. Solve for the variable:
   a. $\log_b 64 = 3$. b. $\log_{1/3} x = -3$.

8. Write each expression as a sum or difference of simpler logarithmic quantities. Assume all variables and expressions represent real numbers.
   a. $\log_b \frac{x^3y^2}{z}$. b. $\log_{10} \sqrt[4]{\frac{xy^2}{x - y}}$.

9. Write each expression as a single logarithm with coefficient 1:
   a. $3 \log_{10} x + 4 \log_{10} y - \log_{10} z$.
   b. $\frac{1}{2}\left(3 \log_b x + 5 \log_b y - 3 \log_b z\right)$.

10. Simplify: $\log_{10}[\log_2(\log_3 9)]$.

11. Solve $\log_{10}(x + 3) + \log_{10} x = 1$.

In Problems 12-17, use the table of logarithms as necessary.

12. Find each logarithm:

    a. $\log_{10} 21.4$. b. $\log_{10} 0.00214$.

13. Find each antilogarithm:

    a. $\text{antilog}_{10}\ 3.9258$. b. $\text{antilog}_{10}\ 2.8156 - 4$.

In Problems 14-15, use properties of logarithms and a table of logarithms, reading it to the nearest table entry.

14. Compute $\dfrac{0.00214}{.317}$.

15. Compute $\sqrt{\dfrac{(2.85)^3(0.97)}{(0.035)}}$.

16. Find the value of $\ln 23.6$.

17. Solve for x using the base 10 and leave solutions in logarithmic form:

    a. $5^x = 15$. b. $3^{3-x} = 1000$.

18. Solve $E = 2ke^{1+t}$ for t using the base e.

19. To the nearest table entry, find $[H^+]$, given that $\log_{10}\dfrac{1}{[H^+]} = 6.3$.

20. Given that $N = N_o 10^{0.2t}$, find t to the nearest table entry if $N = 180$ and $N_o = 3$.

21. Given that $y = y_o e^{-0.3t}$, find t if $y = 6$ and $y_o = 30$.

10

SYSTEMS OF EQUATIONS

10.1

LINEAR SYSTEMS IN TWO VARIABLES

In Frames 1-6, $A = \{(x,y) \mid a_1x + b_1y + c_1 = 0\}$ *and* $B = \{(x,y) \mid a_2x + b_2y + c_2 = 0\}$.

1. If the graphs of A and B are the same line, the equations defining A and B are said to be ______________.

   dependent

2. If the graphs of A and B are the same line, the solution sets of the equations defining A and B are _______ (equal/ unequal) and the intersection of the two solution sets contains all those ordered pairs in either A or ___.

   equal

   B

3. If the graphs of A and B are parallel but distinct lines, the equations are said to be ______________.

   inconsistent

4. If the graphs of A and B are parallel but distinct lines, the intersection of the solution sets of the equations defining A and B is the ______ set.

   null

5. If the graphs of A and B intersect in one and only one point, the intersection of the solution sets of the equations defining A and B are said to be independent and ______________.

   consistent

6. If the graphs of A and B intersect in one and only one point, the intersection of the solution sets of the equations defining A and B has _____ (how many?) and only _____ (how many?) ordered pair.

   one

   one

7. Any ordered pair (x,y) that satisfies the equations
   $a_1x + b_1y + c_1 = 0$ and $a_2x + b_2y + c_2 = 0$
   will also satisfy the equation
   $A(a_1x + b_1y + c_1) + B(a_2x + b_2y + c_2) =$ ___,
   for every pair of real numbers A and B.

   0

8. The left-hand member of

$A(a_1x + b_1y + c_1) + B(a_2x + b_2y + c_2) = 0$

is called a ________ combination of

$a_1x + b_1y + c_1$ and $a_2x + b_2y + c_2$.

linear

In Frames 9-15, solve the system by linear combinations. Sketch the graphs of these equations.

$$3x + 4y = 10 \quad (1)$$
$$x - 2y = 0 \quad (2)$$

9. Obtain an equivalent system whose y-coefficients are negatives of each other by multiplying both members of Equation (2) by ___.

2

10. The equivalent system is

$3x + 4y = 10$ (1')

_____________. (2')

$2x - 4y = 0$

11. Add the corresponding members of Equations (1') and (2') and obtain _________. (3)

$5x = 10$

12. The solution of Equation (3) is $x =$ ___.

2

13. Substitute 2 for x in Equation (2) and obtain ___ $- 2y = 0$, whose solution is $y =$ ___.

2
1

14. The solution set of the given system is _________.

{(2,1)}

15. The graphs of the equations are

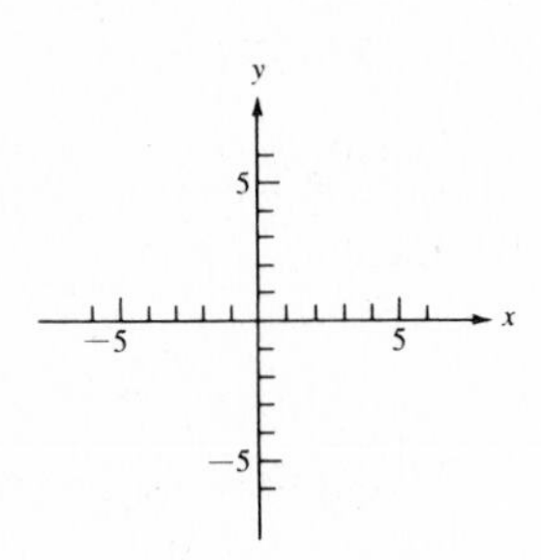

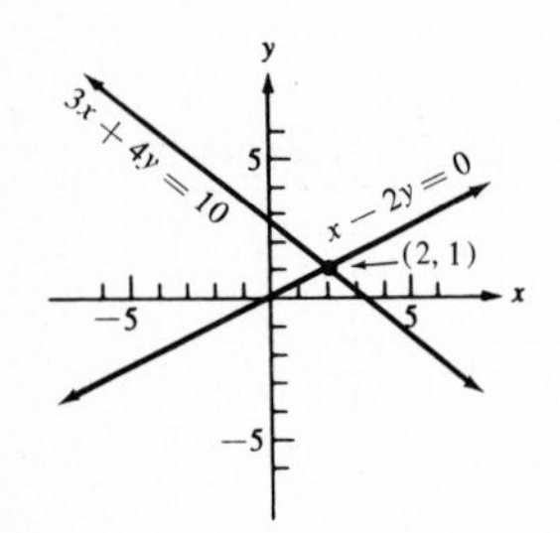

In Frames 16-20, solve the system by linear combinations.

$$\frac{2x}{3} - y = 4 \quad (1)$$
$$x - \frac{3y}{4} = 6 \quad (2)$$

16. Rewrite Equations (1) and (2) equivalently, free of fractions, by multiplying both members of Equations (1) by ___ and (2) by ___ and obtain

_____________ (1')

_____________. (2')

3
4
$2x - 3y = 12$
$4x - 3y = 24$

17. Multiply Equation (1') by -1 and add the left members and the right members of Equations (1') and (2') and obtain ________.

$2x = 12$

18. Solve $2x = 12$ and obtain $x =$ _____.

6

19. Substitute 6 for x in either Equation (1) or (2) and obtain $y =$ ______.

0

20. The solution set of the given system is ________.

$\{(6,0)\}$

21. If, in the system

$$a_1x + b_1y + c_1 = 0; \quad a_2x + b_2y + c_2 = 0,$$

$\frac{a_1}{a_2} \neq \frac{b_1}{b_2}$, the system has a unique solution;

$\frac{a_1}{a_2} = \frac{b_1}{b_2} = \frac{c_1}{c_2}$, the system has ______________ solutions;

$\frac{a_1}{a_2} = \frac{b_1}{b_2} \neq \frac{c_1}{c_2}$, the system has ___ solutions.

infinitely many

no

22. The system

$$9x + 12y = 6$$
$$6x + 8y = 4$$

has infinitely many solutions because, for this system,

$\frac{a_1}{a_2} = \frac{b_1}{b_2} = \frac{c_1}{c_2} =$ ___.

$\frac{3}{2}$

23. The system

$$6x - 8y = -16$$
$$3x - 4y = 8$$

has no solutions because, for this system

$\frac{a_1}{a_2} = \frac{b_1}{b_2} =$ ___, but $\frac{c_1}{c_2} =$ ___.

2; -2

24. The graph of the system of Frame 23 is:

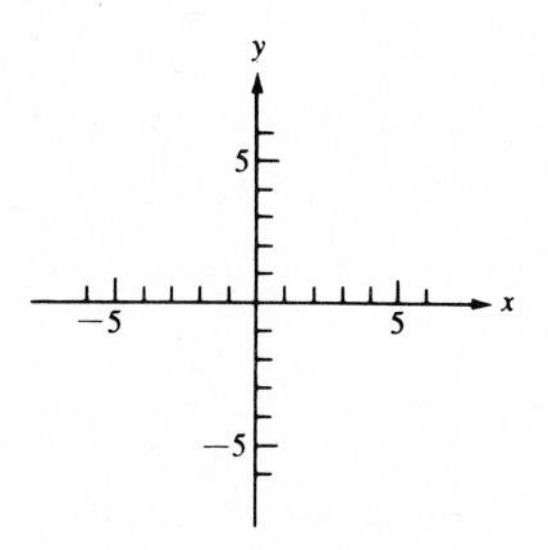

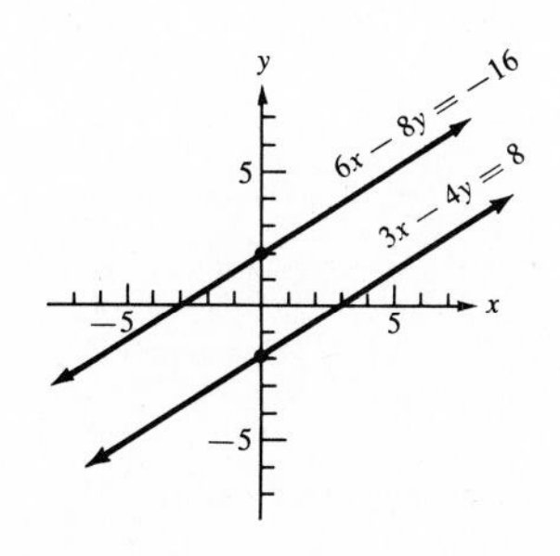

In Frames 25-31, find the solution set of the system

$$\frac{1}{x} + \frac{2}{y} = -\frac{11}{12} \quad (1)$$
$$\frac{1}{x} + \frac{1}{y} = -\frac{7}{12}. \quad (2)$$

25. Set $u = \frac{1}{x}$ and $v = \frac{1}{y}$ and substitute u for $\frac{1}{x}$ and v for $\frac{1}{y}$ in Equations (1) and (2) to obtain

$u + 2v = -\frac{11}{12}$ (1')

____________ . (2')

$u + v = -\frac{7}{12}$

26. To make the coefficients of u negatives of each other, multiply both members of Equation (1') by ____ and obtain the equivalent system

-1

____________ (1")

$u + v = -\frac{7}{12}$. (2")

$-u - 2v = \frac{11}{12}$

27. Add the left and right members of Equations (1") and (2") and obtain

______ . (3)

$-v = \frac{4}{12}$

28. Solve Equation (3) and obtain $v =$ ____ .

$-\frac{1}{3}$

29. Substitute $-\frac{1}{3}$ for v in either Equation (1') or (2'), say Equation (2'), and obtain

$u + (\_\_\_) = -\frac{7}{12}$.

from which $u =$ ____ .

$-\frac{1}{3}$

$-\frac{1}{4}$

30. Since $\frac{1}{x} = u$ and $\frac{1}{y} = v$, we have upon replacing u by $-\frac{1}{4}$ and v by $-\frac{1}{3}$

$\frac{1}{x} =$ ____ and $\frac{1}{y} =$ ____ .

$-\frac{1}{4}$; $-\frac{1}{3}$

31. Solving for x and y we obtain $x =$ ____ and $y =$ ____ and the solution set is __________ .

-4; -3
$\{(-4,-3)\}$

In Frames 32-36, solve the following problem using two variables. A man has $1000.00 more invested at 5 percent than he has invested at 4 percent. If his annual income from the two investments is $698.00, how much does he have invested at each rate?

32. Amount invested at 4 percent : x

Amount invested at _____: y

5 percent

33. The two independent conditions stated in the problem are represented by the equations

____________ (1) and

____________. (2)

$.04x + .05y = 698$

$y = x + 1000$

34. Rewrite Equations (1) and (2) in standard form and free of decimals and obtain

____________ (1')

____________. (2')

$4x + 5y = 69800$

$-x + y = 1000$

7200.00; 8200.00	35. Solve the system of Equations (1') and (2') and obtain $x =$ ________ and $y =$ __________.
yes yes	36. Check: Does .04(7200) + .05(8200) = 698? _____ (yes/no). Does 8200 = 7200 + 1000? _____ (yes/no). Hence he has invested $7200 at 4% and $8200 at 5%.
inconsistent dependent	37. If, upon adding the left members and right members of a system, a statement of the form $0 = k (k \in R$ and $k \neq 0)$ results, the equations are ____________; if the form $0 = 0$ results, they are __________.

10.2

LINEAR SYSTEMS IN THREE VARIABLES

triple; true	38. A solution of an equation in three variables is an ordered ________ of numbers which makes the equation a ______ statement.
solution	39. The intersection of the solution sets of three equations in three variables is the _________ set of the system of equations.

In Frames 40-46, solve the system

$$\begin{aligned} 2x - y + 3z &= -9 \quad (1) \\ x + 3y - z &= 10 \quad (2) \\ 3x + y - z &= 8. \quad (3) \end{aligned}$$

3 $5x + 8y = 21$	40. To make the coefficients of z in Equations (1) and (2) negatives of each other, multiply Equation (2) by ___. Then add the result to Equation (1) and obtain ____________. (4)
-1 $2x - 2y = -2$	41. To make the coefficients of z in Equations (2) and (3) negatives of each other, multiply Equation (2) by ____. Then add the results to Equation (3) and obtain ______________. (5)
$13x = 13$	42. To solve the system made up of Equations (4) and (5), multiply Equation (5) by 4 and add the result to Equation (4) and obtain __________. (6)
1	43. Solve Equation (6) and obtain $x =$ ___.
1 2	44. To obtain y, replace x in Equation (4) or (5), say (5), by 1 and obtain $2(\_\_\_) - 2y = -2$, from which $y =$ ___.
-3	45. To obtain z, replace x by 1 and y by 2 in Equation (1), (2), or (3), say (2), and obtain $1 + 3(2) - z = 10$, from which $z =$ ____.
{(1,2,-3)}	46. From Frames 43, 44, and 45, the solution set of the given system is ___________.

In Frames 47-51, solve the system

$$\begin{aligned} 2x + z &= -4 \quad (1)\\ 3y + 4z &= 11 \quad (2)\\ x + y &= -2. \quad (3) \end{aligned}$$

(2); (3)

47. Since Equation (1) involves only the variables x and z, obtain another equation in x and z, by eliminating the variable y between equations _____ and _____.

-3

$-3x + 4z = 17$

48. To eliminate y between Equations (2) and (3), multiply Equation (3) by ____ and add the result to Equation (2) and obtain

_______________. (4)

-3; 2

49. Now solve the system consisting of Equations (1) and (4) and obtain $x =$ ____ and $z =$ ___.

1

50. To obtain y, substitute $x = -3$ and $z = 2$ into Equation (2) or (3), say (3), and obtain $-3 + y = -2$, from which $y =$ ___.

{(-3,1,2)}

51. From Frames 49 and 50, the solution set of the given system is ____________.

infinitely
no

52. If at any step in the procedure described in the preceding frames the system vanishes or yields a contradiction, the system has ____________ many members in its solution set or ____ members, respectively.

In Frames 53-56, solve the system

$$\begin{aligned} x + y + z &= 4 \quad (1)\\ 2x - 3y - 2z &= -2 \quad (2)\\ 3x - 2y - z &= -1. \quad (3) \end{aligned}$$

2

$4x - y = 6$

53. To make the coefficients of z in Equations (1) and (2) negatives of each other, multiply Equation (1) by ___. Then add the result to Equation (2) and obtain

____________. (4)

-2

$-4x + y = 0$

54. To make the coefficients of z in Equations (2) and (3) negatives of each other, multiply Equation (3) by ____. Then add the result to Equation (2) and obtain

_____________. (5)

$0 = 6$

55. Add the members of the reduced system consisting of Equations (4) and (5) to obtain

________. (6)

no

56. Since Equation (6) is a contradiction, the system has ____ members in its solution set.

In Frames 57-60, solve the following problem. The sum of three numbers is 2. The first number is equal to the sum of the other two, and the third number is the result of subtracting the first from the second. Find the numbers.

the third number

57. the first number: x
the second number: y
__________________: z

58. The statement of the problem yields the equations

$\_\_\_\_\_\_ = 2$

$x = \_\_\_\_$

$z = \_\_\_\_.$

$x + y + z$

$y + z$

$y - x$

59. Rewrite the equations in standard form and obtain

$x + y + z = 2.$

$\_\_\_\_\_\_\_\_.$

$\_\_\_\_\_\_\_\_.$

$x - y - z = 0$

$x - y + z = 0$

60. Solve the system and obtain

$x = \_\_$, $y = \_\_$, $z = \_\_$.

Hence, the first number is 1, the second number is 1, and the third number is 0.

1; 1; 0

10.3

SECOND-DEGREE SYSTEMS IN TWO VARIABLES—SOLUTION BY SUBSTITUTION

61. Systems of two equations in two variables can be solved by graphing both equations on the same axis and estimating the coordinates of any points they have in ____.

common

62. Systems of equations in two variables can also be solved by substitution, a method which involves solving one of the equations explicitly for one variable in terms of the other and substituting this solution into the other equation to obtain an equation that involves ____ (how many?) variable(s).

one

In Frames 63-70, solve the system

$$x^2 + 2y^2 = 12$$
$$2x - y = 2$$

by substitution and check the solutions by graphing and estimating the coordinates of any intersection points.

63. Solve $2x - y = 2$ explicitly for y in terms of x and obtain ______.

$y = 2x - 2$

64. Substitute $2x - 2$ for y in $x^2 + 2y^2 = 12$ and obtain $x^2 + 2(\_\_\_\_)^2 = 12$.

$2x - 2$

65. Write this equation in standard quadratic form and obtain ________.

$9x^2 - 16x - 4 = 0$

66. Then $9x^2 - 16x - 4 = (\_\_\_\_)(\_\_\_\_)$.

$9x + 2$; $x - 2$

67. Therefore, solutions to $9x^2 - 16x - 4 = 0$ are __ and __.

$\frac{-2}{9}$; 2

68. Now, replacing x in the equation of Frame 63 by each of these numbers in turn, obtain

$$y = 2\left(\frac{-2}{9}\right) - 2 = \_\_\_\_$$

and $y = 2(\_\_\_) - 2 = \_\_\_.$

$\frac{-22}{9}$

2; 2

69. Therefore, the solution set of the system is

________________.

$\left\{\left(\frac{-2}{9}, \frac{-22}{9}\right), (2,2)\right\}$

70. The graph is

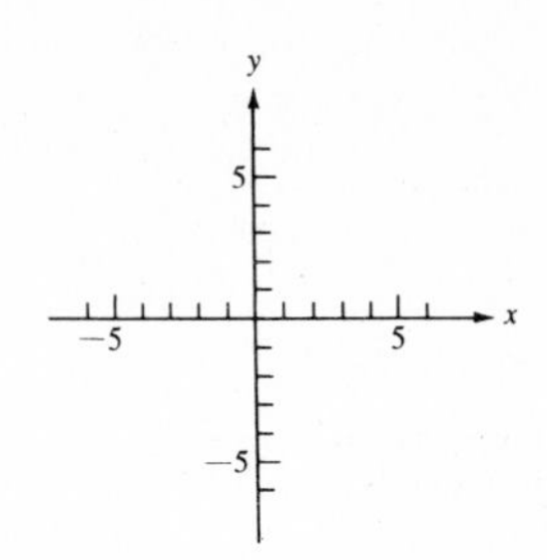

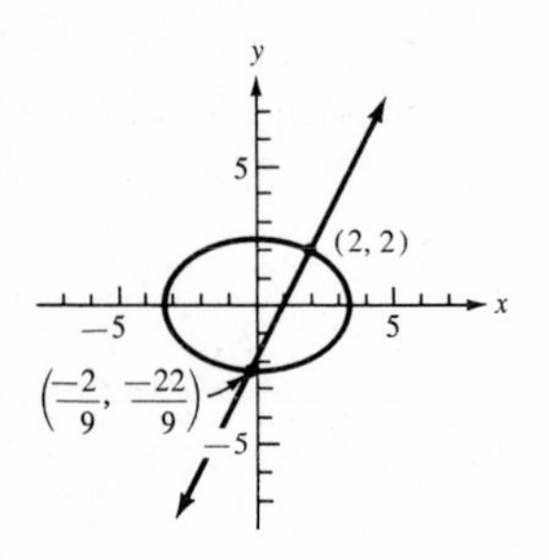

In Frames 71-77, solve the system

$$2x^2 + xy + y^2 = 9$$
$$-x + 3y = 9$$

by substitution.

71. Solve $-x + 3y = 9$ explicitly for x in terms of y and obtain __________. (Note that the solution for y in terms of x would have involved fractions.)

$x = 3y - 9$

72. Substitute $3y - 9$ for x in $2x^2 + xy + y^2 = 9$ and obtain $2(\_\_\_\_\_\_)^2 + (\_\_\_\_\_\_)y + y^2 = 9$.

$3y - 9$; $3y - 9$

73. Write this equation in standard quadratic form and obtain ________________.

$22y^2 - 117y + 153 = 0$

74. Then $22y^2 - 117y + 153 = (y - 3)(\_\_\_\_\_\_\_\_)$.

$22y - 51$

75. So,solutions to $22y^2 - 117y + 153 = 0$ are ___ and ___.

3; $\frac{51}{22}$

76. Now, replacing y in the equation of Frame 71 by each of these numbers in turn, obtain

$$x = 3(3) - 9 = \_\_\_$$

and $x = 3(\_\_\_) - 9 = \_\_\_\_.$

0

$\frac{51}{22}$; $\frac{-45}{22}$

77. Therefore, the solution set of the system is

________________.

$\left\{(0,3), \left(\frac{-45}{22}, \frac{51}{22}\right)\right\}$

In Frames 78-87, solve the following problem. The area of a rectangle is 216 square feet. If the perimeter is 60 feet, find the dimensions of the rectangle.

Answer	Frame
width	78. The length of the rectangle: x The _______ of the rectangle: y
$2x + 2y$	79. Then, $xy = 216$ and _________ $= 60$.
$y = 30 - x$	80. Solve $2x + 2y = 60$ explicitly for y in terms of x and obtain ____________.
$(30 - x)x = 216$	81. Replace y in $xy = 216$ by $30 - x$ and obtain _________________.
$x^2 - 30x + 216 = 0$	82. Write this equation in standard quadratic form and obtain ____________________.
$x - 18$; $x - 12$	83. Then, $x^2 - 30x + 216 =$ (________)(________).
18; 12	84. So, solutions to $x^2 - 30x + 216 = 0$ are ____ and ____.
	85. Now, replacing x in the equation of Frame 80 by each of these numbers in turn, obtain
18; 12	$y = 30 -$ ____ $=$ ____
12; 18	$y = 30 -$ ____ $=$ ____.
$\{(18,12),(12,18)\}$	86. Therefore, the solution set of the system is ___________________.
18 12	87. In both solutions the dimensions are ____ feet by ____ feet.

10.4

SECOND-DEGREE SYSTEMS IN TWO VARIABLES— SOLUTION BY OTHER METHODS

Answer	Frame
second	88. The method of linear combinations of members of the equations of a system can provide a simple means of solution if both equations are of ________ degree in both variables.

In Frames 89-96, solve the system

$$9x^2 + 16y^2 = 100 \quad (1)$$
$$x^2 + y^2 = 8. \quad (2)$$

Answer	Frame
$7y^2 = 28$	89. Multiply Equation (2) by -9 and add the result to Equation (1) and obtain __________. (3)
2; -2	90. Solve Equation (3) and find $y =$ ___ or $y =$ ___.
2	91. Now, substitute 2 for y into either Equation (1) or (2), say (2), and obtain $x^2 + (\_\_\_)^2 = 8.$ (4)
	92. Solve Equation (4) and obtain $x = \pm 2.$ (5)

(2,2); (-2,2)

93. From Frames 90 and 92, when $y = 2$, $x = 2$ or $x = -2$. Hence we have the ordered pairs _______ and ________ in the solution set of the system.

±2

94. Now, substitute the other value of y from Frame 90, namely -2, into Equation (2) and solve for x to obtain $x =$ _____.

(2,-2); (-2,-2)

95. When $y = -2$, $x = 2$ or $x = -2$. Hence we have the additional ordered pairs in the solution set of the system ________ and _________.

{(2,2),(-2,2),(2,-2),(-2,-2)}

96. From Frames 93 and 95, the solution set of the system is ______________________________.

In Frames 97-107, solve the system

$$2x^2 + xy - 2y^2 = 16 \quad (1)$$
$$x^2 + 2xy - y^2 = 17. \quad (2)$$

$-3xy = -18$

97. Multiply Equation (2) by -2 and add the result to Equation (1) and obtain

____________. (3)

$y = \frac{6}{x}$ $(x \neq 0)$

98. Solve Equation (3) for either x or y, say y, and obtain

_______________. (4)

$\frac{6}{x}$; $\frac{6}{x}$

99. In either Equation (1) or (2), say (2), replace y by $\frac{6}{x}$ and obtain

$x^2 + 2x(\_\_\_) - (\_\_\_)^2 = 17.$ (5)

$x^4 - 5x^2 - 36 = 0$ $(x \neq 0)$

100. Write Equation (5) in descending powers of x, free of fractions, with 0 as the right member and obtain

___________________________. (6)

$(x^2 - 9)(x^2 + 4)$

101. Factoring the left member of Equation (6), write

__________________ = 0. (7)

$x^2 + 4 = 0$

102. The solutions to Equation (7) are the solutions to

$x^2 - 9 = 0$ (8)

or ____________. (9)

±3
$\pm 2i$

103. The solutions to Equation (8) are $x =$ ____ and the solutions to Equation (9) are $x =$ _____.

2
-2
(3,2); (-3,-2)

104. From Equation (4) in Frame 98, when $x = 3$, $y =$ ___ and when $x = -3$, $y =$ ____, which give the ordered pairs in the solution set of the system _______ and _________.

$\frac{3}{i}$
$\frac{-3}{i}$; $-3i$; $3i$
(see below)
$\frac{3}{i} = \frac{3 \cdot i}{i \cdot i} = \frac{3i}{-1} = -3i$

105. Now, with $x = 2i$ in Equation (4), $y =$ ___; and then with $x = -2i$, $y =$ ____, which simplify to _____ and ____, respectively.

$(2i,-3i)$
$(-2i,3i)$

106. Frame 105 yields the ordered pairs __________ and __________ in the solution set of the system.

{(3,2),(-3,-2),
$(2i,-3i)$,$(-2i,3i)$}

107. From Frames 104 and 106, the solution set of the system is _______________________________.

TEST PROBLEMS

In Problems 1-3, solve each system by linear combinations.

1. $x + 4y = -14$
   $3x + 2y = -2.$

2. $\frac{2}{3}x - y = 4$
   $x - \frac{3}{4}y = 6.$

3. $\frac{1}{x} + \frac{2}{y} = \frac{-11}{12}$
   $\frac{1}{x} + \frac{1}{y} = \frac{-7}{12}.$

In Problems 4-6, state how many solutions each system has and whether the equations in each system are dependent, inconsistent, or consistent and independent.

4. $3x + 4y = 4$
   $9x + 12y = 3.$

5. $3x + 4y = 1$
   $9x + 12y = 3.$

6. $3x + 4y = 7$
   $9x + 6y = 7.$

7. Solve by linear combinations:

   $2x + 4y + z = 0$
   $5x + 3y - 2z = 1$
   $4x - 7y - 7z = 6.$

In Problems 8 and 9, solve each system by substitution or linear combinations.

8. $y = x^2 - 2x + 1$
   $x + y = 3.$

9. $9x^2 + 16y^2 = 100$
   $x^2 + y^2 = 8.$

10. The area of a rectangle is 240 square feet. If the perimeter is 64 feet, find the dimensions of the rectangle.

11. The perimeter of a triangle is 155 inches. The side x is 20 inches shorter than the side y, and the side y is five inches longer than the side z. Find the lengths of the sides of the triangle.

11

NATURAL NUMBER FUNCTIONS

11.1

SEQUENCES AND SERIES

sequence

1. A function whose domain is the set of successive natural numbers—for example, the function defined by
$\{(n, s(n)) | s(n) = 2n + 1,\ n \in \{1,2,3, \ldots\}\}$, (1)
is called a ________ function.

sequence

2. The elements in the range of a sequence function, arranged in order, form a ________.

7; 9

3. The sequence associated with (1) in Frame 1 is
$s(1) = 2(1) + 1$, $s(2) = 2(2) + 1$,
$s(3) = 2(3) + 1$, $s(4) = 2(4) + 1$, etc.,
which, when computed, yields 3, 5, ___, ___, etc.

general

4. A formula such as $s(n) = 2n + 1$, which generates the terms of a sequence, is called the nth term or ________ term.

$\frac{1}{4}$; $\frac{2}{11}$

5. The first four terms of the sequence whose general term is
$$s(n) = \frac{2}{3n - 1}$$
are 1, $\frac{2}{5}$, ___, and ___.

s_n

6. Instead of using $s(n)$ to denote the general term of a sequence, ___ is used.

In Frames 7-8, find the first four terms of the sequence whose general term is $s_n = \frac{2}{n^2 + 1}$.

1; 2

3; 4

7. $s_1 = \dfrac{2}{(\_\_)^2 + 1}$, $s_2 = \dfrac{2}{(\_\_)^2 + 1}$

$s_3 = \dfrac{2}{(\_\_)^2 + 1}$, $s_4 = \dfrac{2}{(\_\_)^2 + 1}$.

1; $\frac{2}{5}$; $\frac{1}{5}$; $\frac{2}{17}$

8. Simplify s_1, s_2, s_3, and s_4 and obtain

$s_1 =$ ___, $s_2 =$ ___, $s_3 =$ ___, $s_4 =$ ___.

In Frames 9-10, find the first four terms of the sequence whose general term is $s_n = (-1)^{n+1}2^{n-1}$.

1; 1; 2; 2

3; 3; 4; 4

9. $s_1 = (-1)^{(\_\_)+1}2^{(\_\_)-1}$, $s_2 = (-1)^{(\_\_)+1}2^{(\_\_)-1}$,

$s_3 = (-1)^{(\_\_)+1}2^{(\_\_)-1}$, $s_4 = (-1)^{(\_\_)+1}2^{(\_\_)-1}$.

1; -2
4; -8

10. Simplify s_1, s_2, s_3, and s_4 and obtain $s_1 =$ ___, $s_2 =$ ____, $s_3 =$ ___, $s_4 =$ ____.

series

11. The indicated sum of the terms of a sequence is called a ________.

$1 + 3 + 5 + 7 + \cdots + (2n - 1)$

12. The sequence 1, 3, 5, 7, . . . , $2n - 1$ has associated with it the series $S_n =$ ____________________.

$x^2 + x^4 + x^6 + x^8 + \cdots + x^{2n}$

13. Associated with the sequence x^2, x^4, x^6, x^8, . . . , x^{2n} is the series $S_n =$ ____________________.

$2i - 1$

14. A compact way to denote a series is by a means of sigma, or summation, notation, such as

$$S_n = \sum_{i=1}^{n} (2i - 1),$$

where it is understood that S_n is the series whose terms are obtained by replacing i in the expression ________ by 1, 2, 3, . . . , n, successively.

In Frames 15-19, write $S_4 = \sum_{i=1}^{4} (2i - 1)$ *in expanded form.*

1

15. In $(2i - 1)$, replace i by 1 and obtain ___.

3

16. In $(2i - 1)$, replace i by 2 and obtain ___.

5

17. In $(2i - 1)$, replace i by 3 and obtain ___.

7

18. In $(2i - 1)$, replace i by 4 and obtain ___.

1 + 3 + 5 + 7

19. In expanded form, $S_4 =$ ______________.

In Frames 20-22, write $\sum_{i=3}^{7} x^{2i}$ in expanded form.

3; x^6

20. In x^{2i}, replace i first by ___ and obtain ____.

4; 5; 6; 7
x^8; x^{10}; x^{12}
x^{14}

21. Then replace i successively by ___, ___, ___, ___, obtaining the successive terms ____, ____, ____, and ____.

$x^6 + x^8 + x^{10} + x^{12} + x^{14}$

22. The required expanded form is ________________.

In Frames 23-25, write $\sum_{i=2}^{4} \frac{(-1)^{i+1}}{i-1}$ *in expanded form.*

2; -1

23. In $\frac{(-1)^{i+1}}{i-1}$, first replace i by ___ and obtain ____.

3; 4
$\frac{1}{2}$; $\frac{-1}{3}$

24. Then replace i successively by ___ and ___, obtaining the successive terms ___ and ___.

$-1 + \frac{1}{2} - \frac{1}{3}$

25. The required expanded form is ________.

26. $S_\infty = \sum_{i=2}^{\infty} (2i - 1)$ indicates infinitely many terms of the series whose first term is 2(___) - 1, whose second term is 2(___) - 1, whose third term is 2(___) - 1, etc., indefinitely.

2;
3; 4

3 + 5 + 7 + 9 + · · ·

27. In expanded form, $\sum_{i=2}^{\infty} (2i - 1)$ is ________________.

In Frames 28-30, write $\sum_{j=3}^{\infty} \frac{1}{j+1}$ *in expanded form.*

3; $\frac{1}{4}$

28. Replace j in $\frac{1}{j+1}$ first by ___, obtaining ___.

4; 5; 6
$\frac{1}{5}$; $\frac{1}{6}$; $\frac{1}{7}$

29. Then replace j successively by ___, ___, and ___ and obtain the successive terms ___, ___, and ___.

$\frac{1}{4} + \frac{1}{5} + \frac{1}{6} + \frac{1}{7} + \cdots$

30. The required expanded form is ________________.

In Frames 31-33, write 1 + 4 + 9 + 16 in sigma notation.

3^2; 4^2

31. To find a general term, notice that the series can be written as $1^2 + 2^2 +$ ____ + ____.

i^2

32. So, a general term is ____.

$\sum_{i=1}^{4} i^2$

33. The required sigma notation is ________.

In Frames 34-35, write $-1 + \frac{1}{2} - \frac{1}{3} + \frac{1}{4} - \cdots + \frac{(-1)^n}{n}$ *in sigma notation.*

$\frac{(-1)^i}{i}$

34. Replace n in the general term by i and obtain ______.

$\sum_{i=1}^{4} \frac{(-1)^i}{i}$

35. Then the required sigma notation is ______.

In Frames 36-38, write $1 \cdot 2 + 2 \cdot 3 + 3 \cdot 4 + 4 \cdot 5 + \cdots$ in sigma notation.

term
1

36. To find a general term, observe that in each term the first factor is the same as the number of that ______ and the second factor is greater than the first by ___.

$i(i + 1)$

37. Expressed in terms of i, the general term is ______.

$\sum_{i=1}^{\infty} i(i + 1)$

38. The required sigma notation is ______.

In Frames 39-42, write $\frac{1}{2} + \frac{3}{4} + \frac{9}{6} + \frac{27}{8} + \cdots$ in sigma notation.

3^2; 3^3
3^{i-1}

39. To find a general term, observe that the numerators can be expressed as 3^0, 3^1, ____, ____ and, in general, in terms of i, as ______.

$2 \cdot 3$; $2 \cdot 4$
$2i$

40. Observe also that the denominators can be expressed as $2 \cdot 1$, $2 \cdot 2$, ______, ______ and, in general, in terms of i, as ____.

$\frac{3^{i-1}}{2i}$

41. From Frames 39 and 40, a general term is ______.

$\sum_{i=1}^{\infty} \frac{3^{i-1}}{2i}$

42. The required sigma notation is ______.

11.2

ARITHMETIC PROGRESSIONS

arithmetic

43. A sequence in which all terms but the first are generated by adding a constant to the preceding term to obtain the succeeding term is called an ____________ progression.

equal

44. It can be verified that a finite sequence is an arithmetic progression by subtracting each term from its successor and observing that all of the differences are _______ (equal/unequal).

2

45. The sequence 2, 4, 6, 8, 10 is an arithmetic progression, because the differences between consecutive terms are each ___.

$s_n = a + (n - 1)d$

46. If the first term of an arithmetic progression is denoted by a and d denotes the common difference between consecutive terms, the general term of the progression is given by ________________.

In Frames 47-49, write the next three terms and the general term of the arithmetic progression whose first two terms are 2, 7, ...

5

47. The common difference $d = 7 - 2 =$ ___.

12
17
22

48. Then, adding 5 to 7, obtain the third term, ____, to which 5 is again added to obtain the fourth term, ____. Similarly, find the fifth term to be ____.

2
5
2; 5; $5n - 3$

49. From the given terms, observe that $a =$ ___ and, as has been seen, $d =$ ___. So the general term is $s_n =$ (___) + $(n - 1)$(___) = ________.

In Frames 50-52, write the next three terms and the general term of the arithmetic progression whose first two terms are $a + 3b$, $a - 3b$, ...

$-6b$

50. $d = (a - 3b) - (a + 3b) =$ _____.

$a - 9b$; $a - 15b$; $a - 21b$

51. Using this value of d to continue the sequence, obtain ________, _________, and _________.

$a + 3b$; $-6b$
$a + 9b - 6bn$

52. In the formula for the general term, let the role of a be played by the first term, $a + 3b$, and from Frame 50, $d = -6b$, and find $s_n =$ (________) + $(n - 1)$(_____) = ________________.

In Frames 53-54, find the 12th term of the arithmetic progression 3, 7, 11, ...

3; 4

53. First, observe $a =$ ___ and compute $d =$ ___.

4; 47

54. Then, using $s_n = a + (n - 1)d$ and taking $n = 12$, find $s_{12} = 3 + (12 - 1)$(___) = ____.

In Frames 55-56, find the 8th term of the arithmetic progression $\frac{3}{4}$, 2, $\frac{13}{4}$, ...

$\frac{5}{4}$; $\frac{3}{4}$

55. Compute $d = 2 - \frac{3}{4}$ or $\frac{13}{4} - 2 =$ ___ and observe that $a =$ ___.

7; $\frac{38}{4}$ or $\frac{19}{2}$

56. Using $s_n = a + (n - 1)d$, taking $n = 8$, find $s_8 = \frac{3}{4} +$ (___)$\frac{5}{4} =$ ____________.

In Frames 57-60, find the first term of the arithmetic progression whose fourth term is 13 and whose twelfth term is 61.

ninth

57. Set up the diagram

?, ?, ?, 13, ?, ?, ?, ?, ?, ?, ?, 61

and observe that if 13 is taken as the first term of an arithmetic progression, 61 can be taken as the _______ term.

6

58. Using $s_n = a + (n - 1)d$ to find d, taking $n = 9$, $a = 13$, $s_9 = 61$, write $61 = 13 + 8d$ and find $d =$ ___.

13

59. Now use $d = 6$ to find the first term of the arithmetic progression whose fourth term is ____ (see Frame 57) in $s_n = a + (n - 1)d$.

4; -5

60. So, $13 = a + (\_\_\_ - 1)6$ and find $a =$ ____.

In Frames 61-63, an alternative method is used to solve the above problem (stated between Frames 56 and 57).

$a + 3d$

61. Use $s_n = a + (n - 1)d$ with $n = 4$ and $s_4 = 13$; write the equation $13 =$ ________. (1)

$a + 11d$

62. Use $s_n = a + (n - 1)d$ with $n = 12$ and $s_{12} = 61$; write the equation $61 =$ _________. (2)

-5

63. Solve the system (1) and (2) to find $a =$ ____.

means

64. Terms between given terms in an arithmetic progression are called arithmetic _______.

In Frames 65-67, insert 4 arithmetic means between -7 and 8.

5

65. The diagram -7, ___, ___, ___, ___, 8 suggests that if there are four terms between -7 and 8, then the difference between -7 and 8 must be ___ times d.

-7; 3

66. Therefore, $5d = 8 -$ ___ from which $d =$ ___.

2; 5

67. By four successive additions of 3 we obtain the required arithmetic means -4, -1, ___, and ___.

$n\left(\frac{a + s_n}{2}\right)$

68. The sum of n terms of an arithmetic progression is given in terms of s_n by

$S_n =$ ________.

$\frac{n}{2}\left[2a + (n - 1)d\right]$

69. An alternative form for S_n that does not involve s_n is

$S_n =$ _______________.

In Frames 70-72, find the sum of the finite series $\sum_{i=1}^{19} (3i + 2)$.

5; 8; 11

70. The first three terms in expanded form are ___ + ___ + ____ + • • •.

5; 3

71. Observe $a =$ ___ and $d =$ ___.

5; 18; 608

72. Use $S_n = \frac{n}{2}\left[2a + (n - 1)d\right]$ with $n = 19$ and write $S_{19} = \frac{19}{2}\left[2(\_\_\_) + (\_\_\_\_)3\right]$ and find $S_{19} =$ _____.

In Frames 73-75, find the sum of the finite series $\sum_{i=1}^{150} i$.

1; 2; 3; 150

73. The first three terms of this progression are ___, ___, and ___ and the last term is s_{150} = _____.

1; 1

74. So, a = ___, d = ___, and $s_{150} = 150$.

150; 11,325

75. Use $S_n = n\frac{(a + s_n)}{2}$, and find

$S_{150} = (\_\_\_\_\_)\frac{(1 + 150)}{2} = \_\_\_\_\_\_\_\_$.

In Frames 76-79, find the sum of the finite series $\sum_{j=7}^{15} (2j - 1)$.

13; 15; 17

76. The first three terms are ____, ____, and ____.

13; 2

77. So, a = ____, and d = ___.

6; 9

78. Observe that if the series had been generated by taking $j = 1, 2, 3, \ldots$, the progression would then have had 15 terms; but since we are taking $j = 7, 8, 9, \ldots$, we will have ___ fewer terms, so n = ___.

189

79. Now, use $a = 13$, $d = 2$, and $n = 9$ in

$S_n = \frac{n}{2}\left[2a + (n - 1)d\right]$ and find S_9 = _____.

11.3

GEOMETRIC PROGRESSIONS

geometric

80. Any sequence in which each term except the first is obtained by multiplying the preceding term by a common multiplier is called a __________ progression.

2

81. 2, 4, 8, 16, . . . , is a geometric progression whose common multiplier is ___.

$\frac{1}{3}$

82. $1, \frac{1}{3}, \frac{1}{9}, \frac{1}{27}, \ldots$, is a geometric progression whose common multiplier is ___.

multiplier (or ratio)

83. In general, the ratio of any term to its predecessor gives the common __________, which is denoted by r.

In Frames 84-85, find the common ratio and write the next three terms in the geometric progression $\frac{1}{5}, \frac{3}{5}, \frac{9}{5}, \ldots$.

$\frac{3}{5}$; 3

84. To find the common ratio, write

$r = \frac{\_\_\_}{\frac{1}{5}} = \_\_\_$.

$3;\ \frac{27}{5}$

$3;\ \frac{81}{5}$

$\frac{243}{5}$

85. The fourth term is equal to $\frac{9}{5}$(___) = ____.
The fifth term is equal to $\frac{27}{5}$(___) = ____.
The sixth term is equal to ____.

In Frames 86-87, find the common ratio and write the next three terms in the geometric progression $\frac{a}{c}, \frac{a^2}{c^2}, \frac{a^3}{c^3}, \ldots$.

$\frac{a^2}{c^2};\ \frac{a}{c}$

86. $r = \frac{\_\_\_}{\frac{a}{c}} = \_\_\_$.

$\frac{a^4}{c^4};\ \frac{a^5}{c^5};\ \frac{a^6}{c^6}$

87. Multiplying successively by $\frac{a}{c}$, find the next three terms to be ____, ____, and ____.

ar^{n-1}

88. If the first term and the common ratio of a geometric progression are denoted by a and r, respectively, the nth or general term is given by s_n = ______.

In Frames 89-90, find the general term of the geometric progression 2, 4, 8, ...

2; 2

89. a = ___ and r = ___.

2; 2; n

90. With these values for a and r,
$s_n = (\_\_\_)(\_\_\_)^{n-1} = 2^{(\_\_\_)}$.

In Frames 91-92, find the general term of the geometric progression $1, \frac{1}{3}, \frac{1}{9}, \ldots$.

$1;\ \frac{1}{3}$

91. a = ___ and r = ___.

$1;\ \frac{1}{3}$

92. With these values for a and r,
$s_n = (\_\_\_)(\_\_\_)^{n-1}$.

$\frac{a}{c};\ \frac{a}{c}$

93. To find the general term of the geometric progression $\frac{a}{c}, \frac{a^2}{c^2}, \frac{a^3}{c^3}, \ldots$, let the roles of a and r in $s_n = ar^{n-1}$ be played by ___, and ___, respectively, and find

$\frac{a}{c};\ \frac{a}{c};\ n$

$s_n = (\_\_\_)(\_\_\_)^{n-1} = \left(\frac{a}{c}\right)^{(\_\_\_)}$.

8; 7; 256

94. To find the eighth term of the geometric progression 2, 4, 8, . . . , use $s_n = ar^{n-1}$ with $a = 2$, $r = 2$, and n = ___, and obtain $s_8 = 2(2)^{(\_\_\_)}$ = ____.

95. To find the ninth term of the geometric progression $1, \frac{1}{3}, \frac{1}{9}, \ldots,$ use $s_n = ar^{n-1}$ with $a = 1$, $r = \frac{1}{3}$, and $n =$ ___, and obtain

$$s_9 = (1)\left(\frac{1}{3}\right)^{(\_\_\_)} = \_\_\_\_\_ .$$

9

^{8}; $\frac{1}{6561}$

96. The sum of n terms of a geometric progression is given in terms of a and r by

$$S_n = \_\_\_\_\_\_\_\_ \quad (r \neq 1).$$

$\frac{a - ar^n}{1 - r}$

97. The sum of n terms of a geometric progression is given in terms of a, s_n, and r as

$$S_n = \_\_\_\_\_\_\_\_ \quad (r \neq 1).$$

$\frac{a - rs_n}{1 - r}$

In Frames 98-101, compute $\sum_{i=1}^{5} (-2)^i$.

98. Write the first few terms in expanded form and obtain ________________.

$-2 + 4 - 8 + \cdots$

99. By inspection,

$a =$ ____, $r =$ _____ and $n = 5$.

-2; -2

100. Use $S_n = \frac{a - ar^n}{1 - r}$ $(r \neq 1)$ and write

$$S_5 = \frac{(-2) - (-2)(-2)^{(\_\_\_)}}{1 - (-2)} = \frac{-2(\_\_\_ - \_\_\_\_\_)}{3}.$$

^{5}; 1; $(-2)^5$

101. Therefore, $S_5 = \frac{-2(\_\_\_\_)}{3} = \_\_\_\_$.

^{33}; -22

In Frames 102-106, compute $\sum_{j=3}^{7} \left(\frac{1}{4}\right)^j$.

102. Write the first few terms in expanded form and obtain ____________________.

$\left(\frac{1}{4}\right)^3 + \left(\frac{1}{4}\right)^4 + \left(\frac{1}{4}\right)^5 + \cdots$

103. By inspection, $a =$ ______, $r =$ _____.

$\left(\frac{1}{4}\right)^3$; $\left(\frac{1}{4}\right)$

104. Since j ranges over the integers from 3 through 7, $n =$ ___.

5

105. Use $S_n = \frac{a - ar^n}{1 - r}$ $(r \neq 1)$ and write

$$S_5 = \frac{\_\_\_\_\_ - \_\_\_\_\_\left(\frac{1}{4}\right)^5}{1 - \frac{1}{4}} = \frac{\left(\frac{1}{4}\right)^3\left[\_\_\_ - \_\_\_\_\_\right]}{\frac{3}{4}}.$$

$\left(\frac{1}{4}\right)^3$; $\left(\frac{1}{4}\right)^3$; 1; $\left(\frac{1}{4}\right)^5$

$\frac{1}{1024}$

106. Continuing, $S_n = \dfrac{\frac{1}{64}\left[1 - (\_\_\_\_)\right]}{\frac{3}{4}}$,

$\frac{1023}{1024}$; $\frac{341}{16,384}$

$S_n = \frac{1}{64} \cdot \frac{4}{3} \cdot \_\_\_\_ = \_\_\_\_\_$.

In Frames 107-111, compute $\sum_{k=2}^{7} 5^k$ *using the formula*

$S_n = \dfrac{a - rs_n}{1 - r}$ $(r \neq 1)$.

25 + 125 + 625 + • • •

107. Write the first few terms and obtain ____________________.

25; 5

108. By inspection, a = ____, r = ___, and n = 6.

5; 7

109. Use $s_n = ar^{n-1}$ to find s_6 and obtain

$s_6 = 25(5)^{(\_\_)} = 5^{(\_\_)}$.

1; 5^6

110. Now use $S_n = \dfrac{a - rs_n}{1 - r}$ $(r \neq 1)$ and obtain

$S_6 = \dfrac{25 - 5 \cdot 5^7}{1 - 5} = \dfrac{5^2 - 5^8}{-4} = \dfrac{5^2[\_\_ - \_\_\_]}{-4}$.

-15,624; 97,650

111. Continuing, $S_6 = \dfrac{25(\_\_\_\_\_)}{-4}$ = ______.

geometric

112. Terms between given terms in a geometric progression are called _________ means.

In Frames 113-116, insert two geometric means between 5 and 135.

3

113. The diagram 5, _?_, _?_, 135 suggests that there are ___ (how many?) multiplications by r between 5 and 135.

r^3

114. Hence, the quotient when 135 is divided by 5 must be ____.

3

115. Thus, we have $r^3 = \frac{135}{5} = 27$ from which r = ___.

3; 45

116. Therefore, the required means are 5(3) = 15 and 15(___) = ____.

11.4 INFINITE GEOMETRIC PROGRESSIONS

$\frac{a}{1 - r}$; 1

117. The sum of an infinite geometric progression is given by S_∞ = ______, provided that $|r| <$ ___.

In Frames 118-119, find the sum of the infinite geometric series $1 + \frac{1}{3} + \frac{1}{9} + \cdots$, if the sum exists.

$\frac{1}{3}$; 1

118. By inspection, $r =$ ___. Therefore, the series has a sum, since $|r| <$ ___.

1; $\frac{3}{2}$

119. Use $a =$ ___ and $r = \frac{1}{3}$ in $S_\infty = \frac{a}{1 - r}$ and obtain $S_\infty = \frac{1}{1 - \frac{1}{3}} = \frac{1}{\frac{2}{3}} =$ ___.

2; 1

120. The series $\frac{1}{4} + \frac{1}{2} + \cdots$ does not have a sum because $r =$ ___; that is, $|r| \not< $ ___.

In Frames 121-123, find the sum of the infinite geometric series $\frac{2}{3} - \frac{4}{15} + \frac{8}{75} - \cdots$, if the sum exists.

$\frac{-2}{5}$

121. Divide the second term by the first term and obtain $r = \frac{-4}{15} \div \frac{2}{3} =$ ___.

$\frac{2}{5}$; 1

122. $|r| =$ ___, and so the sum exists because $|r| <$ ___.

$\frac{-2}{5}$; $\frac{7}{5}$; $\frac{10}{21}$

123. Using $S_\infty = \frac{a}{1 - a}$, obtain $S_\infty = \frac{\frac{2}{3}}{1 - (\_\_)} = \frac{\frac{2}{3}}{(\_\_)} =$ ___.

In Frames 124-127, compute $\sum_{i=1}^{\infty} \left(-\frac{1}{3}\right)^i$, if it exists.

$-\frac{1}{3} + \frac{1}{9} - \frac{1}{27} + \cdots$

124. Write the first few terms in expanded form and obtain ___________.

$-\frac{1}{3}$; $-\frac{1}{3}$

125. By inspection, $a =$ ___ and $r =$ ___.

$\frac{1}{3}$; less

126. Since $|r| =$ ___, which is ______ (less/greater) than 1, the sum exists.

$\frac{a}{1 - r}$; $\frac{-1}{4}$

127. Substitute $a = \frac{-1}{3}$, $r = \frac{-1}{3}$ into the formula $S_\infty =$ ______ and obtain $S_\infty =$ ___.

In Frames 128-132, find a fraction equivalent to $.3535\overline{35}$, where the bar indicates the sequence of digits that endlessly repeats.

.000035

128. Write the decimal numeral as the series $.35 + .0035 +$ ______ $+ \cdots$.

.01

129. Observe that this is a geometric progression with $r = \frac{.0035}{.35} =$ ___.

1

130. Since $|r| <$ ___, the series has a sum.

$\frac{a}{1-r}$; .35

131. Substitute into the formula $S_\infty =$ ______, taking $a =$ _____ and $r = .01$.

.99
$\frac{35}{99}$

132. So, $S_\infty = \frac{.35}{1 - .01} = \frac{.35}{\_\_\_}$, which, when expressed in simple form is ___.

In Frames 133-136, find a fraction equivalent to $1.3252\overline{525}$.

.00025; .0000025

133. Write the decimal numeral as the series
$1 + .3 + .025 +$ ________ $+$ __________ $+ \cdots$.

.025; .01

134. Observe that this series is geometric, beginning with the term ______ with $r =$ _____.

.025; .025; 25; $\frac{5}{198}$

135. The series $.025 + .00025 + .0000025 + \cdots$ has a sum, because $|r| < 1$ and this sum is given by
$S_\infty = \frac{\_\_\_}{1 - .01} = \frac{\_\_\_}{.99} = \frac{\_\_\_}{990} =$ _____.

2624; $\frac{656}{495}$

136. Therefore, $1.3252\overline{525} = 1 + \frac{3}{10} + \frac{5}{198} = \frac{\_\_\_}{1980} =$ _____.

11.5

THE BINOMIAL EXPANSION

$n!$

137. If n is a positive integer,
$n(n - 1)(n - 2) \ldots (1) =$ ____.

7

138. $7 \cdot 6 \cdot 5 \cdot 4 \cdot 3 \cdot 2 \cdot 1 =$ ___! .

24

139. $4! = 4 \cdot 3 \cdot 2 \cdot 1 =$ ____.

In Frames 140-142, simplify the given expression.

5; 42; $\frac{1}{7}$

140. $\frac{3!5!}{7!} = \frac{3!5!}{7 \cdot 6 \cdot (\_\_)!} = \frac{3!}{\_\_\_} =$ ___.

$12 \cdot 11 \cdot 10$; 1320

141. $\frac{12!}{9!} = \frac{(\_\_\_\_\_\_\_\_) \cdot 9!}{9!} =$ ______.

3

142. $\frac{6! + 3!}{6! - 3!} = \frac{6 \cdot 5 \cdot 4 \cdot 3! + 3!}{6 \cdot 5 \cdot 4 \cdot (\_\_)! - 3!} =$

3!; $\frac{121}{119}$

$= \frac{3!(6 \cdot 5 \cdot 4 + 1)}{(\_\_\_)(6 \cdot 5 \cdot 4 - 1)} =$ _____.

In Frames 143-146, write the first three factors and the last three factors for $(4n)!$.

$4n - 1$

143. From the largest factor, which is $4n$, subtract 1 to obtain the next factor, which is ________.

$4n - 2$

144. From $4n - 1$, subtract 1 to obtain the third factor, ________.

3, 2, 1

145. For the factorial of any number ≥ 3, the last three factors will always be ________.

$4n - 1$; $4n - 2$

146. Therefore, $(4n)! = 4n(\_\_\_\_)(\_\_\_\_) \cdot\cdot\cdot 3 \cdot 2 \cdot 1$.

In Frame 147, write the product $10 \cdot 9 \cdot 8 \cdot 7$ in factorial notation.

$6!$; $\frac{10!}{6!}$

147. $10 \cdot 9 \cdot 8 \cdot 7 = \frac{10 \cdot 9 \cdot 8 \cdot 7 \cdot (6!)}{\_\_\_\_} = \_\_\_\_$.

1

148. By definition, $0! = \_\_\_$.

In Frames 149-153, a sequence of steps is given that yields the expansion of $(a + b)^n$, n a positive integer.

a^n

149. The first term of the expansion is ____.

1

150. The second term can be written as $\frac{na^{n-1}b}{(\_\_)!}$.

2

$2!$

151. The third term can be written as $\frac{n(n - 1)a^{n-2}b^{(\_\_)}}{\_\_\_\_}$.

$n-3$; 3

$3!$

152. The fourth term can be written as

$\frac{n(n - 1)(n - 2)a^{(\_\_\_\_)}b^{\_\_}}{\_\_\_\_}$.

b^n

153. The pattern of development of the above terms continues until the last term of the expansion, which is ____, is obtained.

In Frames 154-159, expand $(x + 2y)^4$.

x^4

154. The first term is ____.

4

155. The second term is $\frac{(\_\_)x^3(2y)}{1!}$.

3; 2

$2!$

156. The third term is $\frac{(4)(\_\_)x^2(2y)^{(\_\_)}}{\_\_\_\_}$.

2; 1

157. The fourth term is $\frac{(4)(3)(\_\_)x^{(\_\_)}(2y)^3}{3!}$.

1; 0

158. The last term is $\frac{(4)(3)(2)(\_\_)x^{(\_\_)}(2y)^4}{4!}$.

$32xy^3$; $16y^4$

159. Simplify each of the above terms and obtain

$(x + 2y)^4 = x^4 + 8x^3y + 24x^2y^2 + \_\_\_\_\_\_ + \_\_\_\_\_\_$.

In Frames 160-167, expand $\left(\frac{x}{3} - 3\right)^5$.

-3

160. Write $\left(\frac{x}{3} - 3\right)^5$ in the form $(a + b)^n$ and obtain $\left[\frac{x}{3} + (\_\_\_\_)\right]^5$.

5

161. The first term of the expansion is $\left(\frac{x}{3}\right)^{(\_\_\_)}$.

^{4} -3 ;

162. The second term is $\dfrac{(5)\left(\frac{x}{3}\right)^{(\_\_\_)}(\_\_\_\_)}{1!}$.

4; ^{2} 2!

163. The third term is $\dfrac{(5)(\_\_\_)\left(\frac{x}{3}\right)^{3}(-3)^{(\_\_\_)}}{\_\_\_\_}$.

3; ^{3} 3!

164. The fourth term is $\dfrac{(5)(4)(\_\_\_)\left(\frac{x}{3}\right)^{2}(-3)^{(\_\_\_)}}{\_\_\_\_}$.

^{1}; ^{4} 4!

165. The fifth term is $\dfrac{(5)(4)(3)(2)\left(\frac{x}{3}\right)^{(\_\_\_)}(-3)^{(\_\_\_)}}{\_\_\_\_}$.

^{0}; ^{5} 5!

166. The sixth term is $\dfrac{5!\left(\frac{x}{3}\right)^{(\_\_\_)}(-3)^{(\_\_\_)}}{\_\_\_\_}$.

$\frac{10x^3}{3}$; $135x$; 243

167. Simplify each of the above terms and obtain

$$\left(\frac{x}{3} - 3\right)^5 = \frac{x^5}{243} - \frac{5x^4}{27} + \_\_\_\_\_ - 30x^2 + \_\_\_\_\_ - \_\_\_\_\_.$$

In Frames 168-171, write the first four terms for the expansion of $\left(\frac{x}{2} - 2\right)^{12}$. *Do not simplify the terms.*

$\left(\frac{x}{2}\right)^{12}$

168. The first term is ______.

12 ; ^{11}

169. The second term is $\dfrac{(\_\_\_)\left(\frac{x}{2}\right)^{(\_\_\_)}(-2)}{1!}$.

$12 \cdot 11\left(\frac{x}{2}\right)^{10}$

170. The third term is $\dfrac{[\_\_\_\_\_\_\_\_](-2)^2}{2!}$.

$\dfrac{12 \cdot 11 \cdot 10\left(\frac{x}{2}\right)^{9}(-2)^{3}}{3!}$

171. The fourth term is ____________.

$r - 1$; $^{r-1}$

172. The rth term in a binomial expansion is given by

$$\frac{n(n-1)(n-2)\cdots(n-r+2)}{(\_\_\_\_)!}a^{n-r+1}b^{(\_\_\_\_)}.$$

In Frames 173-179, write the ninth term in the expansion of $(2x + y)^{13}$ *in simple form.*

9

173. In this expansion, $n = 13$ and, since the ninth term is being written, $r =$ ___.

8!

174. First write the denominator, which is $(r - 1)!$, and obtain ____.

8

175. Then, since the exponent of b in the formula of Frame 172 is also $r - 1$, the exponent of the symbol playing the role of b, namely y, will be ___.

5

176. Because in each term the sum of the exponents must equal 13 and one exponent is known to be 8 from Frame 175, the exponent for $(2x)$ must be $13 - 8 =$ ___.

6
13
$6 \cdot 7 \cdot 8 \cdot 9 \cdot 10 \cdot 11 \cdot 12 \cdot 13$

177. To write the factors that form the coefficient of $\frac{(2x)^5y^8}{8!}$, which is the factor of the required term thus far developed, observe that the smallest factor must be one greater than the exponent on $2x$; so it is ___; and then build by integers to n, which is ____, making the coefficient ______________.

$\frac{13 \cdot 12 \cdot 11 \cdot 10 \cdot 9 \cdot 8 \cdot 7 \cdot 6}{8!}$

178. From Frames 173-177, the ninth term of $(2x + y)^{13}$ is $\frac{\_\_\_\_\_\_\_\_\_\_(2x)^5y^8}{\_\_\_\_}$.

$41{,}184x^5y^8$

179. Simplifying, obtain:

$$\frac{\overset{1}{\cancel{6 \cdot 7 \cdot 8}} \cdot 9 \cdot \overset{2}{\cancel{10}} \cdot 11 \cdot \overset{1}{\cancel{12}} \cdot 13 \cdot \overset{4}{\cancel{2^5}}x^5y^8}{\underset{1}{\cancel{8 \cdot 7 \cdot 6}} \cdot \underset{1}{\cancel{5}} \cdot \underset{1}{\cancel{4 \cdot 3}} \cdot \underset{1}{\cancel{2}} \cdot 1} = \_\_\_\_\_\_\_\_.$$

In Frames 180-186, write the fifth term in the expansion of $(x - 2)^{12}$ in simple form.

$x + (-2)$

180. Express the binomial as a sum and obtain $(\_\_\_\_\_\_\_\_)^{12}$.

12
5

181. In this expansion, $n =$ ____ and, since the fifth term is being written, $r =$ ___.

4!

4

182. The denominator is $(r - 1)! =$ ____, and the number $(r - 1)$ is also the exponent of the last factor of the term and so this last factor is $(-2)^{(\_\_)}$.

8

183. Because the sum of the exponents in each term is always n, the next-to-last factor must be $x^{(\_\_)}$.

9
12

184. To write the factors which form the coefficient of $\frac{x^8(-2)^4}{4!}$, which is the factor of the required term thus far developed, first write the smallest one that is one greater than the exponent on x; it is ___; then build by integers to ____.

$\frac{9 \cdot 10 \cdot 11 \cdot 12x^8(-2)^4}{4!}$

185. From Frames 180-184, the required fifth term is ______________.

$7920x^8$

186. Simplifying, obtain:

$$\frac{9 \cdot 10 \cdot 11 \cdot \overset{1}{\cancel{12}} \cdot \overset{2^3}{\cancel{2^4}}x^8}{\underset{1}{\cancel{4 \cdot 3}} \cdot \underset{1}{\cancel{2}} \cdot 1} + \_\_\_\_\_\_.$$

11.6

PERMUTATIONS

permutation

187. An arrangement, in a specified order, of the elements of a set is called a __________ of the elements.

ba

188. The permutations of the elements of $\{a,b\}$ are ab and ____.

zxy; *zyx*

189. The permutations of the elements of $\{x,y,z\}$ are xyz, xzy, yzx, yxz, _____, and _____.

$P(n,n)$; $n!$

190. The number of permutations of n things taken n at a time, denoted by ________, is given by $P(n,n) =$ _____.

120

191. $P(5,5) = 5! =$ _____.

$6! = 720$

192. The number of ways six books can be arranged on a shelf is $P(6,6) =$ __________.

$8! = 40,320$

193. The number of ways 8 students can be seated at 8 desks is $P(8,8) =$ _____________.

$m \cdot n \cdot p \cdot \cdot \cdot$

194. A general counting principle states that if one thing can be done in m ways, another thing in n ways, another thing in p ways, etc., then the number of ways all the things can be done is ____________________.

4; 3

195. If there are four routes from town A to town B and three routes from town B to town C, then it is possible to go from town A to town C via town B in ___ $\cdot$ ___ ways.

$n - r + 1$

196. The number of permutations of n things taken r at a time, denoted by $P(n,r)$ is given by $P(n,r) = n(n-1)(n-2) \cdot \cdot \cdot ($__________$)$.

In Frames 197-199, determine how many different four-digit numerals can be formed from the digits 0, 1, 2, 3, 4, if no digit can be used more than once.

4

197. Let the four cells [ ] [ ] [ ] [ ] represent positions of the four digits. Then, since no numeral has its first digit 0, a digit can be chosen for the first cell in ___ ways and the cell diagram looks like [4] [ ] [ ] [ ].

4
3; 2

198. Having fixed the first digit, a digit can be chosen for the next cell from among ___ digits, for the third cell from among ___ digits and for the last cell from among ___ digits.

96

199. Now the cell diagram looks like [4] [4] [3] [2] and, from the general counting principle, the act of choosing digits for the cells can be done in $4 \times 4 \times 3 \times 2 =$ ____ ways.

In Frames 200-202, do the problem stated for Frames 197-199 under the condition that no digit may be used more than once and the number named is odd.

1; 3
2

2

200. Since the number named must be odd, the last digit should be chosen first. It must be odd and so can be chosen from the digits ___ or ___. Therefore, a last digit can be chosen in ___ ways and the cell diagram so far is

[ ] [ ] [ ] [_].

3; 3
2

201. Now, since 0 cannot be used as a first digit, a first digit can be chosen in ___ ways, a second digit in ___ ways, then a third digit in ___ ways.

[3] [3] [2] [2]

36

202. The cell diagram is now ________________ and, by the general counting principle, the cells can be filled in $3 \cdot 3 \cdot 2 \cdot 2 =$ ____ ways, which is the answer to the question.

In Frames 203-206, do the problem stated for Frames 197-199, given that no restriction is placed on the repetition of digits and the number named is even.

3

203. Since the last digit must be even, the last cell can be filled in ___ ways. (Recall that zero is even.)

4

204. Then, the first cell can be filled in ___ ways.

5

205. Continuing, since any of the five digits can be chosen for the second and third cells, they can be filled in ___ ways.

[4] [5] [5] [3]
300

206. The final cell diagram is ________________ and the solution to the problem is _____.

In Frames 207-209, determine the number of distinguishable permutations of the letters of the word "cheese."

6

207. If all of the six letters were different, there would be $P(6,6) = (\_\_\_)!$.

6!

208. Since "cheese" contains 3 letters that are the same (the e's), each distinguishable permutation contains 3! permutations that are not distinguishably different. So, if P denotes the number of distinguishable permutations, then $3!P =$ _____.

120

209. Solve this equation for P and obtain $P = \frac{6!}{3!} =$ _____.

In Frames 210-212, determine the number of distinguishable permutations of the letters of the word "hallelujah."

10!

210. First observe that there are 2 h's, 2 a's, and 3 l's. Then, extending the ideas of the preceding problem, letting P denote the number of distinguishable permutations, write $2!2!3!P =$ ______.

$\frac{10!}{2!2!3!}$

211. Solve this equation for P and obtain $P =$ _________.

151,200

212. Then, $\frac{10!}{2!2!3!} = \frac{10 \cdot 9 \cdot 8 \cdot 7 \cdot \overset{1}{\cancel{6}} \cdot 5 \cdot \overset{1}{\cancel{4}} \cdot 3 \cdot 2 \cdot 1}{\underset{1}{\cancel{2}} \cdot 1 \cdot \underset{1}{\cancel{2}} \cdot 1 \cdot \underset{1}{\cancel{3}} \cdot \underset{1}{\cancel{2}} \cdot 1} =$ __________.

3!; 3!; 8!; 1120

213. The number of distinguishable permutations, P, of the letters in the word "nineteen" is given by the equation $(\_\_\_\_)(\_\_\_\_)P = (\_\_\_\_)$, and from this, $P =$ ______.

11.7

COMBINATIONS

combination

214. If the order in which the elements of a set are considered is not of importance, then the elements of the set are called a ____________.

$\binom{n}{r}$
r

215. The number of r-element subsets of an n-element set is denoted by the symbol ____ and is read "the number of combinations of n things taken ___ at a time."

bcd

4

216. From $\{a,b,c,d\}$, a listing of the combinations of its 4 elements taken 3 at a time is *abc*, *abd*, *acd*, and _____. That is,

$$\binom{4}{3} = \_\_\_.$$

$n - r + 1$

217. In general, $\binom{n}{r} = \dfrac{n(n-1)(n-2)\cdots(\_\_\_\_\_\_)}{r!}$.

$\dfrac{n!}{r!(n-r)!}$

218. $\binom{n}{r}$ is also given by: $\binom{n}{r} = \_\_\_\_\_\_\_\_$.

4

219. $\binom{4}{3} = \dfrac{4 \cdot 3 \cdot 2}{3!} = \_\_\_$ or

$$\binom{4}{3} = \frac{4!}{3!(4-3)!} = \frac{4 \cdot 3 \cdot 2 \cdot 1}{3 \cdot 2 \cdot 1 \cdot 1} = \_\_\_.$$

4; 70

220. The number of committees consisting of four people that can be chosen from a group of eight people is

$$\binom{8}{4} = \frac{8 \cdot 7 \cdot 6 \cdot 5}{(\_\_)!} = \_\_\_.$$

In Frames 221-224, determine how many different amounts of money can be formed from a dime, a quarter, and a half-dollar.

3

221. Since there are 3 coins, using 1 coin, there are $\binom{3}{1} = \_\_\_$ different amounts.

3

222. Using 2 coins, there are $\binom{3}{2} = \dfrac{3 \cdot 2}{2 \cdot 1} = \_\_\_$ different amounts.

1

223. Using all 3 coins, there is $\binom{3}{3} = \_\_\_$ amount.

7

224. From Frames 221-223, the total number of different amounts is ___.

$\binom{52}{13}$; 39

225. The number of ways a set of 13 cards can be selected from a deck of 52 cards is

$$\binom{\_\_}{\_\_} = \frac{52!}{13!(\_\_\_)!}.$$ (See Frame 218.)

In Frames 226-229, determine the number of ways a hand consisting of 4 aces and one card that is not an ace can be selected from a deck of 52 cards.

226. If m denotes the number of ways four aces can be selected from the deck and n denotes the number of ways the non-ace card can be selected, then, by the general counting principle (see Frame 194), four aces and another card can be selected in _______ ways.

$m \cdot n$

227. Since a deck of cards contains only 4 aces, the number of combinations of these 4 aces that can be selected from the 4 aces in the deck is given by $\binom{4}{4}$; so $m = \binom{4}{4} =$ ___.

1

228. Since the deck of cards contains 48 non-ace cards, the number of ways one other card can be chosen is given by $\binom{48}{1}$; so $n = \binom{48}{1} =$ ____.

48

229. Then, from Frame 226, the solution to the stated problem is ____ • ___ = ____.

48; 1; 48

In Frames 230-233, determine how many ways a committee of 8 men and 8 women can be selected from a club with 10 men and 11 women.

230. If m denotes the number of combinations of 8 men that can be selected from the 10 men and n denotes the number of combinations of 8 women that can be selected from the 11 women, then the selection of 8 men and 8 women can be made in _______ ways.

$m \cdot n$

231. m is given by $\binom{\_\_}{\_\_}$ and equals ____.

$\binom{10}{8}$; 45

232. n is given by $\binom{\_\_}{\_\_}$ and equals _____.

$\binom{11}{8}$; 165

233. From Frame 230, the solution to the problem is ____ • _____ = ______.

45; 165; 7425

TEST PROBLEMS

1. Find the first four terms in a sequence if the general term is $s_n = \dfrac{n}{n^2 + 1}$.

2. Write $\displaystyle\sum_{j=0}^{4} \frac{(-1)^j 3^j}{j + 1}$ in expanded form.

3. Write the series 1 - 8 + 27 - 64 + 125 using sigma notation.

4. Given that $x, x + 3, \cdots$ is an arithmetic progression:

   a. write the next three terms;

   b. find the expression for the general term.

5. Find the seventeenth term in the progression $-5, -2, 1, \cdots$.

6. If the twentieth term of an arithmetic progression is -46 and the twelfth term is -30, find the fifth term.

7. Find the sum of the series $\sum_{j=10}^{20} (2j - 3)$.

8. Given that $\frac{x}{a}, -1, \frac{a}{x}, \cdots$ is a geometric progression:

   a. write the next three terms;

   b. find an expression for the general term.

9. Find the ninth term in the geometric progression $-81, -27, -9, \cdots$.

10. Find the second term of a geometric progression if the sixth term is 60 and the ratio is 3.

11. Find the sum $\sum_{j=5}^{8} (2j - 5)$.

12. Insert three geometric means between 3 and 243.

13. Find the sum of the infinite geometric series $2 - \frac{3}{2} + \frac{9}{8} + \cdots$.

14. Write $\frac{(12!)(8!)}{16!}$ in expanded form and simplify.

15. Write the first four terms in the expansion of $(x - 2y)^8$ and simplify.

16. Write the seventh term in the expansion of $(a^3 - b)^9$.

17. Find a fraction equivalent to the repeating decimal $1.027\overline{027}$.

18. In how many ways can 5 people be seated in a row of 5 chairs?

19. In how many ways can 12 boys be assigned to the 11 positions on a football team if there is one center, one fullback, and one quarterback in the group?

20. How many different license plates can be made if each plate consists of 3 letters chosen from the first 8 letters of the alphabet and these are then followed by 3 digits?

21. How many distinguishable permutations are there of the letters of the word "succeeded"?

22. How many ways can a committee of 3 men and 3 women be chosen from a group of 6 men and 8 women?

23. How many different amounts of money can be formed from a dollar bill, a half-dollar, a quarter, and a dime?

24. An urn contains 5 red and 4 black balls. In how many ways can 2 red and 3 black balls be drawn from the urn?

APPENDIX A

SYNTHETIC DIVISION; POLYNOMIAL FUNCTIONS

A.1

SYNTHETIC DIVISION

$x - c$

1. Synthetic division is an efficient method of dividing a polynomial by a binomial of the form ______.

remainder (r)

quotient $Q(x)$

2. In the format used for synthetic division, the last number in the bottom row is the ______ and the other numbers are the coefficients in the ______.

In Frames 3-8, use synthetic division to write each quotient $\frac{P(x)}{D(x)}$ *in the form* $Q(x)$ *or* $Q(x) + \frac{r}{D(x)}$.

3. $\frac{2x^4 - x^3 - 1}{x - 2} \quad (x \neq 2)$.

24

23

$\frac{23}{x - 2}$

2	2	-1	0	0	-1
		4	6	12	___
	2	3	6	12	___
	↓	↓	↓	↓	

so, $\frac{2x^4 - x^3 - 1}{x - 2} = 2x^3 + 3x^2 + 6x + 12 +$ ______.

4. $\frac{3x^3 + x^2 - 7}{x + 2} = \frac{3x^3 + x^2 - 7}{x - (-2)} \quad (x \neq -2)$.

-20

-27

$3x^2 - 5x + 10 - \frac{27}{x + 2}$

-2	3	1	0	-7
		-6	10	___
	3	-5	10	___

so, $\frac{3x^3 + x^2 - 7}{x + 2} =$ ______.

5. $\dfrac{x^6 + x^4 + x^2}{x + 1} = \dfrac{x^6 + x^4 + x^2}{x - (-1)} \quad (x \neq 1).$

$$\begin{array}{r|rrrrrrr} -1 & 1 & 0 & 1 & 0 & 1 & 0 & 0 \\ & & -1 & 1 & -2 & 2 & \_\_ & \_\_ \\ \hline & 1 & -1 & 2 & -2 & 3 & \_\_ & \_\_ \end{array}$$

-3; 3

-3; 3

$3x - 3 + \dfrac{3}{x + 1}$

so, $\dfrac{x^6 + x^4 + x^2}{x + 1} = x^5 - x^4 + 2x^3 - 2x^2 +$ ________________ .

6. $\dfrac{x^6 + 1}{x - 1}$.

$$\begin{array}{r|rrrrrrr} 1 & 1 & 0 & 0 & 0 & 0 & 0 & 1 \\ & & 1 & 1 & 1 & 1 & \_\_ & \_\_ \\ \hline & 1 & 1 & 1 & 1 & 1 & \_\_ & \_\_ \end{array}$$

1; 1

1; 2

$x^2 + x + 1 + \dfrac{2}{x - 1}$

so, $\dfrac{x^6 + 1}{x - 1} = x^5 + x^4 + x^3 +$ ________________ $(x \neq 1)$.

$x^2 - 10x + 29 - \dfrac{84}{x + 3}$

(detail below)

$$\begin{array}{r|rrrr} -3 & 1 & -7 & -1 & 3 \\ & & -3 & 30 & -87 \\ \hline & 1 & -10 & 29 & -84 \end{array}$$

7. $\dfrac{x^3 - 7x^2 - x + 3}{x + 3} =$ ________________ $(x \neq 3)$.

$2x^2 + 2x + 3$

(detail below)

$$\begin{array}{r|rrrr} 1 & 2 & 0 & 1 & -3 \\ & & 2 & 2 & 3 \\ \hline & 2 & 2 & 3 & 0 \end{array}$$

8. $\dfrac{2x^3 + x - 3}{x - 1} =$ ______________ $(x \neq 1)$.

A.2

GRAPHING POLYNOMIAL FUNCTIONS

9. **If $P(x)$ is a polynomial of degree $n \geq 1$ with real coefficients and $Q(x)$ is a polynomial of degree $n - 1$ with real coefficients, then the remainder, when $P(x)$ is divided by $x - a$ is the value $P($____$)$. This is called the ____________ theorem.**

a

remainder

In Frames 10-12, use synthetic division and the remainder theorem to find $P(1)$, $P(2)$, and $P(-3)$ if $P(x) = 2x^3 - 3x^2 + x - 1$.

10. To find $P(1)$:

$$\begin{array}{r|rrrr} 1 & 2 & -3 & 1 & -1 \\ & & 2 & -1 & \_\_ \\ \hline & 2 & -1 & 0 & \_\_ \end{array}$$

and so, $P(1) =$ ____.

0

-1

-1

11. To find $P(2)$:

$$\begin{array}{r|rrrr} 2 & 2 & -3 & 1 & -1 \\ & & 4 & 2 & \_\_ \\ \hline & 2 & 1 & 3 & \_\_ \end{array}$$

and so, $P(2) =$ ____.

6

5

5

Answers: -84; -85; -85

12. To find $P(-3)$:

-3	2	-3	1	-1
		-6	27	___
	2	-9	28	___

and so, $P(-3) =$ _____.

In Frames 13-21, graph $\{x, P(x) \mid P(x) = 3x^3 + 2x^2 - x + 1\}$.

Answers: 1; (0,1)

13. We need to find ordered pairs $(x, P(x))$ that name points of the graph. By inspection, $P(0) =$ ___. Therefore, the graph includes the point _______.

Answers: 4; 5; 5; 1; 5

14. To obtain other points on the graph, we shall use synthetic division and the remainder theorem. For $x = 1$, we have

1	3	2	-1	1
		3	5	___
	3	5	4	___

and $P(1) =$ ___, so that (___,___) is on the graph.

Answers: 30; 31; 31; 2; 31

15. For $x = 2$, we have

2	3	2	-1	1
		6	16	____
	3	8	15	____

and $P(2) =$ ____, so that (___,____) is on the graph.

Answers: 0; 1; (-1,1)

16. We now investigate values of $P(x)$ for negative values of x. For $x = -1$, we have

-1	3	2	-1	1
		-3	1	___
	3	-1	0	___

and $P(-1) = 1$, so that the point ________ is on the graph.

Answers: -14; -13; (-2,-13)

17. For $x = -2$, we have

-2	3	2	-1	1
		-6	8	_____
	3	-4	7	_____

and $P(-2) = -13$, so that the point __________ is on the graph.

Answers: -60; -59; (-3,-59)

18. For $x = -3$, we have

-3	3	2	-1	1
		-9	21	_____
	3	-7	20	_____

and $P(-3) = -59$, so that the point __________ is on the graph.

19. From Frames 13-18, we have the ordered pairs (0,1), (1,5), (2,31), (-1,1), (-2,-13), and (-3,-59), which when graphed suggest the graph below. Note that (-3, -59) was not convenient to graph.

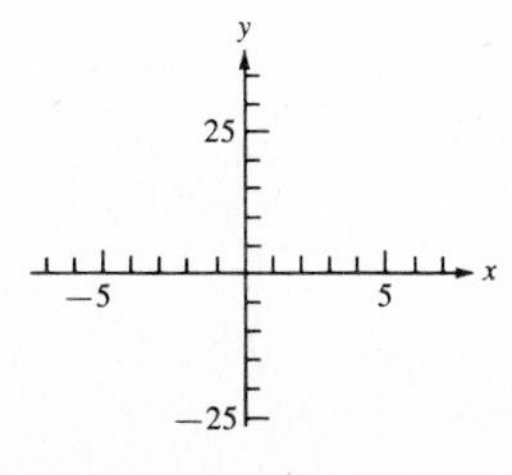

20. It is not clear whether the graph should "hump up" between $-1 < x < 0$. To investigate this, we find $P(x)$ for $x = -1/2$ below.

$$\begin{array}{r|rrrr} -\frac{1}{2} & 3 & 2 & -1 & 1 \\ & & -\frac{3}{2} & -\frac{1}{4} & \_\_\_ \\ \hline & 3 & \frac{1}{2} & -\frac{5}{4} & \_\_\_ \end{array}$$

Since $P\left(-\frac{1}{2}\right) = \frac{13}{8}$, the point ________ is on the graph.

$\frac{5}{8}$

$\frac{13}{8}$

$\left(-\frac{1}{2}, \frac{13}{8}\right)$

21. Including the point $\left(-\frac{1}{2}, \frac{13}{8}\right)$ in the graph shows that it does "hump up" for $-1 < x < 0$ as shown in the graph below:

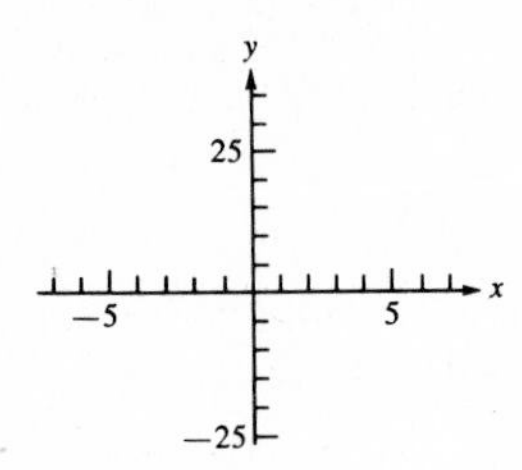

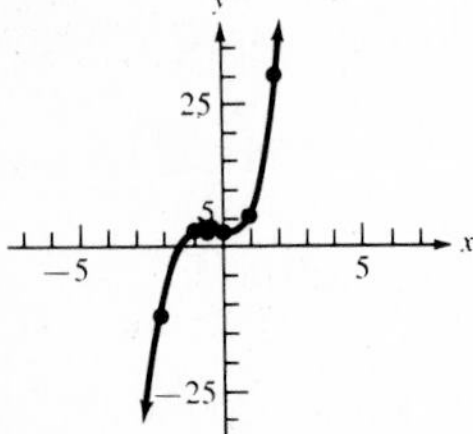
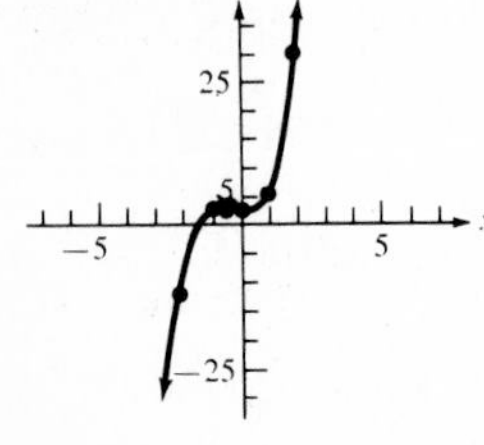

In Frames 22-28, graph $\{ x, P(x) \mid P(x) = -x^4 - x\}$.

22. We need to find ordered pairs $(x, P(x))$ that name points on the graph. By inspection, $P(0) =$ ___. Therefore, the graph includes the point _______.

0
(0,0)

23. To obtain other points on the graph, we shall use synthetic division and the remainder theorem. The synthetic division will not be shown here, but you should do it on scratch paper. For $x = 1$, we find $P(1) =$ ____ and the point ________ lies on the graph.

-2
(1,-2)

24. Continuing, we have $P(2) =$ _____ and the point _________ lies on the graph. $P(3) =$ _____ and the point _________ lies on the graph.

-18; (2,-18)
-84; (3,-84)

25. It seems clear that for $x > 3$, $P(x)$ will __________ (increase/decrease) steadily. Hence we look for no more ordered pairs with $x > 0$.

decrease

26. Proceeding as above, we now investigate values of $P(x)$ for negative values of x. We find $P(-1) =$ ___, so the point ________ is on the graph. $P(-2) =$ _____, so the point __________ is on the graph. $P(-3) =$ _____, so the point __________ is on the graph.

0
(-1,0); -14
(-2,-14); -78
(-3,-78)

27. It seems clear that for $x < -3$, $P(x)$ will __________ (increase/decrease) steadily. Hence, we look for no more ordered pairs with $x < 0$.

decrease

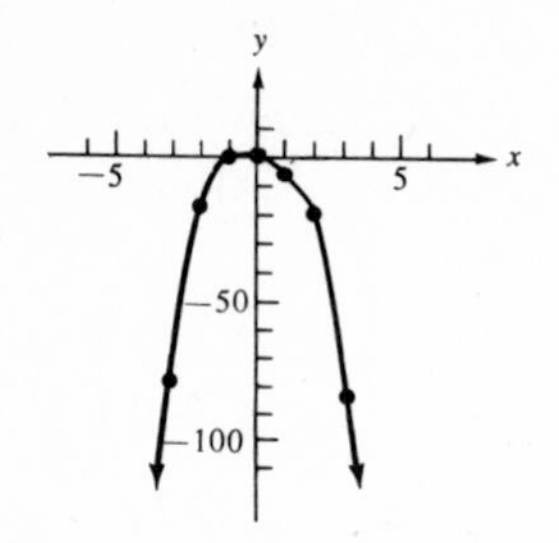

28. The graph is

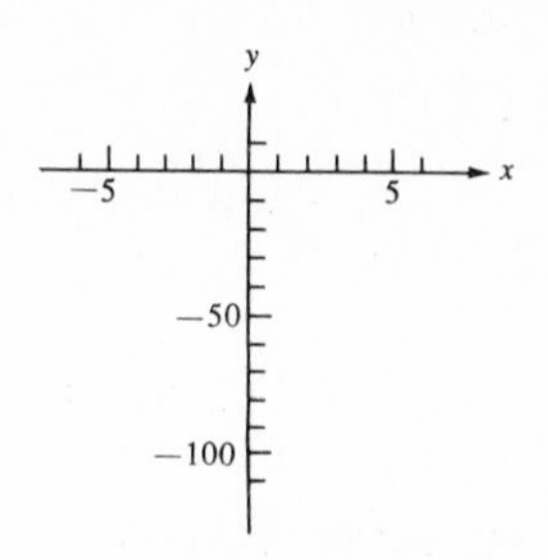

0
factor

29. Let $P(x)$ be a polynomial. Then, $x - a$ is a factor of $P(x)$ if $P(a) =$ ___. This statement is called the _______ theorem.

In Frames 30-32, use the remainder in synthetic division to determine whether $x - 1$ is a factor of $x^4 + 2x^3 - 7x^2 + 5x - 1$.

1

30. By the factor theorem, we need to show that $P($___$) = 0$.

1
0

31. Using synthetic division, we obtain

1⌋	1	2	-7	5	-1
		1	3	-4	___
	1	3	-4	1	___

0; $x - 1$

32. Therefore, $P(1) =$ ___ and we conclude that _______ is a factor of $P(x)$.

In Frames 33-36, verify that 2 is a solution of $x^3 - 2x^2 - x + 2 = 0$ and find the other solutions.

0

33. Let $P(x) = x^3 - 2x^2 - x + 2$. Then 2 is a solution if $P(2) =$ ___.

0
0

34. To find $P(2)$, divide $P(x)$ by $x - 2$ synthetically:

2⌋	1	-2	-1	2
		2	0	-2
	1	0	-1	___

Hence, $P(2) =$ ___ and 2 is a solution.

$x - 2$
$x^2 - 1$

35. By the factor theorem, since $P(2) = 0$, _______ is a factor of $P(x)$ and $P(x) = (x - 2) \cdot Q(x)$ where $Q(x) =$ __________.

$x^2 - 1$
1; -1

36. Hence, the given equation can be written as $(x - 2)(x^2 - 1) = 0$ and it is seen that the other solutions are replacements for x such that _______ $= 0$. By inspection, these are ___ and ___.

TEST PROBLEMS

In Problems 1 and 2, use synthetic division to write the given quotient in the form $Q(x) + \frac{r}{x - a}$ (r a constant) or as a polynomial in simplest form.

1. $\frac{x^3 - 3x^2 + 2x}{x - 3}$.

2. $\frac{y^5 + 1}{y + 1}$.

3. If $P(x) = 2x^3 - 3x^2 + x + 1$, use synthetic division to find $P(1)$ and $P(-1)$.

4. Use synthetic division and the remainder theorem to find solutions to $y = x^3 - x^2 + 3x$ and then graph the equation.

5. a. Is $x + 1$ a factor of $x^3 + x^2 - 4x - 4$?
   b. Is $x - 1$ a factor of $x^3 + x^2 - 4x - 4$?

6. Verify that $x = -1$ is a solution of $x^3 + x^2 - 4x - 4$ and find the other solutions.

APPENDIX B

MATRICES AND DETERMINANTS

B.1

MATRICES

matrix
elements

1. A rectangular array of real numbers is called a ________. The real numbers are called entries or __________ of the matrix.

order

2. The ordered pair having as first component the number of rows (horizontal) and as second component the number of columns (vertical) in the matrix names the dimension or _______ of the matrix.

3

3. $\begin{bmatrix} 1 & 3 & 5 \\ 2 & 4 & 6 \end{bmatrix}$ is a matrix of order 2 × ___.

3 × 2

4. $\begin{bmatrix} 1 & 4 \\ 2 & 5 \\ 3 & 6 \end{bmatrix}$ is a matrix of order _______.

1 × 3

5. $[1 \quad 2 \quad 3]$ is a matrix of order _______.

rows

real

row

6. Two matrices, A and B, are row-equivalent if each can be obtained from the other by a finite number of transformations of the following kinds:

   1. Interchanging any two _______.
   2. Multiplying each entry in a row by a nonzero __________ number.
   3. Adding the same multiple of each entry in any row to the corresponding entry in another _______.

~
row-equivalent

7. The transformations described in Frame 6 are called elementary transformations. The symbol used to denote row equivalence is ___. $A \sim B$ is read, "A is _______________ to B."

In Frames 8-15, show that $\begin{vmatrix} 2 & 2 & 4 \\ 1 & -3 & 0 \\ 2 & -4 & 0 \end{vmatrix} \sim \begin{bmatrix} 1 & 0 & 0 \\ 0 & 1 & 0 \\ 0 & 0 & 1 \end{bmatrix}$. *This is being done to demonstrate how to change a matrix into a row-equivalent matrix.*

8. We first observe that interchanging row 1 and row 2 will achieve two things. It will make the first and third entries of row 1 a 1 and a 0, respectively. Acting on the above observation, we have

$$\begin{bmatrix} 2 & 2 & 4 \\ 1 & -3 & 0 \\ 2 & -4 & 0 \end{bmatrix} \sim \underline{\hspace{3cm}}.$$

$$\begin{bmatrix} 1 & -3 & 0 \\ 2 & 2 & 4 \\ 2 & -4 & 0 \end{bmatrix}$$

9. To obtain a 0 as the first entry of row 2 in $\begin{bmatrix} 1 & -3 & 0 \\ 2 & 2 & 4 \\ 2 & -4 & 0 \end{bmatrix}$, we multiply each entry of row 1 by -2 and add each product to the corresponding entry of row 2 and we have

$$\begin{bmatrix} 1 & -3 & 0 \\ 2 & 2 & 4 \\ 2 & -4 & 0 \end{bmatrix} \sim \begin{bmatrix} 1 & -3 & 0 \\ \_\_ & \_\_ & \_\_ \\ 2 & -4 & 0 \end{bmatrix}.$$

0 8 4

10. To obtain a 0 as the first entry of row 3 in $\begin{bmatrix} 1 & -3 & 0 \\ 0 & 8 & 4 \\ 2 & -4 & 0 \end{bmatrix}$, we multiply each entry of row 1 by -2 and add each product to the corresponding entry of row 3 and we have

$$\begin{bmatrix} 1 & -3 & 0 \\ 0 & 8 & 4 \\ 2 & -4 & 0 \end{bmatrix} \sim \begin{bmatrix} 1 & -3 & 0 \\ 0 & 8 & 4 \\ \_\_ & \_\_ & \_\_ \end{bmatrix}.$$

0 2 0

11. To obtain a 1 as the third entry of row 3 in the last matrix, we interchange rows 2 and 3 and then multiply each entry of row 3 by 1/4 and obtain

$$\begin{bmatrix} 1 & -3 & 0 \\ 0 & 8 & 4 \\ 0 & 2 & 0 \end{bmatrix} \sim \begin{bmatrix} 1 & -3 & 0 \\ 0 & 2 & 0 \\ 0 & 8 & 4 \end{bmatrix} \sim \underline{\hspace{3cm}}.$$

$$\begin{bmatrix} 1 & -3 & 0 \\ 0 & 2 & 0 \\ 0 & 2 & 1 \end{bmatrix}$$

12. To obtain a 1 as the second entry of row 2 in $\begin{bmatrix} 1 & -3 & 0 \\ 0 & 2 & 0 \\ 0 & 2 & 1 \end{bmatrix}$, we multiply each entry of row 2 by 1/2 and we have

$$\begin{bmatrix} 1 & -3 & 0 \\ 0 & 2 & 0 \\ 0 & 2 & 1 \end{bmatrix} \sim \underline{\hspace{3cm}}.$$

$$\begin{bmatrix} 1 & -3 & 0 \\ 0 & 1 & 0 \\ 0 & 2 & 1 \end{bmatrix}$$

13. To obtain a 0 as the second entry of row 1 in the last matrix, we multiply each entry of row 2 by 3 and add each product to the corresponding entry of row 1 and obtain

$$\begin{bmatrix} 1 & -3 & 0 \\ 0 & 1 & 0 \\ 0 & 2 & 1 \end{bmatrix} \sim \underline{\hspace{3cm}}.$$

$$\begin{bmatrix} 1 & 0 & 0 \\ 0 & 1 & 0 \\ 1 & 2 & 1 \end{bmatrix}$$

14. To obtain a 0 as the second entry of row 3 in the last matrix, we multiply each entry of row 2 by -2 and add each product to the corresponding entry of row 3 and obtain

$$\begin{bmatrix} 1 & 0 & 0 \\ 0 & 1 & 0 \\ 0 & 2 & 1 \end{bmatrix} \sim \underline{\hspace{3cm}}.$$

$$\begin{bmatrix} 1 & 0 & 0 \\ 0 & 1 & 0 \\ 0 & 0 & 1 \end{bmatrix}$$

$\begin{bmatrix} 1 & 0 & 0 \\ 0 & 1 & 0 \\ 0 & 0 & 1 \end{bmatrix}$

15\. From Frames 8-14, we have $\begin{bmatrix} 2 & 2 & 4 \\ 1 & -3 & 0 \\ 2 & -4 & 0 \end{bmatrix} \sim$ ____________ as required.

coefficient

augmented

16\. In a system of linear equations of the form

$$a_1x + b_1y + c_1z = d_1$$
$$a_2x + b_2y + c_2z = d_2$$
$$a_3x + b_3y + c_3z = d_3\,,$$

the matrix $\begin{bmatrix} a_1 & b_1 & c_1 \\ a_2 & b_2 & c_2 \\ a_3 & b_3 & c_3 \end{bmatrix}$ is called the ____________ matrix and the matrix $\begin{bmatrix} a_1 & b_1 & c_1 & d_1 \\ a_2 & b_2 & c_2 & d_2 \\ a_3 & b_3 & c_3 & d_3 \end{bmatrix}$ is called the ____________ matrix.

$\begin{bmatrix} 3 & -2 & 1 \\ 1 & 1 & 1 \\ 2 & -1 & 3 \end{bmatrix}$

$\begin{bmatrix} 3 & -2 & 1 & 1 \\ 1 & 1 & 1 & 0 \\ 2 & -1 & 3 & 2 \end{bmatrix}$

17\. In $\begin{aligned} 3x - 2y + z &= 1 \\ x + y + z &= 0 \\ 2x - y + 3z &= 2 \end{aligned}$, the coefficient matrix is ____________ and the augmented matrix is ____________.

Frames 18-22 demonstrate the validity of a matrix method of solving the linear system

$$2x + y = 8 \quad (1)$$
$$4x - 3y = 21. \quad (2)$$

8
21

18\. The augmented matrix A of the given system is

$$A = \begin{bmatrix} 2 & 1 & \_\_ \\ 4 & -3 & \_\_ \end{bmatrix}.$$

5

19\. Multiplying both members of Equation (1) by -2 and adding each product to the corresponding terms of Equation (2) yields the equation $0x - 5y =$ ___, which together with Equation (1) forms the equivalent system:

$$2x + y = 8 \quad (1)$$
$$-5y = 5. \quad (2')$$

0; -5; 5

20\. The augmented matrix, A', of the linear system of Frame 19 is row-equivalent to matrix A of the given system because it can be obtained from matrix A by the elementary row-transformation of multiplying each entry of the first row of matrix A by -2 and adding each product to the corresponding entry of row 2. Thus, we have

$$A = \begin{bmatrix} 2 & 1 & 8 \\ 4 & -3 & 21 \end{bmatrix} \sim \begin{bmatrix} 2 & 1 & 8 \\ \_\_ & \_\_ & \_\_ \end{bmatrix} = A'.$$

-1

$\frac{9}{2}$;

$\left\{\left(\frac{9}{2}, -1\right)\right\}$

21\. Since, from Equation (2') of Frame 19, $y =$ ____, we can substitute -1 for y in Equation (1) to obtain $2x - 1 = 8$. From this $x =$ ___ and the solution of the system is ____________.

22. In terms of matrices, the solution above suggests that augmented matrix A' be obtained from augmented matrix A. From this the system of Frame 19 can be written and can then be solved as in Frame 21.

In Frames 23-26, use the method described in Frame 22 to solve the system

$$2x + 6y = 14$$
$$3x - 4y = -5.$$

$\begin{bmatrix} 2 & 6 & 14 \\ 3 & -4 & -5 \end{bmatrix}$

23. The augmented matrix, A, of the given system is

$A =$ __________.

0

24. We now wish to obtain a matrix, A', which is row-equivalent to A and whose first entry of the second row is ___. This can be done by multiplying each entry of row 1 by $-\frac{3}{2}$ and adding each product to the corresponding member of row 2. Thus,

-13; -26

$$A = \begin{bmatrix} 2 & 6 & 14 \\ 3 & -4 & -5 \end{bmatrix} \sim \begin{bmatrix} 2 & 6 & 14 \\ 0 & \_\_\_ & \_\_\_ \end{bmatrix} = A'.$$

25. Matrix A' corresponds to the system

$2x + 6y = 14$

__________.

$-13y = -26$

2
12
1
{(1,2)}

26. From $-13y = -26$ we find $y =$ ___ and substituting 2 for y in the first equation of Frame 25 we obtain $2x +$ ___ $= 14$, from which $x =$ ___. Hence, the solution set of the system is _______.

In Frames 27-30, use the augmented matrix method used in the above frames to solve the system

$$4x + y = 5$$
$$-3x + 2y = -1.$$

$\begin{bmatrix} 4 & 1 & 5 \\ -3 & 2 & -1 \end{bmatrix}$

27. The augmented matrix, A, of the system is $A =$ __________.

$\frac{11}{4}$; $\frac{11}{4}$ (see below)

Obtain row 2 of A' by multiplying row 1 of A by $\frac{3}{4}$ and adding each product to the corresponding entry of row 2.

28. Then, $A = \begin{bmatrix} 4 & 1 & 5 \\ -3 & 2 & -1 \end{bmatrix} \sim \begin{bmatrix} 4 & 1 & 5 \\ 0 & \_\_\_ & \_\_\_ \end{bmatrix} = A'.$

29. A' corresponds to the system

$4x + y = 5$

__________.

$\frac{11}{4}y = \frac{11}{4}$

1

1; 1
{(1,1)}

30. From $\frac{11}{4}y = \frac{11}{4}$ we find $y =$ ___, and substituting $y = 1$ into $4x + y = 5$, we obtain $4x +$ ___ $= 5$, from which $x =$ ___. Hence, the solution set of the system is _______.

In Frames 31-37, use the augmented matrix method to solve the system

$$\begin{aligned} x + y + z &= 6 \\ 2x - y + z &= 3 \\ 4x + 2y - z &= 5. \end{aligned}$$

31. The augmented matrix, A, of the system is

$A =$ ________________ .

Answer: $\begin{bmatrix} 1 & 1 & 1 & 6 \\ 2 & -1 & 1 & 3 \\ 4 & 2 & -1 & 5 \end{bmatrix}$

32. We seek a matrix that is row-equivalent to A of the form $\begin{bmatrix} 1 & 1 & 1 & 6 \\ \_\_ & a & b & c \\ \_\_ & \_\_ & d & e \end{bmatrix}$.

Answer: 0; 0; 0

33. To obtain a matrix that is row-equivalent to A with a 0 as the first entry of row 2, we multiply the entries of row 1 by ____ and add each product to the corresponding entries of row 2, thus obtaining

$$A = \begin{bmatrix} 1 & 1 & 1 & 6 \\ 2 & -1 & 1 & 3 \\ 4 & 2 & -1 & 5 \end{bmatrix} \sim \begin{bmatrix} 1 & 1 & 1 & 6 \\ \_\_ & \_\_ & \_\_ & \_\_ \\ 4 & 2 & -1 & 5 \end{bmatrix} = A'.$$

Answer: -2; 0; -3; -1; -9

34. Now obtain a matrix, A'', which is row-equivalent to A' with a 0 as the first entry of row ___.

$$A' = \begin{bmatrix} 1 & 1 & 1 & 6 \\ 0 & -3 & -1 & -9 \\ 4 & 2 & -1 & 5 \end{bmatrix} \sim \begin{bmatrix} 1 & 1 & 1 & 6 \\ 0 & -3 & -1 & -9 \\ \_\_ & \_\_ & \_\_ & \_\_ \end{bmatrix} = A''.$$

Answer: 3; 0 -2 -5 -19

35. Next, obtain a matrix, A''', which is row-equivalent to A'' with the second entry of row 3 a ___. To do this, multiply the entries of row ___ of A by $-\frac{2}{3}$ and add each product to the corresponding entries of row 3. Thus:

$$A'' = \begin{bmatrix} 1 & 1 & 1 & 6 \\ 0 & -3 & -1 & -9 \\ 0 & -2 & -5 & -19 \end{bmatrix} \sim \begin{bmatrix} 1 & 1 & 1 & 6 \\ 0 & -3 & -1 & -9 \\ \_\_ & \_\_ & \_\_ & \_\_ \end{bmatrix} = A'''.$$

Answer: 0; 2; 0 0 $-\frac{13}{3}$ -13

36. A''' corresponds to the system

$x + y + z = 6$ (1)

________________ (2)

________________ (3)

Answer: $-3y - z = -9$; $\frac{-13}{3}z = -13$

37. Solving Equation (3) for z we find $z =$ ___. Substitute 3 for z in Equation (2) and obtain $-3y - (\_\_) = -9$ from which $y =$ ___. Then substitute 3 for z and 2 for y in Equation (___) and obtain $x +$ ___ $+$ ___ $= 6$ from which $x =$ ___. Hence, the solution set is __________.

Answer: 3; 3; 2; 1; 2; 3; 1; {(1,2,3)}

B.2

LINEAR SYSTEMS IN TWO VARIABLES—SOLUTION BY DETERMINANTS

38. The value of the determinant $\begin{vmatrix} a_1 & b_1 \\ a_2 & b_2 \end{vmatrix}$ is defined to be __________.

Answer: $a_1b_2 - a_2b_1$

3; 1; -7

39. $\begin{vmatrix} 2 & 1 \\ 3 & -2 \end{vmatrix} = 2(-2) - (\_\_)(\_\_) = \_\_\_\_.$

-12; 2

40. $\begin{vmatrix} 5 & 4 \\ -3 & -2 \end{vmatrix} = -10 - (\_\_\_\_) = \_\_\_.$

0; $\frac{-3}{8}$; $\frac{3}{8}$

41. $\begin{vmatrix} \frac{2}{3} & -\frac{1}{2} \\ \frac{3}{4} & 0 \end{vmatrix} = (\_\_) - \left(\_\_\_\right) = \_\_\_.$

a_1 b_1
a_2 b_2

a_1 c_1
a_2 c_2

42. The solution of the system of equations

$$a_1x + b_1y = c_1$$
$$a_2x + b_2y = c_2$$

by Cramer's rule is

$$x = \frac{D_x}{D}$$

and $$y = \frac{D_y}{D},$$

where $D = \begin{vmatrix} \_\_\_ \end{vmatrix}$, $D_x = \begin{vmatrix} c_1 & b_1 \\ c_2 & b_2 \end{vmatrix}$,

and $D_y = \begin{vmatrix} \_\_\_ \end{vmatrix}$.

In Frames 43-46, solve the system

$$3x + 4y = 10$$
$$x - 2y = 0$$

by Cramer's rule.

-10

43. $D = \begin{vmatrix} 3 & 4 \\ 1 & -2 \end{vmatrix} = \_\_\_\_.$

-20

44. $D_x = \begin{vmatrix} 10 & 4 \\ 0 & -2 \end{vmatrix} = \_\_\_\_.$

-10

45. $D_y = \begin{vmatrix} 3 & 10 \\ 1 & 0 \end{vmatrix} = \_\_\_\_.$

2; 1

46. $x = \frac{D_x}{D} = \_\_\_$ and $y = \frac{D_y}{D} = \_\_\_.$

In Frames 47-50, solve the system

$$\frac{2}{3}x - y = 4$$
$$x - \frac{3}{4}y = 6$$

by Cramer's rule.

$2x - 3y = 12$
$4x - 3y = 24$

47. Rewrite each equation equivalently, free of fractions, and obtain

$\_\_\_\_\_\_\_\_$
$\_\_\_\_\_\_\_\_.$

$\begin{vmatrix} 2 & -3 \\ 4 & -3 \end{vmatrix}$; $\begin{vmatrix} 2 & 12 \\ 4 & 24 \end{vmatrix}$

48. From the equations of Frame 47,

$D = \begin{vmatrix} \_\_\_\_ \end{vmatrix}$, $D_x = \begin{vmatrix} 12 & -3 \\ 24 & -3 \end{vmatrix}$, $D_y = \begin{vmatrix} \_\_\_\_ \end{vmatrix}$

6; 36; 0

49. Evaluating D, D_x, and D_y, obtain

$D =$ ___, $D_x =$ ___, $D_y =$ ___.

6; 0

50. $x = \frac{D_x}{D} =$ ___ and $y = \frac{D_y}{D} =$ ___.

In Frames 51-56, solve the system

$$3x = 2y + 12$$
$$x = 4$$

by Cramer's rule.

$3x - 2y = 12$
$x + 0y = 4$

51. Rewrite the equations in standard form and obtain

___
___.

$\begin{matrix} 3 & -2 \\ 1 & 0 \end{matrix}$; 2

52. $D = \begin{vmatrix} & \\ & \end{vmatrix} =$ ___.

$\begin{matrix} 12 & -2 \\ 4 & 0 \end{matrix}$; 8

53. $D_x = \begin{vmatrix} & \\ & \end{vmatrix} =$ ___.

$\begin{matrix} 3 & 12 \\ 1 & 4 \end{matrix}$; 0

54. $D_y = \begin{vmatrix} & \\ & \end{vmatrix} =$ ___.

4; 0

55. $x =$ ___ and $y =$ ___.

dependent; inconsistent

56. If $D = 0$ in Cramer's rule, then the system is either ___ or ___.

B.3

THIRD-ORDER DETERMINANTS

three

57. A determinant of the form $\begin{vmatrix} a_1 & b_1 & c_1 \\ a_2 & b_2 & c_2 \\ a_3 & b_3 & c_3 \end{vmatrix}$ is a determinant of order ___.

row; column

58. The minor of an element in a determinant is defined to be the determinant that remains after deleting the ___ and ___ in which the element appears.

$\begin{matrix} b_1 & c_1 \\ b_3 & c_3 \end{matrix}$

59. In $\begin{vmatrix} a_1 & b_1 & c_1 \\ a_2 & b_2 & c_2 \\ a_3 & b_3 & c_3 \end{vmatrix}$ the minor of the element a_2 is found by deleting a_2's row and column as shown by the shading to obtain $\begin{vmatrix} & \\ & \end{vmatrix}$.

$\begin{matrix} a_1 & c_1 \\ a_2 & c_2 \end{matrix}$

60. In $\begin{vmatrix} a_1 & b_1 & c_1 \\ a_2 & b_2 & c_2 \\ a_3 & b_3 & c_3 \end{vmatrix}$ the minor of the element b_3 is found by deleting b_3's row and column as shown by the shading to obtain $\begin{vmatrix} & \\ & \end{vmatrix}$.

61. The value of the third-order determinant can be found by expanding it by minors about any _____ or _________.

row; column

62. The sign of each term of the expansion of the third-order determinant can be found by choosing from the sign array

___ ___ ___
___ ___ ___
___ ___ ___

that sign whose position corresponds to the position of the element whose minor is being written.

+ - +
- + -
+ - +

63. The expansion of $\begin{vmatrix} a_1 & b_1 & c_1 \\ a_2 & b_2 & c_2 \\ a_3 & b_3 & c_3 \end{vmatrix}$ by minors about the second row is

$$-a_2\begin{vmatrix} b_1 & c_1 \\ b_3 & c_3 \end{vmatrix} + \_\_\_\begin{vmatrix} a_1 & c_1 \\ a_3 & c_3 \end{vmatrix} \_\_\_\begin{vmatrix} a_1 & b_1 \\ a_3 & b_3 \end{vmatrix}.$$

b_2; $-c_2$

64. In Frame 63, the sign in front of the first term of the expansion is negative because a_2's position in the sign array is __________. The sign in front of the second term is positive because b_2's position in the sign array is __________. Similarly, the sign in front of the third term is negative because of the position of ____ in the sign array.

negative

positive

c_2

In Frames 65-67, expand the determinant $\begin{vmatrix} -1 & 2 & -1 \\ -2 & 0 & -2 \\ 4 & 3 & -3 \end{vmatrix}$ *about the first column and find its value.*

65. The expansion about the first column is

$$+(-1)\begin{vmatrix} 0 & -2 \\ 3 & -3 \end{vmatrix} \_\_\_\_\begin{vmatrix} 2 & -1 \\ 3 & -3 \end{vmatrix} \_\_\_\_\begin{vmatrix} 2 & -1 \\ 0 & -2 \end{vmatrix}.$$

-(-2); +4

66. Evaluating each second-order determinant, obtain

(-1)[0 - (____)] + 2[(____) - (____)] + 4[(____) - (___)].

-6; -6; -3; -4; 0

67. Therefore,

$$\begin{vmatrix} -1 & 2 & -1 \\ -2 & 0 & -2 \\ 4 & 3 & -3 \end{vmatrix} = (-1)(6) + 2(\_\_\_\_) + 4(\_\_\_\_) = (\_\_\_\_\_).$$

-3; -4; -28

In Frames 68-70, expand the determinant $\begin{vmatrix} -1 & 2 & -1 \\ -2 & 0 & -2 \\ 4 & 3 & -3 \end{vmatrix}$ *about the second row and find its value.*
(Note: This is the same determinant evaluated in Frames 65-67.)

68. The expansion about the second row is

$$-(-2)\begin{vmatrix} 2 & -1 \\ 3 & -3 \end{vmatrix} + \_\_\_\_\_\_\_\_\_\_ - \_\_\_\_\_\_\_\_\_\_\_\_ .$$

$0\begin{vmatrix} -1 & -1 \\ 4 & -3 \end{vmatrix}$; $(-2)\begin{vmatrix} -1 & 2 \\ 4 & 3 \end{vmatrix}$

-3; 7; -11

69. Evaluating each second-order determinant, obtain
2(____) + 0(___) + 2(_____).

-28

70. Therefore, $\begin{vmatrix} -1 & 2 & -1 \\ -2 & 0 & -2 \\ 4 & 3 & -3 \end{vmatrix} =$ _____.

In Frames 71-73, solve the equation

$$\begin{vmatrix} x & 0 & 0 \\ 1 & 2 & 3 \\ 0 & -2 & 3 \end{vmatrix} = 36 \text{ for } x.$$

$\begin{matrix} 2 & 3 \\ -2 & 3 \end{matrix}$

71. Expand the left member about the first row and obtain

$$x\begin{vmatrix} \_\_ & \_\_ \\ \_\_ & \_\_ \end{vmatrix} - 0\begin{vmatrix} 1 & 3 \\ 0 & 3 \end{vmatrix} + 0\begin{vmatrix} 1 & 2 \\ 0 & -2 \end{vmatrix} = 36.$$

$12x = 36$

72. Evaluate the left member and obtain the equation _________.

3

73. Therefore, $x =$ ___.

B.4

LINEAR SYSTEMS IN THREE VARIABLES—SOLUTION BY DETERMINANTS

$\begin{vmatrix} a_1 & b_1 & c_1 \\ a_2 & b_2 & c_2 \\ a_3 & b_3 & c_3 \end{vmatrix}$; $\begin{vmatrix} d_1 & b_1 & c_1 \\ d_2 & b_2 & c_2 \\ d_3 & b_3 & c_3 \end{vmatrix}$

$\begin{vmatrix} a_1 & d_1 & c_1 \\ a_2 & d_2 & c_2 \\ a_3 & d_3 & c_3 \end{vmatrix}$; $\begin{vmatrix} a_1 & b_1 & d_1 \\ a_2 & b_2 & d_2 \\ a_3 & b_3 & d_3 \end{vmatrix}$

74. The linear system in three variables

$$a_1x + b_1y + c_1z = d_1$$
$$a_2x + b_2y + c_2z = d_2$$
$$a_3x + b_3y + c_3z = d_3$$

has the solution $x = \frac{D_x}{D}$, $y = \frac{D_y}{D}$, $z = \frac{D_z}{D}$,

where $D =$ __________, $D_x =$ __________

$D_y =$ __________, $D_z =$ __________.

In Frames 75-77, solve the following system by Cramer's rule:

$$2x - 6y + 3z = -12$$
$$3x - 2y + 5z = -4$$
$$4x + 5y - 2z = 10.$$

$\begin{matrix} 2 & -6 & 3 \\ 3 & -2 & 5 \\ 4 & 5 & -2 \end{matrix}$; $\begin{matrix} -12 \\ -4 \\ 10 \end{matrix}$

$\begin{matrix} -12 \\ -4 \\ 10 \end{matrix}$; $\begin{matrix} -12 \\ -4 \\ 10 \end{matrix}$

75. $D = \begin{vmatrix} \_\_ \end{vmatrix}$, $D_x = \begin{vmatrix} \_\_ & -6 & 3 \\ \_\_ & -2 & 5 \\ \_\_ & 5 & -2 \end{vmatrix}$,

$D_y = \begin{vmatrix} 2 & \_\_ & 3 \\ 3 & \_\_ & 5 \\ 4 & \_\_ & -2 \end{vmatrix}$, $D_z = \begin{vmatrix} 2 & -6 & \_\_ \\ 3 & -2 & \_\_ \\ 4 & 5 & \_\_ \end{vmatrix}$.

-129; 0; -258; 0

76. Evaluate D, D_x, D_y, D_z and obtain

$D =$ ______, $D_x =$ ____, $D_y =$ ______, $D_z =$ ____.

0; 2; 0

77. $x = \frac{D_x}{D} =$ ____, $y = \frac{D_y}{D} =$ ____, $z = \frac{D_z}{D} =$ ____.

In Frames 78-80, solve the following system by Cramer's rule:

$$2x + y = 18$$
$$y + z = -1$$
$$3x - 2y - 5z = 38.$$

78. The determinants D, D_x, D_y, and D_z are

$D =$ ______, $D_x =$ ______,

$D_y =$ ______, $D_z =$ ______.

$\begin{vmatrix} 2 & 1 & 0 \\ 0 & 1 & 1 \\ 3 & -2 & -5 \end{vmatrix}$; $\begin{vmatrix} 18 & 1 & 0 \\ -1 & 1 & 1 \\ 38 & -2 & -5 \end{vmatrix}$

$\begin{vmatrix} 2 & 18 & 0 \\ 0 & -1 & 1 \\ 3 & 38 & -5 \end{vmatrix}$; $\begin{vmatrix} 2 & 1 & 18 \\ 0 & 1 & -1 \\ 3 & -2 & 38 \end{vmatrix}$

79. Evaluate D, D_x, D_y, and D_z and obtain

$D =$ ___, $D_x =$ ___, $D_y =$ ___, $D_z =$ ___.

-3; -21; -12; 15

80. $x =$ ___, $y =$ ___, $z =$ ___.

7; 4; -5

81. If $D = 0$ for a linear system in three variables, the system has ____ (how many?) unique solutions.

no

TEST PROBLEMS

In Problems 1-2, solve each system by the augmented matrix method.

1. $x + 4y = -14$
   $3x + 2y = -2.$

2. $2x + 4y + z = 0$
   $5x + 3y - 2z = 1$
   $4x - 7y - 7z = 6.$

3. Evaluate:
   $\begin{vmatrix} 10 & 3 \\ -10 & -2 \end{vmatrix}.$

4. Solve by Cramer's rule:
   $3x - 4y = -2$
   $x - 2y = 0.$

5. Evaluate:
   $\begin{vmatrix} 2 & 3 & 1 \\ 0 & 1 & 0 \\ -4 & 2 & 1 \end{vmatrix}.$

6. Solve for x:
   $\begin{vmatrix} x & 1 & 1 \\ 0 & x & 1 \\ 0 & x & 0 \end{vmatrix} = -4.$

7. Solve by Cramer's rule:
   $3x - 2y + 5z = 6$
   $4x - 4y + 3z = 0$
   $5x - 4y + z = -5.$

APPENDIX C

LINEAR INTERPOLATION

LINEAR INTERPOLATION

In Frames 1-7, find $\log_{10}$ 34.67.

1
$\log_{10}$ 3.467

1. The characteristic of $\log_{10}$ 34.67 is ___, and the mantissa is given by ____________.

0.5391

0.5403

2. Use consecutive x's in the table to bound 3.467 and set up the following tabular arrangement:

x	$\log_{10} x$
3.460	______
3.467	?
3.470	______

.0012

3. Compute the differences indicated by the pairings below and obtain:

$$10\left\{\begin{array}{l} 7\left\{\begin{array}{l}3.460\\3.467\end{array}\right.\\ \quad 3.470\end{array}\right. \quad \begin{array}{|l} \left.\begin{array}{l}0.5391\\ ?\end{array}\right\} y \\ 0.5403\end{array}\Bigg\} \text{______.}$$

Note that for convenience 7 and 10 have been written for 0.007 and 0.010 respectively since only their ratios will be involved.

$\frac{y}{.0012}$

4. Equate the ratios of corresponding differences and obtain the proportion

$$\frac{7}{10} = \text{______} .$$

.0008 (detail below)
(First obtain $y = \frac{7}{10}(.0012)$)

5. Solve for y, rounding off to four decimal places and obtain y = ______.

0.5399

6. Add this value of y to 0.5391 to obtain $\log_{10}$ 3.467 = ______.

1.5399

7. So $\log_{10} 34.67 = 1 + 0.5399 =$ ________.

In Frames 8-14, find $\log_{10}$ 0.08753.

(8 - 10)
8.753

8. The characteristic is __________ and the mantissa is $\log_{10}$ ______.

8.750

8.760

9. Use consecutive x's in the table to bound 8.753 and set up the following tabular arrangement:

x	$\log_{10} x$
______	0.9420
8.753	?
______	0.9425

.0005

10. Compute the differences indicated by the pairings below and obtain:

$$10\left\{\begin{array}{l} 3\left\{\begin{array}{l} 8.750 \\ 8.753 \end{array}\right. \\ \quad 8.760 \end{array}\right. \quad \begin{array}{l} \left.\begin{array}{l} 0.9420 \\ ? \end{array}\right\} y \\ 0.9425 \end{array}\Bigg\} \text{\_\_\_\_\_\_}$$

$\frac{3}{10} = \frac{y}{.0005}$

11. Equate the ratios of corresponding differences and obtain the proportion

__________.

.0002 (detail below)
$y = \frac{3}{10}(.0005)$

12. Solve for y, rounding off to four decimal places, and obtain $y =$ ______.

0.9422

13. Add this value of y to 0.9420 to obtain $\log_{10} 8.753 =$ ______.

8.9422 - 10

14. So $\log_{10} 0.08753 =$ __________.

In Frames 15-21, find antilog$_{10}$ 3.7288.

.7284; .7292

15. From the table, find consecutive mantissa entries that bound the given mantissa 0.7288 to be ______ and ______.

5.350

5.360

16. With these, set up the following tabular arrangement:

x	antilog$_{10}$ x
.7284	______
.7288	?
.7292	______

.010

17. Compute the differences indicated by the pairings below and obtain:

$$.0008\left\{\begin{array}{l} .0004\left\{\begin{array}{l} .7784 \\ .7288 \end{array}\right. \\ \quad .7292 \end{array}\right. \quad \begin{array}{l} \left.\begin{array}{l} 5.350 \\ ? \end{array}\right\} y \\ 5.360 \end{array}\Bigg\} \text{\_\_\_\_\_\_}.$$

$\frac{4}{8} = \frac{y}{.010}$

18. Equate the ratios of corresponding differences and obtain the proportion

__________.

.005 (detail below)

$$y = \frac{(.010)(4)}{8}$$

$$y = \frac{.04}{8} = .005$$

19. Solve for y, and obtain y = ______.

5.355

3

5355

20. Add this value of y to 5.350 and obtain antilog$_{10}$ 0.7288 = _______.

21. Hence, antilog$_{10}$ 3.7288 = 5.355 x 10——, from which we obtain antilog$_{10}$ 3.7288 = _______.

In Frames 22-27, find antilog$_{10}$ (7.1630 - 10).

.1614; .1644

22. The given mantissa is bounded by the successive mantissa entries in the table, _______ and _______.

.1614; 1.450

.1644; 1.460

$$\frac{16}{30} = \frac{y}{.010}$$

23. Set up the tabular arrangement:

x	antilog$_{10}$ x
_______	_______
.1630	?
_______	_______

24. Obtain the proportion _____________.

.005 (detail below)

$$y = \frac{16(.010)}{30} = \frac{.16}{30},$$

then .16 ÷ 30.

25. Solve for y, rounding off to three decimal places and obtain y = ______.

1.455

-3

.001455

26. Add this value of y to 1.450 and obtain antilog$_{10}$ 0.1630 = _______.

27. Hence, antilog$_{10}$(7.1630 - 10) is 1.455 × 10——, from which we obtain antilog$_{10}$(7.1630 - 10) = _________.

TEST PROBLEMS

Use linear interpolation to find each logarithm or antilogarithm.

1. log$_{10}$ 243.6
2. log$_{10}$ 0.03762
3. antilog$_{10}$ 3.1237
4. antilog$_{10}$ 8.9979 - 10

SOLUTIONS TO TEST PROBLEMS

CHAPTER 1

1. a. 7, 6. b. 0,6.

2. a. $x \nless y$ b. $-2 < x < 1$

3. a.

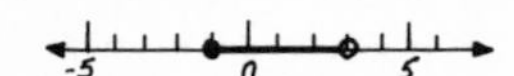

b.

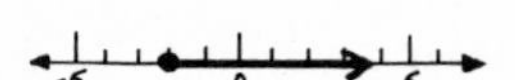

4. a. Negative or additive-inverse property.
   b. Commutative property for multiplication.
   c. Identity element for addition.
   d. Distributive property.

5. a. II b. I c. IV d. III

6. a. 8 b. $-(x - 2)$

7. a. 8 b. undefined c. -7 d. -6

8. a. 2 b. -2 c. 12 d. 16

9. a. $5\left(\frac{1}{9}\right)$ b. $\frac{7}{11}$

10. a. -3 b. $\frac{1}{3}$

CHAPTER 2

1. a. Trinomial. b. Binomial.

2. a. $P(0) = 3(0)^2 + 2(0) - 1 = \boxed{-1}$. b. $P(-1) = 3(-1)^2 + 2(-1) - 1 = \boxed{0}$.

3. a. $Q(1) = (1)^4 - 2(1)^2 = -1$. $P[Q(1)] = P(-1) = (-1)^2 - 3(-1) = \boxed{4}$.
   b. $P(1) = (1)^2 - 3(1) = -2$. $Q[P(1)] = Q(-2) = (-2)^4 - 2(-2)^2 = \boxed{8}$.

4. $(2x - 5y) - (5x - 2y) = 2x - 5y - 5x + 2y = (2x - 5x) + (2y - 5y) = \boxed{-3x - 3y}$.

5. $3c - [2c - (3c - 1) + 1] = 3c - [2c - 3c + 1 + 1] = 3c - [-c + 2] = 3c + c - 2 = \boxed{4c - 2}$.

6. $(-3xy^2)(2x^2y)(xy) = (-3 \cdot 2)(x \cdot x^2 \cdot x)(y^2 \cdot y \cdot y) = \boxed{-6x^4y^4}$.

7. $-3[(x - 3) - (x + 2)^2] = -3[x - 3 - (x^2 + 4x + 4)] = -3[x - 3 - x^2 - 4x - 4]$
   $= -3[(-x^2 - 3x - 7)] = \boxed{3x^2 + 9x + 21}$.

8. $\dfrac{-12x^3y^2z^3}{3xy^2z^2} = \dfrac{-12}{3} \cdot \dfrac{x^3}{x} \cdot \dfrac{y^2}{y^2} \cdot \dfrac{z^3}{z^2} = \boxed{-4x^2z}$.

9. $y^{2n-1} \cdot y^{n+1} = y^{(2n-1)+(n+1)} = \boxed{y^{3n}}$.

10. $\dfrac{a^{3n}b^{4n+2}}{a^{2n}b^2} = \dfrac{a^{3n}}{a^{2n}} \cdot \dfrac{b^{4n+2}}{b^2} = a^{3n-2n} \cdot b^{4n+2-2} = \boxed{a^nb^{4n}}$.

11. $5p^3(p^2 - 2p + 1) = \boxed{5p^5 - 10p^4 + 5p^3}$. 12. $(2x^n + 1)(2x^n - 1) = \boxed{4x^{2n} - 1}$.

13. a. $2 \cdot 2 \cdot 2 \cdot 2 \cdot 3$ b. $2 \cdot 2 \cdot 2 \cdot 2 \cdot 7$

14. $5m^3 + 2m^2 - m = \boxed{m(5m^2 + 2m - 1)}$. 15. $y^{n+3} + 4y^n = y^3 \cdot y^n + 4y^n = \boxed{y^n(y^3 + 4)}$.

16. $3(x^2 + xy - 6y^2) = \boxed{3(x + 3y)(x - 2y)}$.

17. $8a^3 - b^3 = (2a)^3 - b^3 = \boxed{(2a - b)(4a^2 + 2ab + b^2)}$.

18. $3m^2 - 5m + 3mn - 5n = m(3m - 5) + n(3m - 5) = \boxed{(3m - 5)(m + n)}$.

19. $9x^4 - 16x^2y^2 = x^2(9x^2 - 16y^2) = \boxed{x^2(3x - 4y)(3x + 4y)}$.

CHAPTER 3

1. a. $-\dfrac{-4}{7} = \boxed{\dfrac{4}{7}}$. b. $\dfrac{4 - 3x}{-x} = \dfrac{-1(4 - 3x)}{-1(-x)} = \boxed{\dfrac{3x - 4}{x}}$.
   c. $\dfrac{-5a^2b^2}{-2c^2} = \boxed{\dfrac{5a^2b^2}{2c^2}}$. d. $-\dfrac{m - 2n}{-n} = \boxed{\dfrac{m - 2n}{n}}$.

2. $\dfrac{8r^4 - 12r^3}{4r^2 - 6r} = \dfrac{4r^3(2r - 3)}{2r(2r - 3)} = \dfrac{\overset{1}{\cancel{2r(2r - 3)}} \cdot 2r^2}{\underset{1}{\cancel{2r(2r - 3)}} \cdot 1} = \boxed{2r^2}$.

3. $\dfrac{8a^3 - 27b^3}{4a - 9b} = \dfrac{\overset{1}{\cancel{(2a - 3b)}}(4a^2 + 6ab + 9b^2)}{\underset{1}{\cancel{(2a - 3b)}}(2a + 3b)} = \boxed{\dfrac{4a^2 + 6ab + 9b^2}{2a + 3b}}$.

4. $$\begin{array}{r} x + 5 \\ x - 2 \overline{\big) x^2 + 3x - 4} \\ \underline{x^2 - 2x} \\ 5x - 4 \\ \underline{5x - 10} \\ 6 \end{array}$$ Therefore, $\dfrac{x^2 + 3x - 4}{x - 2} = \boxed{x + 5 + \dfrac{6}{x - 2}}$.

5. $$\frac{6x^2 - 2xy + 9x - 3y}{2x + 3} = \frac{2x(3x - y) + 3(3x - y)}{2x + 3} = \frac{(2x + 3)(3x - y)}{2x + 3} = 3x - y.$$

Therefore, $\dfrac{6x^2 - 2xy + 9x - 3y}{2x + 3} = \boxed{3x - y}$.

6. $$\frac{4}{3x} + \frac{5}{2y} - \frac{2}{xy} = \frac{8y}{2 \cdot 3xy} + \frac{15x}{2 \cdot 3xy} - \frac{12}{2 \cdot 3xy} = \boxed{\frac{8y + 15x - 12}{6xy}}.$$

7. $$\frac{m}{m^2 - 1} - \frac{m}{m^2 - 2m + 1} = \frac{m}{(m + 1)(m - 1)} - \frac{m}{(m - 1)(m - 1)}$$
$$= \frac{m(m - 1)}{(m + 1)(m - 1)^2} - \frac{m(m + 1)}{(m + 1)(m - 1)^2}$$
$$= \frac{m(m - 1) - m(m + 1)}{(m + 1)(m - 1)^2} = \frac{m^2 - m - m^2 - m}{(m + 1)(m - 1)^2} = \boxed{\frac{-2m}{(m + 1)(m - 1)^2}}.$$

8. $$\frac{3p}{4pq - 6q^2} \cdot \frac{2p - 3q}{12p} = \frac{\overset{1}{\cancel{3p}}}{2p\underset{1}{\cancel{(2p - 3q)}}} \cdot \frac{\overset{1}{\cancel{2p - 3q}}}{\underset{1}{\cancel{3p}} \cdot 4} = \boxed{\frac{1}{8q}}.$$

9. $$\frac{4y^2 + 8y + 3}{2y^2 - 5y + 3} \cdot \frac{6y^2 - 9y}{1 - 4y^2} = \frac{(2y + 3)\overset{1}{\cancel{(2y + 1)}}}{\underset{1}{\cancel{(2y - 3)}}(y - 1)} \cdot \frac{3y\overset{1}{\cancel{(2y - 3)}}}{\underset{1}{\cancel{(1 + 2y)}}(1 - 2y)} = \boxed{\frac{3y(2y + 3)}{(y - 1)(1 - 2y)}}.$$

10. $$\frac{m^2 + m - 2}{m^2 + 2m - 3} \div \frac{m^2 + 7m + 10}{m^2 - 2m - 15} = \frac{\overset{1}{\cancel{(m + 2)}}\overset{1}{\cancel{(m - 1)}}}{\underset{1}{\cancel{(m + 3)}}\underset{1}{\cancel{(m - 1)}}} \cdot \frac{(m - 5)\overset{1}{\cancel{(m + 3)}}}{(m + 5)\underset{1}{\cancel{(m + 2)}}} = \boxed{\frac{m - 5}{m + 5}}.$$

11. $$\frac{8x^3 - 1}{x - 2} \div \frac{4x^2 + 2x + 1}{x^2 - 4x + 4} = \frac{(2x - 1)\overset{1}{\cancel{(4x^2 + 2x + 1)}}}{\underset{1}{\cancel{x - 2}}} \cdot \frac{\overset{1}{\cancel{(x - 2)}}(x - 2)}{\underset{1}{\cancel{4x^2 + 2x + 1}}} = \boxed{(2x - 1)(x - 2)}.$$

12. $$\frac{\dfrac{m - 2}{m}}{\dfrac{m^2 - 4}{m^2}} = \frac{\dfrac{m - 2}{m}(m^2)}{\dfrac{m^2 - 4}{m^2}(m^2)} = \frac{m(m - 2)}{m^2 - 4} = \frac{m\overset{1}{\cancel{(m - 2)}}}{\underset{1}{\cancel{(m - 2)}}(m + 2)} = \boxed{\frac{m}{m + 2}}.$$

13. $$\frac{\dfrac{a - 1}{a + 1} - \dfrac{a + 1}{a - 1}}{\dfrac{a - 1}{a + 1} + \dfrac{a + 1}{a - 1}} = \frac{\left[\left(\dfrac{a - 1}{a + 1}\right) - \left(\dfrac{a + 1}{a - 1}\right)\right](a + 1)(a - 1)}{\left[\left(\dfrac{a - 1}{a + 1}\right) + \left(\dfrac{a + 1}{a - 1}\right)\right](a + 1)(a - 1)} = \frac{(a - 1)(a - 1) - (a + 1)(a + 1)}{(a - 1)(a - 1) + (a + 1)(a + 1)}$$
$$= \frac{a^2 - 2a + 1 - a^2 - 2a - 1}{a^2 - 2a + 1 + a^2 + 2a + 1} = \frac{-4a}{2a^2 + 2}$$
$$= \frac{\overset{1}{\cancel{2}}(-2a)}{\underset{1}{\cancel{2}}(a^2 + 1)} = \boxed{\frac{-2a}{a^2 + 1}}.$$

CHAPTER 4

1.
$$5[2 + 3(x - 2)] + 20 = 0$$
$$5[2 + 3x - 6] + 20 = 0$$
$$5[3x - 4] + 20 = 0$$
$$15x - 20 + 20 = 0$$
$$15x = 0$$
$$\boxed{x = 0}.$$

2.
$$(2x + 1)(x - 3) = (x - 2) + 2x^2$$
$$2x^2 - 5x - 3 = x - 2 + 2x^2$$
$$-6x = 1$$
$$\boxed{x = -\frac{1}{6}}.$$

3.
$$\frac{x}{5} - 9 = \frac{x}{2}$$
$$10\left(\frac{x}{5} - 9\right) = 10\left(\frac{x}{2}\right)$$
$$2x - 90 = 5x$$
$$-90 = 3x$$
$$\boxed{x = -30}.$$

4.
$$\frac{x}{x + 2} - \frac{x^2 + 8}{x^2 - 4} = \frac{3}{x - 2}$$
$$(x + 2)(x - 2)\left[\frac{x}{x + 2} - \frac{x^2 + 8}{x^2 - 4}\right] = (x + 2)(x - 2)\left[\frac{3}{x - 2}\right]$$
$$x(x - 2) - (x^2 + 8) = 3(x + 2)$$
$$x^2 - 2x - x^2 - 8 = 3x + 6$$
$$-2x - 8 = 3x + 6$$
$$-5x = 14$$
$$\boxed{x = \frac{-14}{5}}.$$

5.
$$\frac{x}{a} - \frac{a}{b} = \frac{b}{c}$$
$$abc\left[\frac{x}{a} - \frac{a}{b}\right] = abc\left[\frac{b}{c}\right]$$
$$bcx - a^2c = ab^2$$
$$bcx = a^2c + ab^2$$
$$\boxed{x = \frac{a^2c + ab^2}{bc}}.$$

6.
$$S = \frac{a}{1 - r}$$
$$S(1 - r) = a$$
$$S - Sr = a$$
$$-Sr = a - S$$
$$r = \frac{a - S}{-S} = \boxed{\frac{S - a}{S}}.$$

7.
$$\frac{1}{C_n} = \frac{1}{C_1} + \frac{1}{C_2}$$
$$\frac{1}{5} = \frac{1}{15} + \frac{1}{C_2}$$
$$(15C_2)\frac{1}{5} = 15C_2\left(\frac{1}{15} + \frac{1}{C_2}\right)$$
$$3C_2 = C_2 + 15$$
$$2C_2 = 15$$
$$\boxed{C_2 = 7.5}$$

8.
$$\frac{2x + y}{2} = \frac{4x - y}{3}$$
$$6\left(\frac{2x + y}{2}\right) = 6\left(\frac{4x - y}{3}\right)$$
$$3(2x + y) = 2(4x - y)$$
$$6x + 3y = 8x - 2y$$
$$-2x + 3y = -2y$$
$$-2x = -5y$$
$$x = \frac{-5y}{-2}$$

In standard form $\boxed{x = \frac{5y}{2}}$.

9. $\dfrac{2x - 3}{2} \le 5$

$2x - 3 \le 10$

$2x \le 13$

$\boxed{x \le \dfrac{13}{2};}$.

10. $\dfrac{1}{2}(x + 2) \ge \dfrac{2x}{3}$

$6\left[\dfrac{1}{2}(x + 2)\right] \ge 6\left[\dfrac{2x}{3}\right]$

$3(x + 2) \ge 2(2x)$

$3x + 6 \ge 4x$

$-x \ge -6$

$\boxed{x \le 6;}$.

11. $2 \le 3x - 4 \le 8$

$6 \le 3x \le 12$

$\boxed{2 \le x \le 4;}$.

12. $\{x|2x - 3 < 5\} \cap \{x|2x - 3 > -5\}$

$\{x|2x < 8\} \cap \{x|2x > -2\}$

$\{x|x < 4\} \cap \{x|x > -1\}$

$\{x|x < 4 \text{ and } x > -1\} = \boxed{\{x|-1 < x < 4\};}$.

13. $|2x + 5| = 2$

$2x + 5 = 2$ or $-(2x + 5) = 2$

$2x = -3$ or $-2x = 7$

$\boxed{x = -\dfrac{3}{2} \quad \text{or} \quad x = -\dfrac{7}{2}}$.

14. $\left|2x + \dfrac{1}{2}\right| = \dfrac{1}{4}$

$2x + \dfrac{1}{2} = \dfrac{1}{4}$ or $-\left(2x + \dfrac{1}{2}\right) = \dfrac{1}{4}$

$8x + 2 = 1$ or $-8x - 2 = 1$

$8x = -1$ or $-8x = 3$

$\boxed{x = -\dfrac{1}{8} \quad \text{or} \quad x = -\dfrac{3}{8}}$.

15. $|x + 5| > 2$

$x + 5 > 2$ or $-(x + 5) > 2$

$x > -3$ or $-x - 5 > 2$

$x > -3$ or $-x > 7$

$x > -3$ or $x < -7$

$\boxed{\{x|x > -3\} \cup \{x|x < -7\};}$.

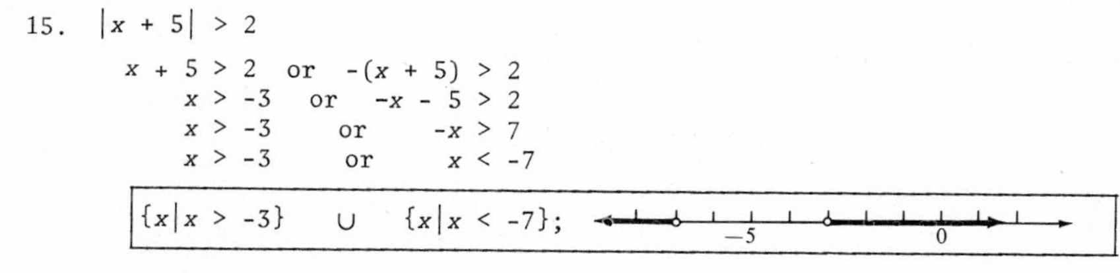

16. $|2x + 4| < 6$

$-6 < 2x + 4 < 6$

$-10 < 2x < 2$

$-5 < x < 1$

$\boxed{\{x|-5 < x < 1\};}$.

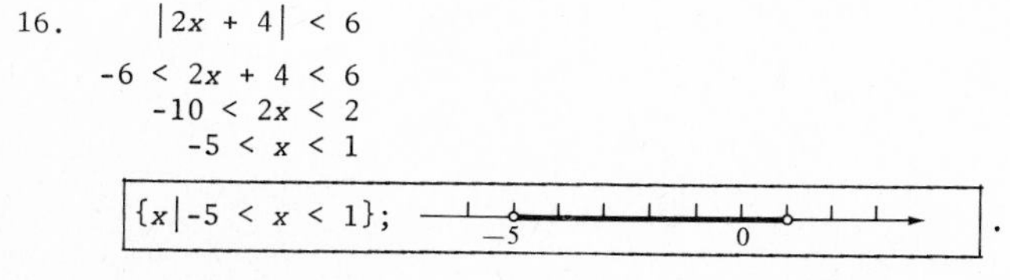

17. $\dfrac{x}{8} = 4 + \dfrac{5}{8}$

$x = 32 + 5$

$\boxed{x = 37}$.

18. The length: x

The width: $\frac{3}{4}x$

New area = original area + 96

$$(x + 8)\left(\frac{3}{4}x + 6\right) = x\left(\frac{3}{4}x\right) + 96$$

$$\cancel{\frac{3}{4}x^2} + 12x + 48 = \cancel{\frac{3}{4}x^2} + 96$$

$$12x = 48$$

$$x = 4, \ \frac{3}{4}x = 3.$$

4 feet × 3 feet.

19. The number of gallons of water to be added: x

[amount of acid in original 2 gallons] = [amount of acid in the final 10% solution.]

$$2 = .1(2 + x)$$
$$2 = .2 + .1x$$
$$20 = 2 + x$$
$$x = 18.$$

18 gallons should be added.

20. The number of nickels: x

The number of dimes: $x + 5$

The number of quarters: $2(x + 5) - 25$

$$5x + 10(x + 5) + 25[2(x + 5) - 25] = 455$$
$$5x + 10x + 50 + 50x + 250 - 625 = 455$$
$$65x - 325 = 455$$
$$65x = 780$$
$$x = 12$$
$$x + 5 = 17$$
$$2(x + 5) - 25 = 9$$

Hence, there are **12 nickels, 17 dimes, and 9 quarters**.

21. Rate of the second plane: x

	r	t	$d = rt$
1st plane	300	5	1500
2nd plane	x	4	$4x$

[distance 2nd plane travels] = [distance 1st plane travels]

$$4x = 1500$$
$$x = 375$$

The rate of the 2nd plane is **375 miles per hour**.

CHAPTER 5

1. a. $(x^3y^2)^3 = (x^3)^3(y^2)^3 = \boxed{x^9y^6}$.

   b. $(-2x^3)^2(-2y^2)^2 = (-2)^2(x^3)^2(-2)^2(y^2)^2 = (-2)^2(-2)^2(x^3)^2(y^2)^2 = \boxed{16x^6y^4}$.

2. a. $\left(\dfrac{x^3}{y^4}\right)^3 = \dfrac{(x^3)^3}{(y^4)^3} = \boxed{\dfrac{x^9}{y^{12}}}$.

   b. $\left(\dfrac{x^2}{y^2}\right)^2\left(-\dfrac{2xy}{5}\right)^3 = \dfrac{(x^2)^2}{(y^2)^2} \cdot \dfrac{(-2)^3x^3y^3}{5^3} = \dfrac{x^4}{y^4} \cdot \dfrac{-8x^3y^3}{125} = \dfrac{-8x^7y^3}{125y^4} = \boxed{\dfrac{-8x^7}{125y}}$.

3. a. $\dfrac{(-x^3)^2(-x)^3}{(x^2)^3} = \dfrac{x^6(-x^3)}{x^6} = \boxed{-x^3}$.

   b. $(x^{2n+1} \cdot x^{n-2}) = (x^{2n+1+n-2})^2 = (x^{3n-1})^2 = x^{2(3n-1)} = \boxed{x^{6n-2}}$.

4. a. $\left(\dfrac{x^{2n}x^{3n}}{x^{5n-2}}\right)^3 = \left(\dfrac{x^{5n}}{x^{5n-2}}\right)^3 = (x^{5n-(5n-2)})^3 = (x^2)^3 = \boxed{x^6}$.

   b. $\left(\dfrac{x^{2n-2}}{x^{2n}}\right)^{2n} = (x^{2n-2-2n})^{2n} = (x^{-2})^{2n} = x^{-4n} = \boxed{\dfrac{1}{x^{4n}}}$.

5. $\left(\dfrac{x^{3n}y^{4n}}{x^{2n}y^n}\right)^{1/3} = (x^{3n-2n}y^{4n-n})^{1/3} = (x^ny^{3n})^{1/3} = x^{n(1/3)}y^{3n(1/3)} = \boxed{x^{n/3}y^n}$.

6. $\left[\left(\dfrac{x^{4n}}{y^{3n}}\right)^{1/4n}\right]^{1/3} = \left[\dfrac{(x^{4n})^{1/4n}}{(y^{3n})^{1/4n}}\right]^{1/3} = \left[\dfrac{x^{4n/4n}}{y^{3n/4n}}\right]^{1/3} = \left[\dfrac{x}{y^{3/4}}\right]^{1/3} = \dfrac{x^{1/3}}{(y^{3/4})^{1/3}} = \boxed{\dfrac{x^{1/3}}{y^{1/4}}}$.

7. a. $x^{-5}x^2 = x^{-5+2} = x^{-3} = \boxed{\dfrac{1}{x^3}}$.

   b. $(x^4y^{-2})^{-1/3} = (x^4)^{-1/3}(y^{-2})^{-1/3} = x^{-4/3}y^{2/3} = \boxed{\dfrac{y^{2/3}}{x^{4/3}}}$.

8. a. $\dfrac{x^{-3}y^{-2}z^{-1}}{x^0y^{-4}z^2} = x^{-3}y^{-2-(-4)}z^{-1-2} = x^{-3}y^2z^{-3} = \dfrac{1}{x^3} \cdot y^2 \cdot \dfrac{1}{z^3} = \boxed{\dfrac{y^2}{x^3z^3}}$.

   b. $\left(\dfrac{3^0x^{-2}y^{-3}}{x^0y^{-2}}\right)^{-2} = (1 \cdot x^{-2-0}y^{-3-(-2)})^{-2} = (x^{-2}y^{-1})^{-2} = (x^{-2})^{-2}(y^{-1})^{-2} = \boxed{x^4y^2}$.

9. $\dfrac{(2 \times 10^{-4})(4 \times 10^6)(5 \times 10^2)}{8 \times 10^2} = \dfrac{2 \times 4 \times 5}{8} \times \dfrac{10^{-4} \times 10^6 \times 10^2}{10^2} = 5 \times 10^2 = \boxed{500}$.

10. $\dfrac{.009 \times .0008}{.0036} = \dfrac{9 \times 10^{-3} \times 8 \times 10^{-4}}{36 \times 10^{-4}} = \dfrac{9 \times 8}{36} \times \dfrac{10^{-3} \times 10^{-4}}{10^{-4}} = 2 \times 10^{-3} = \boxed{.002}$.

11. $x^{1/3} - x = \boxed{x(x^{-2/3} - 1)}$.   12. $y^{1/3} - y^{-1/3} = \boxed{y^{-1/3}(y^{2/3} - 1)}$.

13. $y^{2n+2} - y^n = \boxed{y^n(y^{n+2} - 1)}$.

14. a. $(2 - y^2)^{-3/4} = \dfrac{1}{(2 - y^2)^{3/4}} = \boxed{\dfrac{1}{\sqrt[4]{(2 - y^2)^3}}}$.

    b. $\sqrt[3]{2m^2n} = (2m^2n)^{1/3} = (2n)^{1/3}(m^2)^{1/3} = \boxed{(2n)^{1/3}m^{2/3}}$.

15. a. $\sqrt[4]{81x^6y^3} = \sqrt[4]{(3^4x^4)x^2y^3} = \sqrt[4]{(3x)^4}\sqrt[4]{x^2y^3} = \boxed{3x\sqrt[4]{x^2y^3}}$.

b. $\sqrt{\frac{5}{6}} = \sqrt{\frac{5 \cdot 6}{6 \cdot 6}} = \frac{\sqrt{30}}{\sqrt{36}} = \boxed{\frac{\sqrt{30}}{6}}$.

c. $\frac{2m}{\sqrt{2mn}} = \frac{2m\sqrt{2mn}}{\sqrt{2mn}\sqrt{2mn}} = \frac{2m\sqrt{2mn}}{2mn} = \boxed{\frac{\sqrt{2mn}}{n}}$.

16. a. $\sqrt{48} + \sqrt{75} = \sqrt{16}\sqrt{3} + \sqrt{25}\sqrt{3} = 4\sqrt{3} + 5\sqrt{3} = \boxed{9\sqrt{3}}$.

b. $\sqrt{9m} + 3\sqrt{36m} - 2\sqrt{25m} = \sqrt{9}\sqrt{m} + 3\sqrt{36}\sqrt{m} - 2\sqrt{25}\sqrt{m} = 3\sqrt{m} + 3(6)\sqrt{m} - 2(5)\sqrt{m}$

$= 3\sqrt{m} + 18\sqrt{m} - 10\sqrt{m} = \boxed{11\sqrt{m}}$.

c. $\frac{2\sqrt{8}}{3} + \frac{3\sqrt{18}}{2} = \frac{2\sqrt{4}\sqrt{2}}{3} + \frac{3\sqrt{9}\sqrt{2}}{3} = \frac{2(2)\sqrt{2}}{3} + \frac{3(3)\sqrt{2}}{2} = \frac{4\sqrt{2}}{3} + \frac{9\sqrt{2}}{2} = \frac{8\sqrt{2}}{6} + \frac{27\sqrt{2}}{6} = \boxed{\frac{35\sqrt{2}}{6}}$

d. $\sqrt[3]{8x} - \sqrt[3]{-64x^4} = \sqrt[3]{8}\sqrt[3]{x} - \sqrt[3]{-64x^3}\sqrt[3]{x} = 2(\sqrt[3]{x}) + 4x(\sqrt[3]{x}) = \boxed{(2 + 4x)(\sqrt[3]{x})}$.

17. a. $\sqrt{5}(\sqrt{6} - \sqrt{5}) = \sqrt{5}\sqrt{6} - \sqrt{5}\sqrt{5} = \boxed{\sqrt{30} - 5}$.

b. $(\sqrt{2} - \sqrt{2x})(\sqrt{2} + \sqrt{2x}) = (\sqrt{2})^2 - (\sqrt{2x})^2 = \boxed{2 - 2x}$.

c. $\sqrt[4]{4}(\sqrt[4]{4} + \sqrt[4]{64}) = (\sqrt[4]{4})(\sqrt[4]{4}) + (\sqrt[4]{4})(\sqrt[4]{64}) = \sqrt[4]{16} + \sqrt[4]{256} = \sqrt[4]{2^4} + \sqrt[4]{2^8}$

$= \sqrt[4]{2^4} + \sqrt[4]{(2^2)^4} = 2 + 2^2 = \boxed{6}$.

18. a. $\frac{\sqrt{15} + \sqrt{21}}{\sqrt{3}} = \frac{\sqrt{3}\sqrt{5} + \sqrt{3}\sqrt{7}}{\sqrt{3}} = \frac{\sqrt{3}(\sqrt{5} + \sqrt{7})}{\sqrt{3}} = \boxed{\sqrt{5} + \sqrt{7}}$.

b. $\frac{m\sqrt{mn^3} - \sqrt{mn}}{\sqrt{mn}} = \frac{m\sqrt{n^2}\sqrt{mn} - \sqrt{mn}}{\sqrt{mn}} = \frac{mn\sqrt{mn} - \sqrt{mn}}{\sqrt{mn}} = \frac{\sqrt{mn}(mn - 1)}{\sqrt{mn}} = \boxed{mn - 1}$.

19. a. $\frac{2\sqrt{3} - 2}{2\sqrt{3} + 2} = \frac{(2\sqrt{3} - 2)(2\sqrt{3} - 2)}{(2\sqrt{3} + 2)(2\sqrt{3} - 2)} = \frac{4(3) - 8\sqrt{3} + 4}{4(3) - 4} = \frac{16 - 8\sqrt{3}}{8} = \frac{8(2 - \sqrt{3})}{8} = \boxed{2 - \sqrt{3}}$.

b. $\frac{\sqrt{x} + \sqrt{y}}{\sqrt{x} - \sqrt{y}} = \frac{(\sqrt{x} + \sqrt{y})(\sqrt{x} + \sqrt{y})}{(\sqrt{x} - \sqrt{y})(\sqrt{x} + \sqrt{y})} = \boxed{\frac{x + 2\sqrt{xy} + y}{x - y}}$.

CHAPTER 6

1.
$$5x^2 = 9x$$
$$5x^2 - 9x = 0$$
$$x(5x - 9) = 0$$
$$x = 0 \quad \text{or} \quad x = \frac{9}{5}.$$

$\boxed{\left\{0, \frac{9}{5}\right\}}$.

2.
$$2x(x - 2) = x + 3$$
$$2x^2 - 4x = x + 3$$
$$2x^2 - 5x - 3 = 0$$
$$(2x + 1)(x - 3) = 0$$
$$x = -\frac{1}{2} \quad \text{or} \quad 3.$$

$\boxed{\left\{-\frac{1}{2}, 3\right\}}$.

3.
$$\frac{2x^2}{7} = 8$$
$$2x^2 = 56$$
$$x^2 = 28$$
$$x^2 = \pm\sqrt{28} = \pm 2\sqrt{7}.$$

$\boxed{\{2\sqrt{7}, -2\sqrt{7}\}}$.

4.
$$(2x + 1)^2 = 25$$
$$2x + 1 = \pm 5$$
$$2x = -1 \pm 5$$
$$x = \frac{-1 \pm 5}{2}.$$

$\boxed{\{-3, 2\}}$.

5. $$3x^2 = 5x - 1$$
$$3x^2 - 5x + 1 = 0. \qquad a = 3,\ b = -5,\ c = 1.$$
$$x = \frac{-(-5) \pm \sqrt{(-5)^2 - 4(3)(1)}}{2(3)} = \frac{5 \pm \sqrt{25 - 12}}{6} = \frac{5 \pm \sqrt{13}}{6}. \qquad \boxed{\left\{\frac{5 + \sqrt{13}}{6}, \frac{5 - \sqrt{13}}{6}\right\}}.$$

6. $$\frac{x^2 - x}{2} + 1 = 0$$
$$x^2 - x + 2 = 0. \qquad a = 1,\ b = -1,\ c = 2.$$
$$x = \frac{-(-1) \pm \sqrt{(-1)^2 - 4(1)(2)}}{2(1)} = \frac{1 \pm \sqrt{1 - 8}}{2} = \frac{1 \pm \sqrt{-7}}{2} = \frac{1 \pm i\sqrt{7}}{2}$$
$$\boxed{\left\{\frac{1 + i\sqrt{7}}{2}, \frac{1 - i\sqrt{7}}{2}\right\}}.$$

7. $$2x^2 - 3x + 2y = 0. \qquad a = 2,\ b = -3,\ c = 2y.$$
$$x = \frac{-(-3) \pm \sqrt{(-3)^2 - 4(2)(2y)}}{2(2)} = \frac{3 \pm \sqrt{9 - 16y}}{4}. \qquad \boxed{\left\{\frac{3 - \sqrt{9 - 16y}}{4}, \frac{3 + \sqrt{9 - 16y}}{4}\right\}}.$$

8. $$\sqrt{x - 3} + 5 = 0$$
$\sqrt{x - 3} = -5$. For every x, $\sqrt{x - 3}$ is non negative. Therefore, the solution set is $\boxed{\emptyset}$.

9.
$$\begin{aligned} \sqrt{3x + 10} &= x + 4 \\ (\sqrt{3x + 10})^2 &= (x + 4)^2 \\ 3x + 10 &= x^2 + 8x + 16 \\ x^2 + 5x + 6 &= 0 \\ (x + 2)(x + 3) &= 0 \\ x &= -2 \text{ or } -3. \end{aligned}$$
Neither of these is extraneous; therefore, $\boxed{\{-2, -3\}}$.

10.
$$\begin{aligned} 4\sqrt{x} + \sqrt{16x + 1} &= 5 \\ \sqrt{16x + 1} &= 5 - 4\sqrt{x} \\ (\sqrt{16x + 1})^2 &= (5 - 4\sqrt{x})^2 \\ 16x + 1 &= 25 - 40\sqrt{x} + 16x \\ 40\sqrt{x} &= 24 \\ 5\sqrt{x} &= 3 \\ (5\sqrt{x})^2 &= 3^2 \\ 25x &= 9 \\ x &= \frac{9}{25}, \end{aligned}$$
which is not extraneous. Therefore, $\boxed{\left\{\frac{9}{25}\right\}}$.

11.
$$\begin{aligned} x^4 - 2x^2 - 24 &= 0 \\ (x^2 - 6)(x^2 + 4) &= 0 \\ x^2 - 6 = 0 \quad &\text{or} \quad x^2 + 4 = 0 \\ x = \pm\sqrt{6} \qquad & \qquad x = \pm\sqrt{-4} = \pm 2i. \end{aligned}$$
$\boxed{\{\sqrt{6}, -\sqrt{6}, 2i, -2i\}}$.

12. $x^{2/3} - 2x^{1/3} = 35$. Let $u = x^{1/3}$.
$$\begin{aligned} u^2 - 2u - 35 &= 0 \\ (u - 7)(u + 5) &= 0 \\ u = 7 \quad &\text{or} \quad u = -5. \end{aligned}$$
Then $x^{1/3} = 7$ or $x^{1/3} = -5$
$$x = 7^3 \quad \text{or} \quad x = (-5)^3$$
$$x = 343 \quad \text{or} \quad x = -125. \qquad \boxed{\{343, -125\}}.$$

13. $\left(x - \left(-\frac{2}{3}\right)\right)(x - 2) = 0$

$\left(x + \frac{2}{3}\right)(x - 2) = 0$

$x^2 + \frac{2}{3}x - 2x - \frac{4}{3} = 0$

$3x^2 + 2x - 6x - 4 = 0$

$\boxed{3x^2 - 4x - 4 = 0}$.

14. $2x^2 + 3x - 2 = 0$

$x^2 + \frac{3}{2}x = 1$

$x^2 + \frac{3}{2}x + \frac{9}{16} = 1 + \frac{9}{16}$

$\left(x + \frac{3}{4}\right)^2 = \frac{25}{16}$

$x + \frac{3}{4} = \pm\frac{5}{4}$

$x = \frac{-3}{4} \pm \frac{5}{4}$. $\boxed{\left\{-2, \frac{1}{2}\right\}}$.

15. a. $(2 - 3i) - (4 + 5i) = (2 - 4) + (-3 - 5)i = \boxed{-2 - 8i}$.

b. $(2 - 3i)(4 + 5i) = 8 - 12i + 10i - 15i^2 = 8 - 2i - 15(-1) = 8 - 2i + 15 = \boxed{23 - 2i}$.

c. $\frac{4 + 5i}{2 - 3i} = \frac{(4 + 5i)(2 + 3i)}{(2 - 3i)(2 + 3i)} = \frac{8 + 22i + 15i^2}{4 - 9i^2} = \frac{8 + 22i - 15}{4 - (-9)} = \frac{-7 + 22i}{13} = \boxed{\frac{-7}{13} + \frac{22}{13}i}$.

16. a. $(3 + \sqrt{-12}) + (1 - \sqrt{-27}) = (3 + i\sqrt{12}) + (1 - i\sqrt{27}) = (3 + i\sqrt{4}\sqrt{3}) + (1 - i\sqrt{9}\sqrt{3})$

$= (3 + 2i\sqrt{3}) + (1 - 3i\sqrt{3}) = (3 + 1) + (2\sqrt{3} - 3\sqrt{3})i = \boxed{4 - i\sqrt{3}}$.

b. $(3 - \sqrt{-3})(1 - \sqrt{-3}) = (3 - i\sqrt{3})(1 - i\sqrt{3}) = 3 - 4i\sqrt{3} + i^2(3) = 3 - 4i\sqrt{3} - 3 = \boxed{-4i\sqrt{3}}$.

c. $\frac{1}{2 - \sqrt{-3}} = \frac{1}{2 - i\sqrt{3}} = \frac{2 + i\sqrt{3}}{(2 - i\sqrt{3})(2 + i\sqrt{3})} = \frac{2 + i\sqrt{3}}{4 - i^2(3)} = \frac{2 + i\sqrt{3}}{4 + 3} = \frac{2 + i\sqrt{3}}{7} = \boxed{\frac{2}{7} + \frac{\sqrt{3}}{7}i}$.

17. For one solution, $b^2 - 4ac = 0$. $a = 1$, $b = -k$, $c = 9$.

$b^2 - 4ac = (-k)^2 - 4(1)(9) = k^2 - 36 = 0$. $\boxed{k = \pm 6}$.

18. For imaginary solutions, $b^2 - 4ac < 0$. $a = 1$, $b = -1$, $c = k - 2$.

$b^2 - 4ac = (-1)^2 - 4(1)(k - 2) = 1 - 4k + 8 < 0$. $-4k < -9$. $\boxed{k > \frac{9}{4}}$.

19. $x^2 - 5x - 6 \geq 0$

$(x - 6)(x + 1) \geq 0$

$(x - 6 \geq 0 \text{ and } x + 1 \geq 0)$ or $(x - 6 \leq 0 \text{ and } x + 1 \leq 0)$

$(x \geq 6 \text{ and } x \geq -1)$ or $(x \leq 6 \text{ and } x \leq -1)$

$(x \geq 6)$ or $(x \leq -1)$.

$\{x | x \geq 6 \text{ or } x \leq -1\}$; 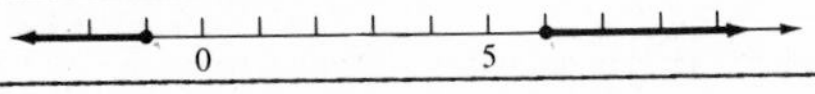

.

20. $\frac{3}{x - 6} > 8$

$(x - 6)^2\left(\frac{3}{x - 6}\right) > (x - 6)^2(8)$

$3(x - 6) > 8(x - 6)^2$

$3(x - 6) - 8(x - 6)^2 > 0$

$(x - 6)[3 - 8(x - 6)] > 0$

$(x - 6)(-8x + 51) > 0$

$(x - 6 > 0 \text{ and } -8x + 51 > 0)$ or $(x - 6 < 0 \text{ and } -8x + 51 < 0)$

$\left(x > 6 \text{ and } x < \frac{51}{8}\right)$ or $\left(x < 6 \text{ and } x > \frac{51}{8}\right)$

$\left(6 < x < \frac{51}{8}\right)$ or $(\emptyset)$

$6 < x < \frac{51}{8}$. $\left\{x | 6 < x < \frac{51}{8}\right\}$; 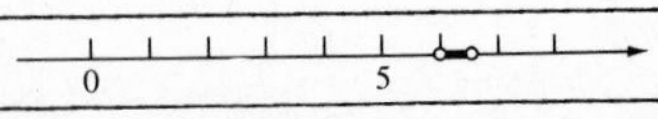

.

21. The first positive integer: x

The next positive integer: $x + 1$

$$x^2 + (x + 1)^2 = 145$$

$$x^2 + x^2 + 2x + 1 = 145$$

$$2x^2 + 2x - 144 = 0$$

$$2(x + 9)(x - 8) = 0.$$

The only positive solution is $x = 8$. Therefore $x + 1 = 9$.

$\boxed{8 \text{ and } 9}$

22. The rate going: x
The rate returning: $x - 2$

	d	r	$t = d/r$
going	6	x	$\frac{6}{x}$
returning	6	$x - 2$	$\frac{6}{x - 2}$

$$\begin{bmatrix}\text{time}\\ \text{going}\end{bmatrix} + \begin{bmatrix}\text{time}\\ \text{returning}\end{bmatrix} = \frac{5}{2}$$

$$\frac{6}{x} + \frac{6}{x - 2} = \frac{5}{2}$$

$$2x(x - 2)\ \frac{6}{x} + \frac{6}{x - 2} = 2(\ \ - 2)\ \frac{5}{2}$$

$$12(x - 2) + 12x = 5x(x - 2)$$
$$12x - 24 + 12x = 5x^2 - 10x$$
$$5x^2 - 34x + 24 = 0$$
$$(5x - 4)(x - 6) = 0$$
$$x = 6 \text{ or } \frac{4}{5}.$$

When $x = \frac{4}{5}$, $x - 2 = \frac{4}{5} - 2 = \frac{4}{5} - \frac{10}{5} = \frac{-6}{5}$, which makes no sense for the stated problem. So we reject $x = \frac{4}{5}$. When $x = 6$, $x - 2 = 4$. Therefore, rate going = $\boxed{6 \text{ mph}}$; rate returning = $\boxed{4 \text{ mph}}$.

CHAPTER 7

1. a. $2(0) - 9y = 18$. $y = -2$. $\boxed{-2}$.

   b. $2x - 9(0) = 18$. $x = 9$. $\boxed{9}$.

   c. $2(2) - 9y = 18$. $4 - 9y = 18$. $-9y = 14$. $y = \frac{-14}{9}$. $\boxed{\frac{-14}{9}}$.

2. $$\frac{2x^4}{3 + 2y^2} = 5$$

$$2x^4 = 15 + 10y^2$$
$$10y^2 = 2x^4 - 15$$

$$y^2 = \frac{2x^4 - 15}{10}$$

$$y = \pm\sqrt{\frac{2x^4 - 15}{10}} = \pm\sqrt{\frac{10(2x^4 - 15)}{10^2}} = \pm\frac{1}{10}\sqrt{10(2x^4 - 15)}.\quad \boxed{y = \pm\frac{1}{10}\sqrt{10(2x^4 - 15)}}.$$

3. $kx - 3y = 9$. $k(1) - 3(-3) = 9$. $k + 9 = 9$. $\boxed{k = 0}$.

4. $\boxed{\{(-2, -5), (-1, -3), (0, -1), (1, 1), (2, 3)\}}$.

The graph is

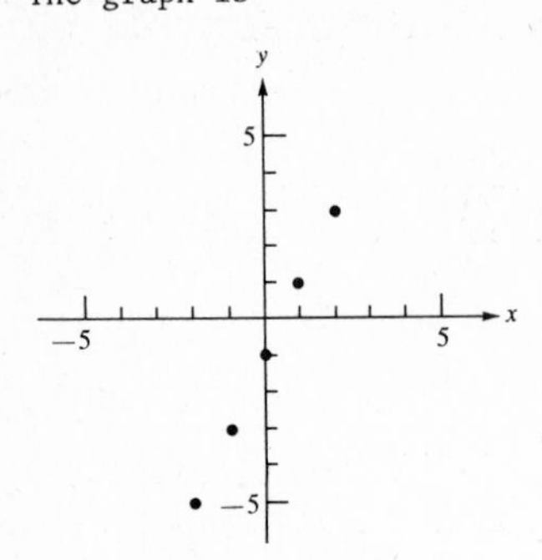

5. x-intercept is $\boxed{\frac{1}{2}}$.

y-intercept is $\boxed{2}$.

The graph is

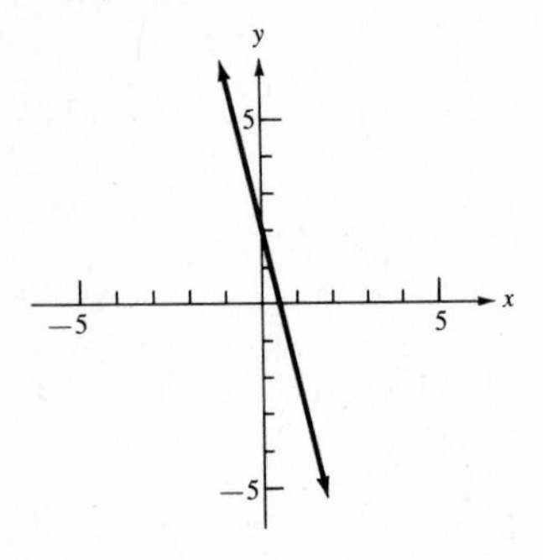

6. x-intercept is $\boxed{2}$.

y-intercept is $\boxed{-3}$.

The graph is

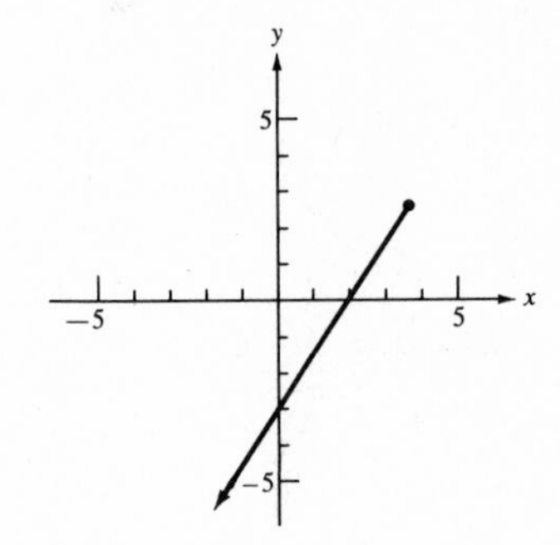

7. $y = |2x - 6|$. For $2x - 6 \geq 0$, that is, for $x \geq 3$, $y = 2x - 6$. For $2x - 6 < 0$, that is, for $x < 3$, $y = -(2x - 6)$. So, for $x \geq 3$, we graph $\boxed{y = 2x - 6}$; for $x < 3$, we graph $\boxed{y = -(2x - 6)}$. The graph is

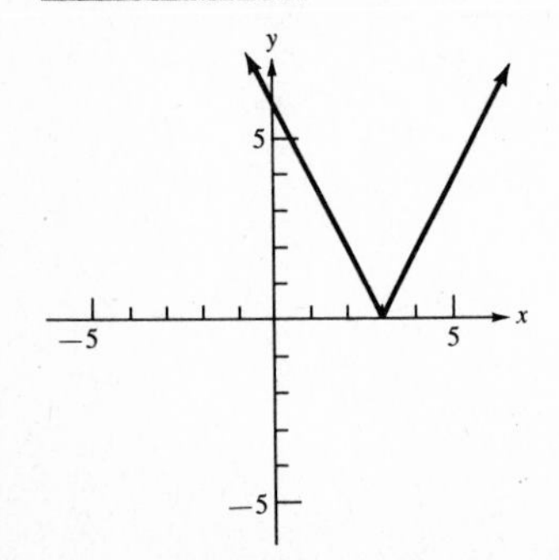

8. a. $\boxed{\text{Domain is } \{3,4,5,6,\}. \text{ Range is } \{1\}. \text{ A function.}}$

b. $\boxed{\text{Domain is } \{1\}. \text{ Range is } \{3,4,5,6\}. \text{ Not a function.}}$

9. a. $h(3) = 2(3)^2 + 3(3) - 2 = 2(9) + 9 - 2 = \boxed{25}$.

b. $h(1) - h(0) = [2(1)^2 + 3(1) - 2] - [2(0)^2 + 3(0) - 2] = 3 - (-2) = \boxed{5}$.

10. a. $f(x + h) = (x + h)^2 + 2(x + h) = \boxed{x^2 + 2hx + h^2 + 2x + 2h}$.

b. $\dfrac{f(x + h) - f(x)}{h} = \dfrac{(x^2 + 2hx + h^2 + 2x + 2h) - (x^2 + 2x)}{h} = \dfrac{2hx + h^2 + 2h}{h}$

$= \dfrac{h(2x + h + 2)}{h} = \boxed{2x + h + 2}$.

11. a. $d = \sqrt{(2 - (-2))^2 + (-3 - (-1))^2} = \sqrt{4^2 + (-2)^2} = \sqrt{20} = \boxed{2\sqrt{5}}$.

b. $m = \dfrac{-3 - (-1)}{2 - (-2)} = \dfrac{-2}{4} = \boxed{-\dfrac{1}{2}}$.

12. $y - y_1 = m(x - x_1)$

$(x_1, y_1) = (2, -1)$, and $m = \dfrac{3}{4}$

$y - (-1) = \dfrac{3}{4}(x - 2)$

$y + 1 = \dfrac{3}{4}(x - 2)$

$4(y + 1) = 3(x - 2)$

$4y + 4 = 3x - 6$

$\boxed{3x - 4y - 10 = 0}$.

13. a. $3x - 2y = 5$

$-2y = -3x + 5$

$\boxed{y = \dfrac{3}{2}x - \dfrac{5}{2}}$.

b. Slope is $\boxed{\dfrac{3}{2}}$.

c. y-intercept is $\boxed{-\dfrac{5}{2}}$.

14. $3x - 2y + 5 = 0$

$-2y = -3x - 5$

$y = \dfrac{3}{2}x + \dfrac{5}{2}$.

Slope of graph of $3x - 2y + 5 = 0$ is $\dfrac{3}{2}$.

Slope of perpendicular is $-\dfrac{2}{3}$.

In point-slope form, we have

$y - 5 = -\dfrac{2}{3}(x - 0)$

$3y - 15 = -2x$

$\boxed{2x + 3y - 15 = 0}$.

15. The graph is

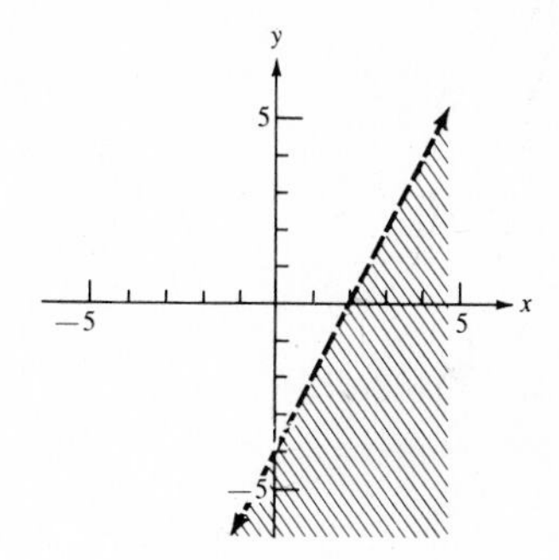

16. The graph is the double-shaded region:

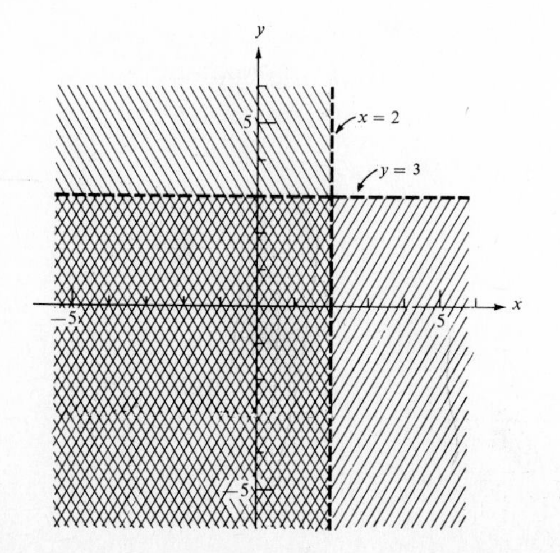

CHAPTER 8

1. $x^2 + x - 6 = 0$

$(x + 3)(x - 2) = 0.$ $\boxed{x\text{-intercepts are -3 and 2}}$.

When $x = 0$, $f(x) = -6$; therefore, the $\boxed{y\text{-intercept is -6}}$.

2. The x-coordinate of the vertex is $\frac{-2}{2(-3)} = \frac{1}{3}$.

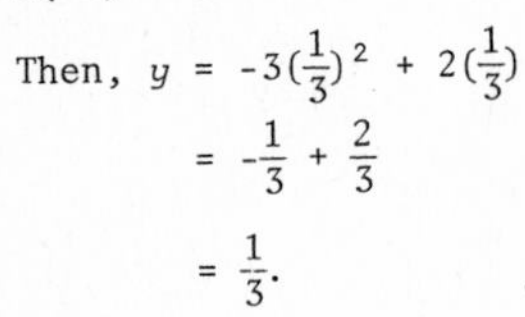

Then, $y = -3(\frac{1}{3})^2 + 2(\frac{1}{3})$

$= -\frac{1}{3} + \frac{2}{3}$

$= \frac{1}{3}.$

Maximum point at $\boxed{\left(\frac{1}{3}, \frac{1}{3}\right)}$.

3. The graph is

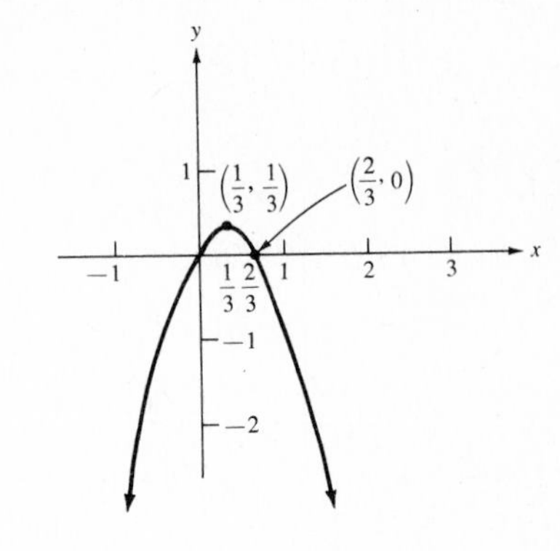

4. $x^2 + y^2 = 49$, $y^2 = 49 - x^2$, $y = \pm\sqrt{49 - x^2}$.
$y \in R$ if and only if $49 - x^2 \geq 0$. Solving this inequality, we obtain $|x| \leq 7$. Hence, the domain is $\boxed{\{x \mid |x| \leq 7\}}$.

5. a. $x^2 - 5y^2 = 3$. a and b have opposite signs and $c \neq 0$. Therefore, the graph is a $\boxed{\text{hyperbola}}$.

b. $y = \frac{2}{3}x^2 - \frac{2}{3}$. This is of the form $y = ax^2 + bx + c$ and so the graph is a $\boxed{\text{parabola}}$.

c. $x^2 + 5y^2 = 9$. a and b and c have the same sign and $a \neq b$. Therefore, the graph is an $\boxed{\text{ellipse}}$.

6. The graph is

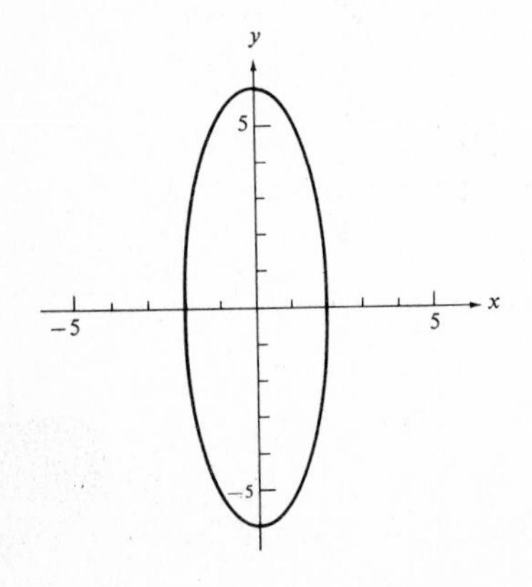

7. The graph is

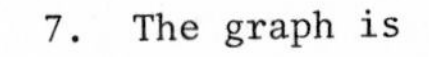

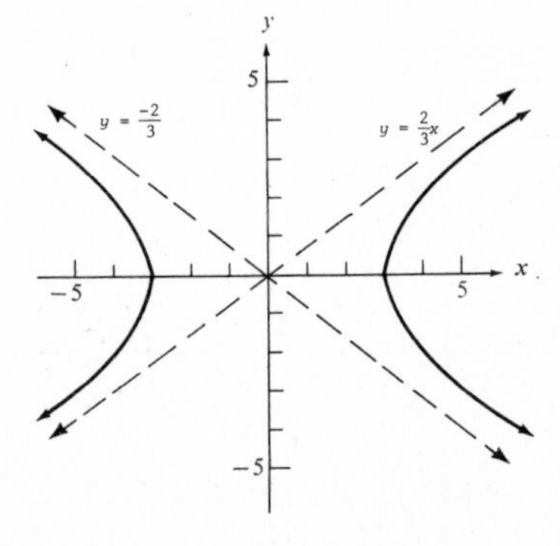

8. The graph is

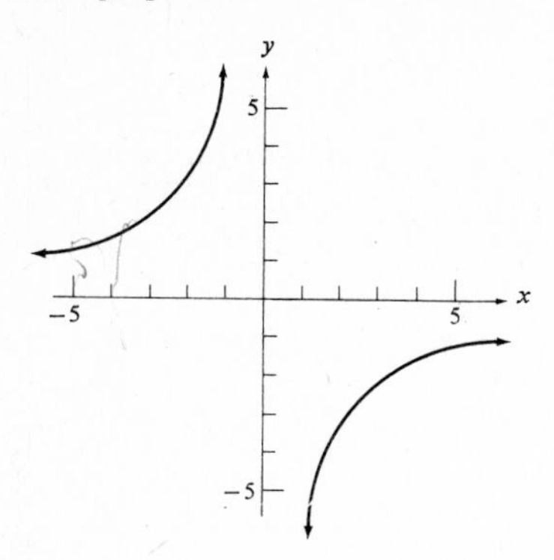

9. The graph is

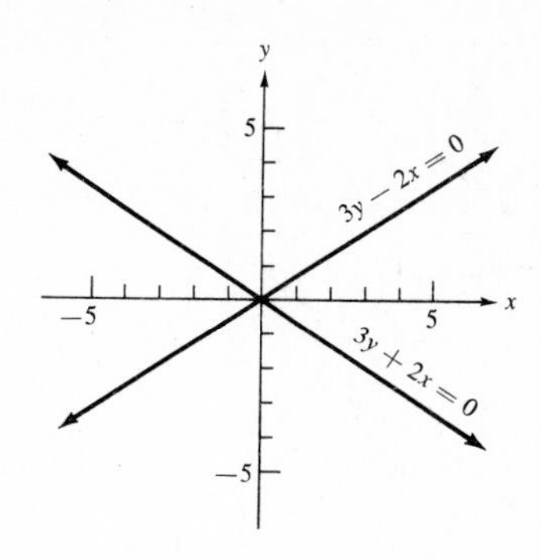

10. $y = \dfrac{k}{t^2}$

$2 = \dfrac{k}{2^2}$

$k = 8$. Then $y = \dfrac{8}{t^2}$

$y = \dfrac{8}{3^2} = \boxed{\dfrac{8}{9}}$.

11. $R = \dfrac{kl}{d^2}$

$100 = \dfrac{k(25)}{(0.006)^2}$

$k = \dfrac{100(0.006)^2}{25} = 4(0.006)^2$.

Then, $R = \dfrac{4(0.006)^2 l}{d^2}$

$R = \dfrac{4(0.006)^2(100)}{(.012)^2} = 4(100)\left(\dfrac{.006}{.012}\right)^2$

$= 400\left(\dfrac{1}{2}\right)^2 = 100.$

Hence, the resistance is $\boxed{100 \text{ ohms}}$.

12. The graph is the two points.

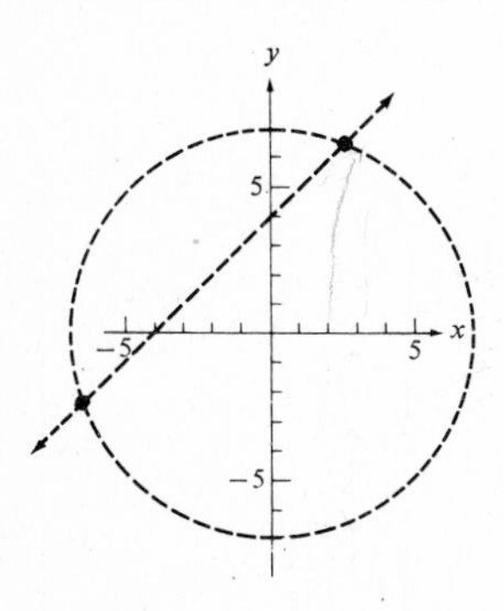

13. The graph is

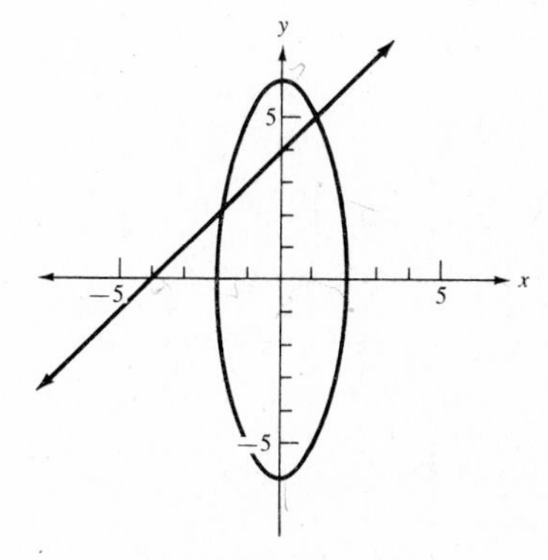

14. The graph is

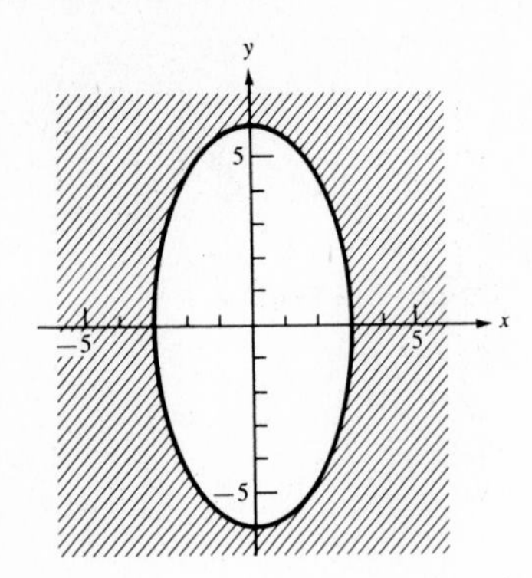

15. a. Replace x by y and y by x in the given equation and obtain $\boxed{2y - 3x = 6}$.

b. The graph is

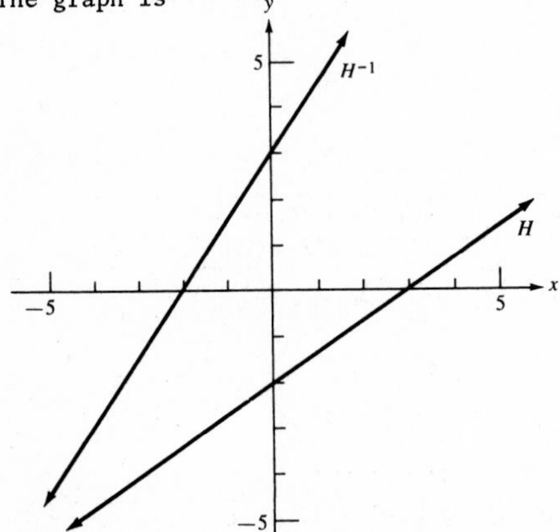

c. Since each x is paired with one and only one y, $\boxed{h^{-1} \text{ is a function}}$.

CHAPTER 9

1. a. $y = 4^{-1} = \boxed{\frac{1}{4}}$. b. $y = 4^0 = \boxed{1}$. c. $y = 4^2 = \boxed{16}$.

2. a. $y = \left(\frac{1}{4}\right)^{-1} = \frac{1}{\frac{1}{4}} = \boxed{4}$. b. $y = \left(\frac{1}{4}\right)^0 = \boxed{1}$. c. $y = \left(\frac{1}{4}\right)^2 = \boxed{\frac{1}{16}}$.

3. The graph is

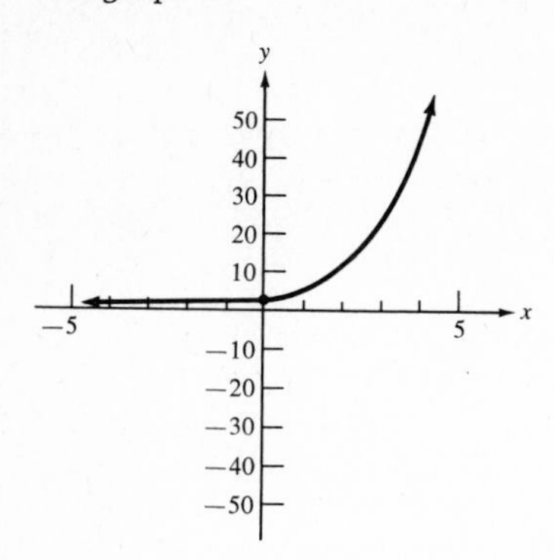

4. The graph is

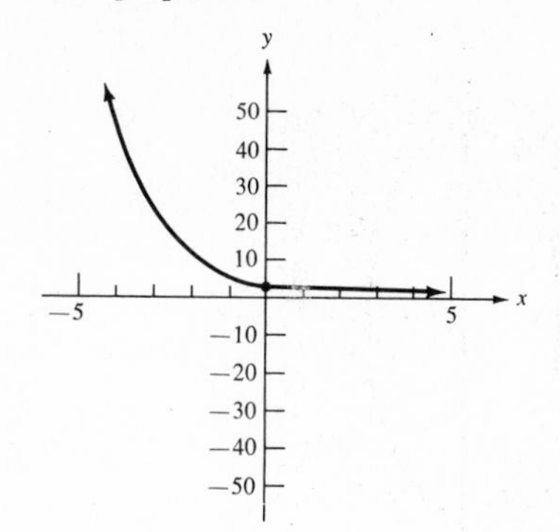

5. a. $\log_2 16 = \boxed{4}$. b. $\log_{10} 0.01 = \boxed{-2}$.

6. a. $10^{-5} = \boxed{0.00001}$. b. $\left(\frac{1}{4}\right)^{-2} = \boxed{16}$.

7. a. $b^3 = 64$
$b^3 = 4^3$
$b = \boxed{4}$.

b. $\left(\frac{1}{3}\right)^{-3} = x$

$x = \dfrac{1}{\left(\frac{1}{3}\right)^3} = \boxed{27}$.

8. a. $\log_b \dfrac{x^3y^2}{z} = \log_b x^3y^2 - \log_b z = \log_b x^3 + \log_b y^2 - \log_b z$
$= \boxed{3 \log_b x + 2 \log_b y - \log_b z}$.

b. $\log_{10}\left(\dfrac{xy^2}{x - y}\right)^{1/4} = \dfrac{1}{4}\log_{10}\left(\dfrac{xy^2}{x - y}\right) = \dfrac{1}{4}\left[\log_{10}(xy^2) - \log_{10}(x - y)\right]$
$= \dfrac{1}{4}\left[\log_{10} x + 2 \log_{10} y - \log_{10}(x - y)\right]$
$= \boxed{\dfrac{1}{4}\log_{10} x + \dfrac{1}{2}\log_{10} y - \dfrac{1}{4}\log_{10}(x - y)}$.

9. a. $3 \log_{10} x + 4 \log_{10} y - \log_{10} z = \log_{10} x^3 + \log_{10} y^4 - \log_{10} z = \boxed{\log_{10} \dfrac{x^3y^4}{z}}$.

b. $\dfrac{1}{2}\left(\log_b x^3 + \log_{10} y^5 - \log_b z^3\right) = \dfrac{1}{2}\left(\log_b \dfrac{x^3y^5}{z^3}\right) = \log_b\left(\dfrac{x^3y^5}{z^3}\right)^{1/2} = \boxed{\log_b\sqrt{\dfrac{x^3y^5}{z^3}}}$.

10. $\log_{10}[\log_2(\log_3 9)] = \log_{10}[\log_2(2)] = \log_{10}[1] = \boxed{0}$.

11. $\log_{10}(x + 3)(x) = 1$. $10^1 = (x + 3)(x)$. $x^2 + 3x - 10 = 0$. $(x + 5)(x - 2) = 0$. $x = -5, 2$. -5 causes $\log_{10}(x + 3)$ or $\log_{10} x$ to be undefined and so must be rejected. $\boxed{\{2\}}$.

12. a. $\boxed{1.3304}$. b. $\boxed{7.3304 - 10}$.

13. a. $\text{antilog}_{10}\ 0.9258 = 8.430$. $\text{antilog}_{10}\ 3.9258 = \boxed{8430}$.
b. $\text{antilog}_{10}\ 0.08156 = 6.54$. $\text{antilog}_{10}\ 2.8156 - 4 = \boxed{0.0654}$.

14. $x = \dfrac{0.00214}{.317}$. $\log_{10} x = \log_{10}(0.00214) - \log_{10}(.317) = (7.3304 - 10) - (9.5011 - 10)$

$= (17.3304 - 20) - (9.5011 - 10) = 7.8293 - 10.$

$x = \text{antilog}_{10}(7.8293 - 10) = \boxed{.00675}$.

15. $x = \dfrac{(2.85)^3(0.97)}{(0.035)}$. $\log_{10} x = \dfrac{1}{2}\left[3 \log_{10} 2.85 + \log_{10}(0.97) - \log_{10}(0.035)\right]$

$= \dfrac{1}{2}\left[3(0.4548) + (9.9868 - 10) - (8.5441 - 10)\right]$

$= \dfrac{1}{2}\left[(11.3512 - 10) - (8.5441 - 10)\right] = \dfrac{1}{2}\left[2.8071\right] = 1.4036.$

$x = \text{antilog}_{10}(1.4036) = \boxed{25.33}$.

16. $\ln 23.6 = 2.303 \log_{10} 23.6$

$= 2.303(1.3729)$

$= \boxed{3.1618}$.

17. a. $x \log_{10} 5 = \log_{10} 15$. $x = \boxed{\dfrac{\log_{10} 15}{\log_{10} 5}}$.

b. $(3 - x)\log_{10} 3 = \log_{10} 1000$

$(3 - x)\log_{10} 3 = 3$

$3 - x = \dfrac{3}{\log_{10} 3}$

$-x = -3 + \dfrac{3}{\log_{10} 3}$

$x = \boxed{3 - \dfrac{3}{\log_{10} 3}}$.

18. $E = 2ke^{1+t}$

$\ln E = \ln 2 + \ln k + (1 + t)\ln e$

$\ln E - \ln 2 - \ln k = (1 + t) \cdot 1$

$\boxed{-1 + \ln E - \ln 2 - \ln k = t}$.

19. $\log_{10} \dfrac{1}{[H^+]} = 6.3$

$\log_{10} 1 - \log_{10}[H^+] = 6.3$

$0 - \log_{10}[H^+] = 6.3$

$\log_{10}[H^+] = -6.3$

$= [10 + (-6.3)] - 10$

$= 3.7 - 10$

$[H^+] = \text{antilog}_{10}(3.7 - 10)$.

To the nearest table entry,

$[H^+] = \boxed{5.01 \times 10^{-7}}$

20.

$$N = N_o 10^{0.2t}$$
$$180 = 3 \cdot 10^{0.2t}$$
$$60 = 10^{0.2t}$$
$$0.2t \ \log_{10} 10 = \log_{10} 60$$
$$(0.2t)(1) = 1.7782$$
$$t = \frac{1.7782}{.2} = \boxed{8.891}$$

21.

$$y = y\ e^{-0.3t}$$
$$6 = 30e^{-0.3t}$$
$$\frac{6}{30} = e^{-0.3t}$$
$$e^{-0.3t} = 0.2$$
$$-0.3t \ \ln e = \ln\ 0.2$$
$$(-0.3t)(1) = (2.303)\log_{10} 0.2$$
$$-0.3t = 2.303(9.3010 - 10) = 2.303\ (-0.6990) = 1.6098$$
$$t = \frac{-1.6098}{-0.3} = \boxed{5.37}$$

CHAPTER 10

1. $x + 4y = -14$ $\quad$ $x + 4y = -14$ $\quad$ $-5x = -10.$ $\quad$ $x = 2.$
   $3x + 2y = -2.$ $\quad$ $-6x - 4y = 4.$

   $2 + 4y = -14$
   $y = -4.$ $\quad$ $\boxed{\{(2,-4)\}}$.

2. $\frac{2}{3}x - y = 4$ $\quad$ $2x - 3y = 12$ $\quad$ $-2x = -12.$ $\quad$ $x = 6.$
   $x - \frac{3}{4}y = 6.$ $\quad$ $-4x + 3y = -24.$

   $2(6) - 3y = 12$
   $-3y = 0$
   $y = 0.$ $\quad$ $\boxed{\{(6,0)\}}$.

3. Let $u = \frac{1}{x}$ and $v = \frac{1}{y}$. Then the system becomes

$u + 2v = -\frac{11}{12}$ $\quad$ $12u + 24v = -11$ $\quad$ $12v = -4.$ $\quad$ $v = -\frac{4}{12} = -\frac{1}{3}.$

$u + v = -\frac{7}{12}.$ $\quad$ $-12u - 12v = 7.$

$u + v = -\frac{7}{12}$

$u + -\frac{4}{12} = -\frac{7}{12}$

$u = -\frac{3}{12} = -\frac{1}{4}.$

Since $u = \frac{1}{x}$, we have $-\frac{1}{4} = \frac{1}{x}$; $x = -4.$

Since $v = \frac{1}{y}$, we have $-\frac{1}{3} = \frac{1}{y}$; $y = -3.$

$\boxed{\{(-4,-3)\}}$.

4. Since $\frac{3}{9} = \frac{4}{12} \neq \frac{4}{3}$, there are no solutions; the equations are $\boxed{\text{inconsistent}}$.

5. Since $\frac{3}{9} = \frac{4}{12} = \frac{1}{3}$, the equations are $\boxed{\text{dependent}}$.

6. Since $\frac{3}{9} \neq \frac{4}{6}$, the equations are $\boxed{\text{consistent and independent}}$.

7. $2x + 4y + z = 0$, $5x + 3y - 2z = 1$. $4x + 8y + 2z = 0$, $5x + 3y - 2z = 1$. $9x + 11y = 1$.

$2x + 4y + z = 0$, $4x - 7y - 7z = 6$. $14x + 28y + 7z = 0$, $4x - 7y - 7z = 6$. $18x + 21y = 6$.

$9x + 11y = 1$, $18x + 21y = 6$. $-18x - 22y = -2$, $18x + 21y = 6$. $-y = 4$. $y = -4$.

$9x + 11y = 1$, $9x + 11(-4) = 1$, $9x = 45$, $x = 5$. $2x + 4y + z = 0$, $2(5) + 4(-4) + z = 0$, $z = 6$. $\boxed{\{(5,-4,6)\}}$.

8. $y = x^2 - 2x + 1$, $x + y = 3$.

$$\begin{aligned} x + (x^2 - 2x + 1) &= 3 \\ x^2 - x - 2 &= 0 \\ (x - 2)(x + 1) &= 0 \\ x &= 2 \text{ or } -1. \end{aligned}$$

From $x + y = 3$, when $x = 2$, we have $2 + y = 3$; $y = 1$. And when $x = -1$, we have $-1 + y = 3$; $y = 4$.

$\boxed{\{(2,1),(-1,4)\}}$.

9. $9x^2 + 16y^2 = 100$, $x^2 + y^2 = 8$. $9x^2 + 16y^2 = 100$, $-16x^2 - 16y^2 = -128$. $-7x^2 = -28$, $x^2 = 4$, $x = \pm 2$.

From $x^2 + y^2 = 8$, when $x = 2$, we have $4 + y^2 = 8$ and $y = \pm 2$. So we have (2,2) and (2,-2). Similarly, when $x = -2$, we obtain (-2,2) and (-2,-2).

$\boxed{\{(2,2),(2,-2),(-2,2),(-2,-2)\}}$.

10. Let x = width. Then $32 - x$ = length.

$$\begin{aligned} x(32 - x) &= 240 \\ x^2 - 32x + 240 &= 0 \\ (x - 12)(x - 20) &= 0 \\ x &= 12 \text{ or } 20. \end{aligned}$$

When $x = 12$, $32 - x = 20$; and when $x = 20$, $32 - x = 12$. In any case, the dimensions are $\boxed{\text{20 feet by 12 feet}}$.

11. $x + y + z = 155$, $x = y - 20$, $y = 5 + z$. $x + y + z = 155$, $x - y = -20$, $y - z = 5$. $x + y + z = 155$, $-x + y = 20$.

$2y + z = 175$. $2y + z = 175$, $y - z = 5$. $3y = 180$, $y = 60$. $x = 60 - 20$, $x = 40$.

$60 - z = 5$, $z = 55$.

$\boxed{\begin{aligned} x &= 40 \text{ inches} \\ y &= 60 \text{ inches} \\ z &= 55 \text{ inches} \end{aligned}}$.

CHAPTER 11

1. $s_n = \dfrac{n}{n^2+1}$. $s_1 = \dfrac{1}{1^2+1} = \dfrac{1}{2}$. $s_2 = \dfrac{2}{2^2+1} = \dfrac{2}{5}$. $s_3 = \dfrac{3}{3^2+1} = \dfrac{3}{10}$. $s_4 = \dfrac{4}{4^2+1} = \dfrac{4}{17}$.

   So we have $\boxed{\dfrac{1}{2}, \dfrac{2}{5}, \dfrac{3}{10}, \dfrac{4}{17}}$.

2. $$\sum_{j=0}^{4} \frac{(-1)^j 3^j}{j+1} = \frac{(-1)^0 3^0}{0+1} + \frac{(-1)^1(3)^1}{1+1} + \frac{(-1)^2(3)^2}{2+1} + \frac{(-1)^3(3)^3}{3+1} + \frac{(-1)^4(3)^4}{4+1}$$
   $$= \boxed{1 - \frac{3}{2} + 3 - \frac{27}{4} + \frac{81}{5}}.$$

3. $1 - 8 + 27 - 64 + 125 = 1^3 - 2^3 + 3^3 - 4^3 + 5^3$
   $= (-1)^2(1^3) + (-1)^3(2^3) + (-1)^4(3^3) + (-1)^5(4^3) + (-1)^6(5^3)$.

   From this, we observe that $s_n = (-1)^{n+1}(n^3)$. Therefore, we have

   $$\boxed{\sum_{j=1}^{5} (-1)^{j+1}(j)^3}.$$

4. a. $d = 3$. Therefore, the next three terms are $\boxed{x + 6, x + 9, x + 12}$.

   b. Use $s_n = a + (n - 1)d$ and obtain $s_n = x + (n - 1)(3) = \boxed{x + 3n - 3}$.

5. $-5, -2, 1, \ldots$ is an arithmetic progression with $a = -5$, $d = 3$. Therefore, using $s_n = a + (n - 1)d$, we have

   $$s_n = -5 + (17 - 1)(3) = \boxed{43}.$$

6. Using $s_n = a + (n - 1)d$, we have $-46 = a + (20 - 1)d$ and $-30 = a + (12 - 1)d$. These can be written equivalently as $a + 19d = -46$ and $a + 11d = -30$. To solve this system of equations, we subtract the second equation from the first and obtain $8d = -16$, from which we have $d = -2$. Using this result in $a + 11d = -30$, we have $a + 11(-2) = -30$ and obtain $a = -8$. Then,

   $$s_5 = -8 + (5 - 1)(-2) = \boxed{-16}.$$

7. $$\sum_{j=10}^{20} (2j - 3) = [2(10) - 3] + [2(11) - 3] + [2(12) - 3] + \cdots + [2(20) - 3]$$
   $$= 17 + 19 + 21 + \cdots + 37.$$

   This series is arithmetic with $a = 17$, $d = 2$, $n = 11$, and $s_{11} = 37$.
   Using $s_n = \frac{n}{2}(a + s_n)$, we find,

   $$s_{11} = \frac{11}{2}(17 + 37) = \frac{11}{2}(54) = \boxed{297}.$$

8. a. To determine r, we divide -1 by $\frac{x}{a}$; that is, we have

$$-1 \div \frac{x}{a} = -1\left(\frac{a}{x}\right) = -\frac{a}{x}. \quad \text{So, } r = -\frac{a}{x}. \quad s_4 = s_3\left(-\frac{a}{x}\right) = \frac{a}{x}\left(-\frac{a}{x}\right) = -\frac{a^2}{x^2}.$$

$$s_5 = s_4\left(-\frac{a}{x}\right) = \left(-\frac{a^2}{x^2}\right)\left(-\frac{a}{x}\right) = \frac{a^3}{x^3}.$$

$$s_6 = s_5\left(-\frac{a}{x}\right) = \left(\frac{a^3}{x^3}\right)\left(-\frac{a}{x}\right) = -\frac{a^4}{x^4}.$$

Therefore, the next three terms are $\boxed{-\frac{a^2}{x^2}, \frac{a^3}{x^3}, -\frac{a^4}{x^4}}$.

b. $s_n = \left(\frac{x}{a}\right)\left(-\frac{a}{x}\right)^{n-1} = \frac{x}{a}(-1)^{n-1}\left(\frac{a^{n-1}}{x^{n-1}}\right) = \boxed{(-1)^{n-1}\frac{a^{n-2}}{x^{n-2}}}$.

9. For this geometric progression, $a = -81$ and $r = \frac{-27}{-81} = \frac{1}{3}$. Using $s_n = ar^{n-1}$ with $n = 9$, we have

$$s_9 = (-81)\left(\frac{1}{3}\right)^8 = -(3)^4\frac{1}{(3)^8} = -\frac{1}{3^4} = \boxed{-\frac{1}{81}}.$$

10. In $s_n = ar^{n-1}$, take $n = 6$, $s_6 = 60$, and $r = 3$ and obtain $60 = a(3)^5$. $60 = 243a$.

$a = \frac{60}{243} = \frac{20}{81}$. Then $s_2 = \frac{20}{81}(3) = \boxed{\frac{20}{27}}$.

11. $$\sum_{j=5}^{8}(2^j - 5) = (2^5 - 5) + (2^6 - 5) + (2^7 - 5) + (2^8 - 5)$$
$$= 2^5 + 2^6 + 2^7 + 2^8 - (5 + 5 + 5 + 5)$$
$$= 2^5 + 2^6 + 2^7 + 2^8 - (20).$$

$2^5 + 2^6 + 2^7 + 2^8$ is a geometric progression with $a = 2^5$, $r = 2$, and $n = 4$.

Using $S_n = \frac{a - ar^n}{1 - r} = \frac{2^5 - 2^5(2)^4}{1 - 2} = \frac{2^5(1 - 2^4)}{-1} = \frac{32(1 - 16)}{-1} = -32(-15) = 480.$

Therefore, $\sum_{j=5}^{8}(2^j - 5) = 480 - 20 = \boxed{460}$.

12. The diagram 3, ___, ___, ___, 243 suggests that there are 4 multiplications by r between 3 and 243. Hence,

$$r^4 = \frac{243}{3}, \quad r^4 = 81, \quad r = 3.$$

Therefore, the required means are $\boxed{9, 27, 81}$.

13. Since $r = -\frac{3}{4}$, that is $|r| < 1$, the series has a sum given by $S_\infty = \frac{a}{1 - r}$. Taking $a = 2$ and $r = -\frac{3}{4}$, we obtain $S_\infty = \frac{2}{1 - \left(-\frac{3}{4}\right)} = \frac{2}{1 + \frac{3}{4}} = \frac{2}{\frac{7}{4}} = \boxed{\frac{8}{7}}$.

14. $$\frac{(12!)(8!)}{16!} = \frac{\overset{1}{\cancel{12!}}8!}{16 \cdot 15 \cdot 14 \cdot 13 \cdot \underset{1}{\cancel{12!}}} = \frac{\overset{1}{\cancel{8}} \cdot \overset{1}{\cancel{7}} \cdot 6 \cdot \overset{1}{\cancel{5}} \cdot \overset{2}{\cancel{4}} \cdot \overset{1}{\cancel{3}} \cdot \overset{1}{\cancel{2}} \cdot 1}{\underset{1}{\cancel{16}} \cdot \underset{1}{\cancel{15}} \cdot \underset{\underset{1}{\cancel{2}}}{\cancel{14}} \cdot 13} = \boxed{\frac{12}{13}}.$$

15. The first four terms of $[x + (-2y)]^8$ are $x^8 + \frac{8x^7(-2y)^1}{1!} + \frac{7 \cdot 8x^6(-2y)^2}{2!} + \frac{6 \cdot 7 \cdot 8x^5(-2y)^3}{3!}$.

These can be simplified to $\boxed{x^8 - 16x^7y + 112x^6y^2 - 448x^5y^3}$.

16. The 7th term of $(a^3 - b)^9$ is $\frac{4 \cdot 5 \cdot 6 \cdot 7 \cdot 8 \cdot 9(a^3)^3(-b)^6}{6!}$.

This can be simplified to $\boxed{84a^9b^6}$.

17. $1.027\overline{027} = 1 + .027 + .000027 + .000000027 + \cdots$. Beginning with the second term, .027, we have an infinite geometric series with $a = .027$ and $r = .001$. So using

$$S_\infty = \frac{a}{1 - r}$$

on this infinite geometric series, we have

$$1.027\overline{027} = 1 + \frac{.027}{1 - .001} = 1 + \frac{.027}{.999} = 1 + \frac{27}{999} = 1 + \frac{1}{37} = \boxed{\frac{38}{37}}.$$

18. $P_{5,5} = 5! = 5 \cdot 4 \cdot 3 \cdot 2 \cdot 1 = \boxed{120}$.

19. Eight positions remain to be filled and nine boys are available to fill them.

$$P_{9,8} = 9 \cdot 8 \cdot 7 \cdot 6 \cdot 5 \cdot 4 \cdot 3 \cdot 2 = \boxed{362,880}.$$

20. Since the letters and digits may be used more than once, the letters may each be chosen in 8 ways and the digits may each be chosen in 10 ways. So, we have

$$8 \cdot 8 \cdot 8 \cdot 10 \cdot 10 \cdot 10 = \boxed{512,000}.$$

21. $$P = \frac{9!}{2!3!2!} = \frac{9 \cdot \overset{4}{\cancel{8}} \cdot 7 \cdot \overset{1}{\cancel{6}} \cdot 5 \cdot 4 \cdot 3 \cdot \overset{1}{\cancel{2}} \cdot 1}{\underset{1}{\cancel{2}} \cdot 1 \cdot \underset{1}{\cancel{3 \cdot 2}} \cdot 1 \cdot \underset{1}{\cancel{2}} \cdot 1} = \boxed{15,120}.$$

22. 3 men can be chosen from 6 men in $\binom{6}{3}$ ways. $\binom{6}{3} = \frac{\overset{1}{\cancel{6}} \cdot 5 \cdot 4}{\underset{1}{\cancel{3 \cdot 2}} \cdot 1} = 20$. 3 women can be chosen from 8 women in $\binom{8}{3}$ ways. $\binom{8}{3} = \frac{8 \cdot 7 \cdot \overset{1}{\cancel{6}}}{\underset{1}{\cancel{3 \cdot 2}} \cdot 1} = 56$. Then, a committee can be chosen in $20 \times 56 = \boxed{1120}$ ways.

23. This is equivalent to being asked to determine how many nonempty subsets there are of a four-member set. This is given by $2^4 - 1 = \boxed{15}$. An alternative solution is obtained by computing $\binom{4}{1} + \binom{4}{2} + \binom{4}{3} + \binom{4}{4} = 4 + 6 + 4 + 1 = \boxed{15}$.

24. 2 red balls can be chosen from 5 red balls in $\binom{5}{2} = 10$ ways. 3 black balls can be chosen from 4 black balls in $\binom{4}{3} = 4$ ways. Each of the choices of a pair of red balls can be paired with 4 choices of a combination of black balls. Therefore, there are $10 \times 4 = \boxed{40}$ ways in total.

APPENDIX A

1. $$\begin{array}{r|rrrr} 3 & 1 & -3 & 2 & 0 \\ & & 3 & 0 & 6 \\ \hline & 1 & 0 & 2 & 6 \end{array}$$

$$\boxed{x^2 + 2 + \frac{6}{x - 3}}.$$

2. $$\begin{array}{r|rrrrrr} -1 & 1 & 0 & 0 & 0 & 0 & 1 \\ & & -1 & 1 & -1 & 1 & -1 \\ \hline & 1 & -1 & 1 & -1 & 1 & 0 \end{array}$$

$$\boxed{y^4 - y^3 + y^2 - y + 1}.$$

3. $P(1)$ is equal to the remainder when $P(x)$ is divided by $x - 1$.

$$\begin{array}{r|rrrr} 1 & 2 & -3 & 1 & 1 \\ & & 2 & -1 & 0 \\ \hline & 2 & -1 & 0 & 1 \end{array}$$

Hence, $\boxed{P(1) = 1}$.

$P(-1)$ is found similarly.

$$\begin{array}{r|rrrr} -1 & 2 & -3 & 1 & 1 \\ & & -2 & 5 & -6 \\ \hline & 2 & -5 & 6 & -5 \end{array}$$

Hence, $\boxed{P(-1) = -5}$.

4. $x^3 - x^2 + 3x$ is divided successively by -3, -2, -1, 0, 1, 2, 3.

$$\begin{array}{r|rrrr} -3 & 1 & -1 & 3 & 0 \\ & & -3 & 12 & -45 \\ \hline & 1 & -4 & 15 & -45 \end{array} \quad \begin{array}{r|rrrr} -2 & 1 & -1 & 3 & 0 \\ & & -2 & 6 & -18 \\ \hline & 1 & -3 & 9 & -18 \end{array} \quad \begin{array}{r|rrrr} -1 & 1 & -1 & 3 & 0 \\ & & -1 & 2 & -5 \\ \hline & 1 & -2 & 5 & -5 \end{array} \quad \begin{array}{r|rrrr} 0 & 1 & -1 & 3 & 0 \\ & & 0 & 0 & 0 \\ \hline & 1 & -1 & 3 & 0 \end{array}$$

$$\begin{array}{r|rrrr} 1 & 1 & -1 & 3 & 0 \\ & & 1 & 0 & 3 \\ \hline & 1 & 0 & 3 & 3 \end{array} \quad \begin{array}{r|rrrr} 2 & 1 & -1 & 3 & 0 \\ & & 2 & 2 & 10 \\ \hline & 1 & 1 & 5 & 10 \end{array} \quad \begin{array}{r|rrrr} 3 & 1 & -1 & 3 & 0 \\ & & 3 & 6 & 27 \\ \hline & 1 & 2 & 9 & 27 \end{array}$$

Hence, the following ordered pairs correspond to points on the graph: (-3,-45), (-2,-18), (-1,-5), (0,0), (1,3), (2,10), (3,27).

The graph is

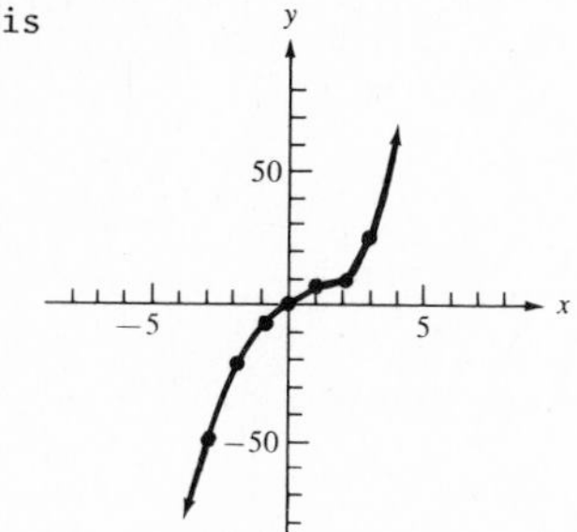

5. a. $$\begin{array}{r|rrrr} -1 & 1 & 1 & -4 & -4 \\ & & -1 & 0 & 4 \\ \hline & 1 & 0 & -4 & 0 \end{array}$$

Since the remainder is 0, $\boxed{x + 1 \text{ is a factor}}$.

b. $$\begin{array}{r|rrrr} 1 & 1 & 1 & -4 & -4 \\ & & 1 & 2 & -2 \\ \hline & 1 & 2 & -2 & -6 \end{array}$$

Since the remainder is not 0, $\boxed{x - 1 \text{ is not a factor}}$.

6. In the solution to 5(a) above, we found that $x - 1$ is a factor of $x^3 + x^2 - 4x + 4$. Hence, from the synthetic division $x^3 + x^2 - 4x + 4 = (x - 1)(x^2 - 4) = (x - 1)(x + 2)(x - 2) = 0$. Then, by inspection the required solutions are $x = 1$, $x = -2$, $x = 2$ and the solution set is $\boxed{\{1,-2,2\}}$.

APPENDIX B

1. $\begin{bmatrix} 1 & 4 & -14 \\ 3 & 2 & -2 \end{bmatrix} \sim \begin{bmatrix} 1 & 4 & -14 \\ 0 & -10 & 40 \end{bmatrix}$. The last matrix corresponds to the system $\begin{aligned} x + 4y &= -14 \\ -10y &= 40 \end{aligned}$. From the second equation, $y = -4$ and when -4 is substituted for y in the first equation, we find $x = 2$. $\boxed{\{(2,-4)\}}$.

2. We start with the augmented matrix $\begin{bmatrix} 2 & 4 & 1 & 0 \\ 5 & 3 & -2 & 1 \\ 4 & -7 & -7 & 6 \end{bmatrix}$. To obtain a row-equivalent matrix with 0 as the first entry in rows 2 and 3 we proceed as follows: multiply each entry of row 1 by $\frac{-5}{2}$ and add each product to the corresponding entry of row 2 to obtain a new row 2; multiply each entry of row 1 by -2 and add each product to the corresponding entry of row 3 to obtain a new row 3. Thus we have $\begin{bmatrix} 2 & 4 & 1 & 0 \\ 5 & 3 & -2 & 1 \\ 4 & -7 & -7 & 6 \end{bmatrix} \sim \begin{bmatrix} 2 & 4 & 1 & 0 \\ 0 & -7 & \frac{-9}{2} & 1 \\ 0 & -15 & -9 & 6 \end{bmatrix}$. Then, to obtain a matrix which is row-equivalent to this last one but with a 0 as the second entry in row 3 (without disturbing the already existing zeros), we multiply each entry of row 2 by $\frac{-15}{7}$ and add each product to the corresponding entry of row 3 to obtain a new row 3. Thus, we have $\begin{bmatrix} 2 & 4 & 1 & 0 \\ 0 & -7 & \frac{-9}{2} & 1 \\ 0 & -15 & -9 & 6 \end{bmatrix} \sim \begin{bmatrix} 2 & 4 & 1 & 0 \\ 0 & -7 & \frac{-9}{2} & 1 \\ 0 & 0 & \frac{9}{14} & \frac{27}{7} \end{bmatrix}$. This last matrix corresponds to

$$\begin{aligned} 2x + 4y + z &= 0 \quad (1) \\ -7y - \frac{9}{2}z &= 1 \quad (2) \\ \frac{9}{14}z &= \frac{27}{7} \quad (3) \end{aligned}$$

From Equation (3), $z = 6$. In Equation (2) substitute 6 for z and solve for y and obtain $y = -4$. Then, in Equation (1) substitute 6 for z and -4 for y and solve for x to obtain $x = 5$. The solution set is $\boxed{\{(5,-4,6)\}}$.

3. $\begin{vmatrix} 10 & 3 \\ -10 & -2 \end{vmatrix} = (10)(-2) - (-10)(3) = -20 - (-30) = \boxed{10}$.

4. $D = \begin{vmatrix} 3 & -4 \\ 1 & -2 \end{vmatrix} = -6 - (-4) = -2.$ $\quad D_x = \begin{vmatrix} -2 & -4 \\ 0 & -2 \end{vmatrix} = 4 - 0 = 4.$

$D_y = \begin{vmatrix} 3 & -2 \\ 1 & 0 \end{vmatrix} = 0 - (-2) = 2.$ $\quad x = \frac{D_x}{D} = \frac{4}{-2} = -2.$ $\quad y = \frac{D_y}{D} = \frac{2}{-2} = -1.$ $\boxed{\{(-2,-1)\}}$.

5. Expanding about the 2nd row, we have

$-0 \begin{vmatrix} 3 & 1 \\ 2 & 1 \end{vmatrix} + 1 \begin{vmatrix} 2 & 1 \\ -4 & 1 \end{vmatrix} - 0 \begin{vmatrix} 2 & 3 \\ -4 & 2 \end{vmatrix}$; $\begin{vmatrix} 2 & 1 \\ -4 & 1 \end{vmatrix} = 2 - (-4) = \boxed{6}$.

6. Expanding the determinant about the first column, we have

$x \begin{vmatrix} x & 1 \\ x & 0 \end{vmatrix} - 0 \begin{vmatrix} 1 & 1 \\ x & 0 \end{vmatrix} + 0 \begin{vmatrix} 1 & 1 \\ x & 1 \end{vmatrix} = -4.$ $\quad x(x \cdot 0 - x \cdot 1) = -4$

$$\begin{aligned} -x^2 &= -4 \\ x^2 &= 4 \\ x &= \pm 2. \end{aligned}$$ $\boxed{\{2,-2\}}$.

7. $D = \begin{vmatrix} 3 & -2 & 5 \\ 4 & -4 & 3 \\ 5 & -4 & 1 \end{vmatrix} = \begin{vmatrix} -22 & 18 & 0 \\ -11 & 8 & 0 \\ -5 & 4 & 1 \end{vmatrix} = \begin{vmatrix} -22 & 18 \\ -11 & 8 \end{vmatrix} = -11\begin{vmatrix} 2 & 18 \\ 1 & 8 \end{vmatrix} = -11(16 - 18) = 22.$

$D_x = \begin{vmatrix} 6 & -2 & 5 \\ 0 & -4 & 3 \\ -5 & -4 & 1 \end{vmatrix} = \begin{vmatrix} 31 & 18 & 0 \\ 15 & 8 & 0 \\ -5 & 4 & 1 \end{vmatrix} = \begin{vmatrix} 31 & 18 \\ 15 & 8 \end{vmatrix} = 248 - 270 = -22.$

$D_y = \begin{vmatrix} 3 & 6 & 5 \\ 4 & 0 & 3 \\ 5 & -5 & 1 \end{vmatrix} = \begin{vmatrix} -22 & 31 & 0 \\ -11 & 15 & 0 \\ 5 & -1 & 1 \end{vmatrix} = \begin{vmatrix} -22 & 31 \\ -11 & 15 \end{vmatrix} = -11\begin{vmatrix} 2 & 31 \\ 1 & 15 \end{vmatrix} = -11(30 - 31) = 11.$

$D_z = \begin{vmatrix} 3 & -2 & 6 \\ 4 & -4 & 0 \\ 5 & -4 & -5 \end{vmatrix} = 4\begin{vmatrix} 3 & -2 & 6 \\ 1 & -1 & 0 \\ 5 & -4 & -5 \end{vmatrix} = 4\begin{vmatrix} 3 & 1 & 6 \\ 1 & 0 & 0 \\ 5 & 1 & -5 \end{vmatrix} = 4(-1)\begin{vmatrix} 1 & 6 \\ 1 & -5 \end{vmatrix} = -4(-5 - 6) = 44.$

$x = \frac{D_x}{D} = \frac{-22}{22} = -1.$ $y = \frac{D_y}{D} = \frac{11}{22} = \frac{1}{2}.$ $z = \frac{D_z}{D} = \frac{44}{22} = 2.$ $\boxed{\left\{\left(-1, \frac{1}{2}, 2\right)\right\}}.$

APPENDIX C

1. $\log_{10} 243.6 = \log_{10}(2.436 \times 10^2)$

Hence the characteristic is 2.

x	$\log_{10} x$
10 { 6 { 2.430	0.3856 } y } 0.0018
2.436	?
2.440	0.3874

$$\frac{6}{10} = \frac{y}{0.0018}$$

$y = \frac{6}{10}(0.0018) = 0.00108 \approx 0.0011.$

Hence the mantissa is $0.3856 + 0.0011 = 0.3867$ and $\log_{10} 253.6 =$ $\boxed{2.3867}$.

2. $\log_{10} 0.03762 = \log_{10}(3.762 \times 10^{-2})$

Hence the characteristic is $-2 = 8 - 10$.

x	$\log_{10} x$
10 { 2 { 3.760	0.5752 } y } 0.0011
3.762	?
3.770	0.5763

$$\frac{2}{10} = \frac{y}{0.0011}$$

$y = \frac{2}{10}(0.0011) = 0.00022 \approx 0.0002.$

Hence the mantissa is $0.5752 + 0.0002 = 0.5754$ and $\log_{10} 0.03762 =$ $\boxed{8.5754 - 10}$.

3. Consecutive mantissa entries that bound the given mantissa 0.1237 are shown below.

x	$\text{antilog}_{10}\, x$
33 { 31 { 0.1206	1.320 } y } 0.010
0.1237	?
0.1239	1.330

$$\frac{31}{33} = \frac{y}{0.010}$$

$y = \frac{31}{33}(0.010) \approx 0.009$ and $\text{antilog}_{10}\, 0.1237 = 1.329.$

Hence $\text{antilog}_{10}\, 3.1237 = 1.329 \times 10^3 =$ $\boxed{1329}$

4.

		x	$\text{antilog}_{10} x$		
5	1	.9978	9.950	y	.010
		.9979	?		
		.9983	9.960		

$$\frac{y}{.010} = \frac{.00011}{.0005}; \quad \frac{y}{.010} = \frac{1}{5}; \quad y = \frac{.010}{5} = .002$$

$\text{antilog}_{10}\ 0.9979 = 9.952$. $\text{antilog}_{10}\ 8.9979 - 10 = \boxed{0.09952}$